Structural Health Monitoring

Structural Health Monitoring

Edited by
Daniel Balageas,
Claus-Peter Fritzen
and
Alfredo Güemes

ISTE Ltd
6 Fitzroy Square
London W1T 5DX
UK

ISTE USA
4308 Patrice Road
Newport Beach, CA 92663
USA

www.iste.co.uk

Library of Congress Cataloging-in-Publication Data

Structural health monitoring / edited by Daniel Balageas ;
[edited by] Daniel Balageas, Claus-Peter Fritzen, Alfredo Güemes.-- 1st ed.
    p. cm.
 Includes bibliographical references and index.
 ISBN-13: 978-1-905209-01-9
 1. Structural analysis (Engineering) 2. Automatic data collection
systems. 3. Detectors. I. Balageas, Daniel. II. Fritzen, Claus-Peter. III. Güemes, Alfredo.

 TA645.S754 2005
 624.1'71--dc22

                                                          2005034291

British Library Cataloguing-in-Publication Data
A CIP record for this book is available from the British Library
ISBN 10: 1-905209-01-0
ISBN 13: 978-1-905209-01-9

Printed and bound in Great Britain by Antony Rowe Ltd, Chippenham, Wiltshire.

Table of Contents

Chapter 6. Low Frequency Electromagnetic Techniques 411
Michel LEMISTRE

**Chapter 7. Capacitive Methods for Structural Health Monitoring in
Civil Engineering** 463
Xavier DÉROBERT and Jean IAQUINTA

Foreword

The origins of this book date back to a pre-conference course given at the First European Workshop on Structural Health Monitoring, which was held at the Ecole Normale Supérieure of Cachan (Paris) in July 2002. In 2004, this course was extended to form a continuing-education short course lasting three and a half days, organized by the Ecole Normal Supérieure of Cachan.

The motivation of the authors has essentially been to make the information collected for this short course more widely available, especially at the present time, which is characterized by the strong emergence of approaches in the technical community to the problems of Structural Health Monitoring.

The book is organized around the various sensing techniques used to achieve the monitoring. For this reason, emphasis is put on sensors, on signal and data reduction methods, and on inverse techniques, allowing the identification of the physical parameters affected by the presence of the damage on which the diagnosis is established. This choice leads to a presentation that is not oriented by the type of applications or linked to special classes of problems, but presents the broad families of techniques: vibration and modal analysis (Chapter 2), optical fibre sensing (Chapter 3), acousto-ultrasonics using piezoelectric transducers (Chapter 4), and electric and electromagnetic techniques (Chapters 5 to 7).

Each chapter has been written by specialists in the domain of the chapter, who have been working in the field for a long time and have wide knowledge and experience. The authors, who come from the academic world or from research centres, have written their contributions in a pedagogical spirit, so that this book can be easily understood by beginners in the field and by students. Nevertheless, the book aims to present an exhaustive overview of present research and development, giving numerous references that will be useful even to experienced researchers and engineers.

The Editors
D.L. Balageas, C.-P. Fritzen, A. Güemes

Chapter 1

Introduction to Structural Health Monitoring

1.1. Definition of Structural Health Monitoring

Structural Health Monitoring (SHM) aims to give, at every moment during the life of a structure, a diagnosis of the "state" of the constituent materials, of the different parts, and of the full assembly of these parts constituting the structure as a whole. The state of the structure must remain in the domain specified in the design, although this can be altered by normal aging due to usage, by the action of the environment, and by accidental events. Thanks to the time-dimension of monitoring, which makes it possible to consider the full history database of the structure, and with the help of Usage Monitoring, it can also provide a prognosis (evolution of damage, residual life, etc.).

If we consider only the first function, the diagnosis, we could estimate that Structural Health Monitoring is a new and improved way to make a Non-Destructive Evaluation. This is partially true, but SHM is much more. It involves the integration of sensors, possibly smart materials, data transmission, computational power, and processing ability inside the structures. It makes it possible to reconsider the design of the structure and the full management of the structure itself and of the structure considered as a part of wider systems. This is schematically presented in Figure 1.1.

Chapter written by Daniel BALAGEAS.

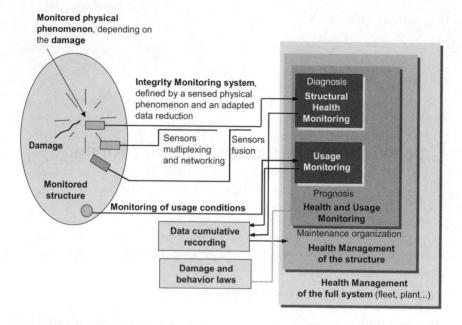

Figure 1.1. *Principle and organization of a SHM system*

In Figure 1.1, the organization of a typical SHM system is given in detail. The first part of the system, which corresponds to the structural integrity monitoring function, can be defined by: i) the type of physical phenomenon, closely related to the damage, which is monitored by the sensor, ii) the type of physical phenomenon that is used by the sensor to produce a signal (generally electric) sent to the acquisition and storage sub-system. Several sensors of the same type, constituting a network, can be multiplexed and their data merged with those from other types of sensors. Possibly, other sensors, monitoring the environmental conditions, make it possible to perform the usage monitoring function. The signal delivered by the integrity monitoring sub-system, in parallel with the previously registered data, is used by the controller to create a diagnostic. Mixing the information of the integrity monitoring sub-system with that of the usage monitoring sub-system and with the knowledge based on damage mechanics and behavior laws makes it possible to determine the prognosis (residual life) and the health management of the structure (organization of maintenance, repair operations, etc.). Finally, similar structure management systems related to other structures which constitute a type of super system (a fleet of aircraft, a group of power stations, etc.) make possible the health management of the super system. Of course, workable systems can be set up even if they are not as comprehensive as described here.

1.2. Motivation for Structural Health Monitoring

Knowing the integrity of in-service structures on a continuous real-time basis is a very important objective for manufacturers, end-users and maintenance teams. In effect, SHM:

– allows an optimal use of the structure, a minimized downtime, and the avoidance of catastrophic failures,

– gives the constructor an improvement in his products,

– drastically changes the work organization of maintenance services: i) by aiming to replace scheduled and periodic maintenance inspection with performance-based (or condition-based) maintenance (long term) or at least (short term) by reducing the present maintenance labor, in particular by avoiding dismounting parts where there is no hidden defect; ii) by drastically minimizing the human involvement, and consequently reducing labor, downtime and human errors, and thus improving safety and reliability. These drastic changes in maintenance philosophy are described in several recent papers, in particular for military air vehicles [DER 03], for Army systems [WAL 03] for civil aircraft [BER 03, GOG 03], and for civil infrastructures [FRA 03].

The improvement of safety seems to be a strong motivation, in particular after some spectacular accidents due to: i) unsatisfactory maintenance, for example, in the aeronautic field, the accident of Aloha Airlines [OTT 88] – see Figure 1.2a) – or, in the civil engineering field, the collapse of the Mianus River bridge; ii) ill-controlled manufacturing process, for example, the Injak bridge collapse (see Figure 1.2b)). In both fields the problem of aging structures was discovered and subsequent programs were established. To pinpoint the importance of the problem of structural aging, the following statistic can be recalled: bridge inspection during the late 1980s revealed that on the 576,000 US highway bridges, 236,000 were rated deficient by present day standards [WAN 97].

Nevertheless, analysis of the various causes of aircraft accidents points to the relatively low influence of maintenance deficiency. Figure 1.3 shows that maintenance is only responsible of 14% of hull loss. Furthermore, it should be noted that only 4% of all accidents are due to structural weakness. It can be concluded that, thanks to the introduction of SHM, even an improvement in maintenance and a decrease of structure-caused accidents by a factor of two would lead to a global reduction of accidents of less than 10%, which is far from what is needed to avoid a significant increase in the number of accidents in the near future if air traffic continues to increase.

The economic motivation is stronger, principally for end-users. In effect, for structures with SHM systems, the envisaged benefits are constant maintenance costs and reliability, instead of increasing maintenance costs and decreasing reliability for classical structures without SHM (see Figure 1.4).

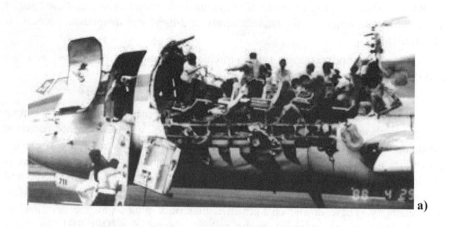

a)

b)

Figure 1.2. *Spectacular accidents have motivated the community to improve safety:*
a) the Aloha Airlines flight 243, April 29, 1988, due to corrosion insufficiently controlled by
maintenance; b) the Injaka bridge collapse, July 1998, due to a poorly controlled
construction process

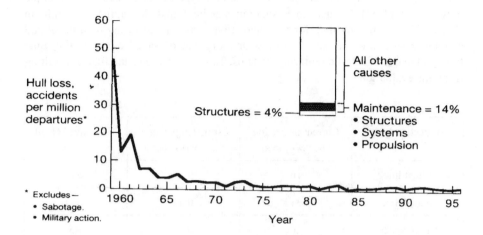

Figure 1.3. *Origin of hull losses: safety record for the worldwide commercial jet fleet, from [GOR 97]*

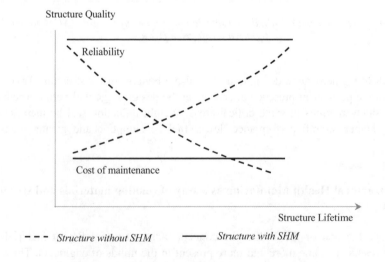

Figure 1.4. *Benefit of SHM for end-users [CHA 02]*

The economic impact of the introduction of SHM for aircraft is not easy to evaluate. It depends on the usage conditions and, furthermore, it is difficult to appreciate the impact on the fabrication cost of the structure. The cost of SHM systems must not be so high as to cancel out the expected maintenance cost savings.

It is easier to evaluate the time saved by the new type of maintenance based on the introduction of SHM. Such an evaluation can be found, for military aircraft, in [BAR 97], who reports that, for a modern fighter aircraft featuring both metal and composite structure, an estimated 40% or more can be saved on inspection time through the use of smart monitoring systems. Table 1.1 presents the figures resulting from this evaluation.

Inspection type	Current inspection time (% of total)	Estimated potential for smart systems	Time saved (% of total)
Flight line	16	0.40	6.5
Scheduled	31	0.45	14.0
Unscheduled	16	0.10	1.5
Service instructions	37	0.60	22.0
	100		44.0

Table 1.1. *Estimated time saved on inspection operations by the use of SHM, for modern fighter aircraft, from [BAR 97]*

Still in the aeronautic domain, there is also a benefit for constructors. Taking into account the permanent presence of sensors at the design stage will permit a reduction in the safety margins in some critical areas. Weight reduction will be then possible, giving higher aircraft performance, lower fuel consumption and greater maximum range.

1.3. Structural Health Monitoring as a way of making materials and structures smart

Since the end of the 1980s, the concept of smart or intelligent materials and structures has become more and more present in the minds of engineers. These new ideas were particularly welcome in the fields of aerospace and civil engineering. In fact, the concept is presently one of the driving forces for innovation in all domains.

The concept of Smart Materials/Structures (SMS) can be considered as a step in the general evolution of man-made objects as shown in Figure 1.5. There is a continuous trend from simple to complex in human production, starting from the use of homogeneous materials, supplied by nature and accepted with their natural properties, followed by multi-materials (in particular, composite materials) allowing us to create structures with properties adapted to specific uses. In fact, composite materials and multi-materials are replacing homogeneous materials in more and more structures. This is particularly true in the aeronautic domain. For instance, composite parts are now currently used or envisaged for modern aircraft (see for instance in Figure 1.6, Boeing's *7E7 Dreamliner* project, which has 50% of its structures made of composites). It is worth noting that this aircraft is the first one in which it is clearly planned to embed SHM systems, in particular systems for impact detection.

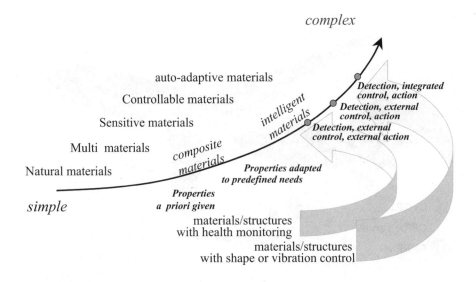

Figure 1.5. *General evolution of materials/structures used by people, and the place of smart structures, including structures with SHM*

The next step consists of making the properties of the materials and structures adapt to changing environmental conditions. This requires making them sensitive, controllable and active. The various levels of such "intelligence" correspond to the existence of one, two or all three qualities. Thus, sensitive, controllable and auto-adaptive materials/structures can be distinguished. Classically, three types of SMS exist: SMS controlling their shape, SMS controlling their vibrations, and SMS controlling their health. It is clear that materials and structures integrating SHM systems belong, at least in the short term, to the less smart type of SMS. In effect, almost all achievements in this field are only intended to make materials/structures sensitive, by embedding sensors. The next step towards smarter structures would be to make self-repairing materials/structures, or at least materials/structures with embedded damage-mitigation properties. For damage mitigation, embedding actuators made of shape memory alloys (SMA) could be a solution that would induce strains in order to reduce the stresses in regions of strain concentration. These SMA actuators could be in the form of wires [YOS 96, CHO 99] or films [TAK 00]. As regards self-healing structures, very few attempts have been made. We could mention, in the field of civil engineering, the existence of self-healing concretes containing hollow adhesive-filled brittle fibers: the adhesive is released when the fibers are broken in the region where cracking occurs [DRY 94]. A similar method can be applied to polymer matrix composites [DRY 96, MOT 99].

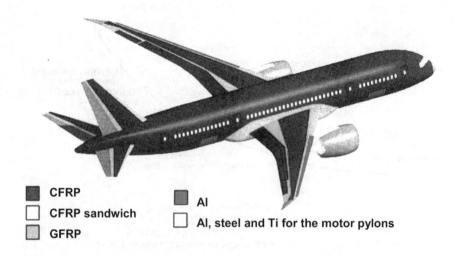

CFRP

CFRP sandwich

GFRP

Al

Al, steel and Ti for the motor pylons

Figure 1.6. *Example of the increasing importance of composites in civil aircraft: the 7E7 Dreamliner has 50% of its structure made of composites. For this aircraft, impact detection monitoring systems are envisaged for outer panels*

As seen above, strong differences exist between structures with SHM and SMS controlling their shape and vibrations. Nevertheless, it is interesting to consider them as part of a whole (see Figure 1.7), since a really smart structure will integrate all three functionalities, and because they all rely on common basic researches aimed at:

– elaborating new sensitive materials to make sensors and actuators,

– developing technologies to miniaturize sensors and actuators, and to embed them without degradation of the host structures,

– conceiving systems for data reduction and diagnostic formulation.

This is the reason why, until recently, works on SHM were often presented at conferences and in journals devoted to the general topic of SMS.

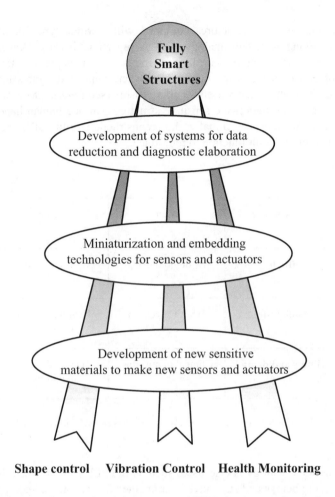

Figure 1.7. *Common basis and complementarity of SHM, shape control and vibration control*

1.4. SHM and biomimetics

The research on SMS in general, and on SHM in particular, is more or less influenced by biomimetics (or bio inspiration). This attitude is a real source of innovation.

Regarding SHM, a strong similarity exists between it and medical activity. This has been well pinpointed in [GAN 92] where a parallelism, given in Table 1.2, is drawn.

Very often, sensitive structures equipped with various types of sensors are compared to living skin. This analogy remains superficial because skin is really an auto-adaptive smart structure controlling its integrity. This is possible thanks to the presence of actuators that can counterbalance environmental aggressions. At the micro scale, the number and variety of skin sensors (see Figure 1.8) is way beyond what is possible with man-made sensitive structures (in one human hand there are more than *100,000 sensors!*). Finally, the reconstruction ability of living tissues is certainly the most difficult function to reproduce.

Phase of life	Man	Structures
Birth	Birth monitoring	Process monitoring
Sound life	Health check-up	Health and usage monitoring
Illness and death	Clinical monitoring	Health (damage) monitoring

Table 1.2. *Parallelism between medical activities and SHM, from [GAN 92]*

Often, another analogy is also used, such as in [BER 03], between the nervous system of living beings and structures instrumented by sensors and equipped with a central processor (see Figure 1.9). The gap between living systems and artefacts is perhaps smaller in this case and study of the functioning of the nervous system and the brain is useful when conceiving control systems (adaptive control influenced by the environment). After detection of the damage by the sensors embedded in the structure, the central processor can build a diagnosis and a prognosis and decide of the actions to undertake (restriction of the operational domain to avoid overloading in the damaged area, and/or scheduling a condition-based inspection possibly followed by a repair).

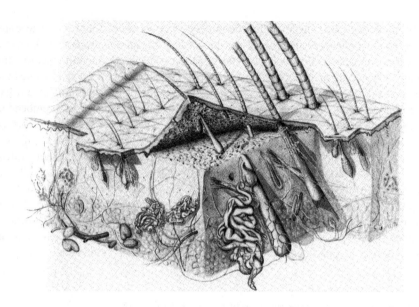

Figure 1.8. *Sketch of human skin showing the variety of sensors and actuators making it a really smart structure, taken from [MON 74]*

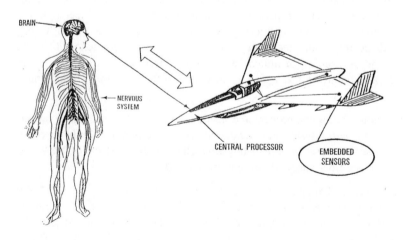

Figure 1.9. *Analogy between the nervous system of man and a structure with SHM, from [ROG 93]*

Biomimetics can help in finding new ideas, but we must avoid trying to copy nature as closely as possible, since we do not use the same materials or the same fabrication processes. For example, bio-inspiration had a strong influence on the strategy adopted by many researchers of supposing that it is mandatory to embed the sensors inside the materials of the structure. Such a choice is important since it has huge consequences for the development of practical systems. The fully embedded solution considerably complicates the technology needed, creating problems of different kinds: higher miniaturization is needed, demonstration of the innocuousness of the embedded sensor for the host structure has to be demonstrated, connectivities complicate the structure design and process, repair ability is problematic, redundancy of sensor networks is needed, the operational life of sensors must be at least as long as that of the structure, etc. The necessity for embedding the sensors is not obvious in most cases, and the drawbacks of surface-mounted sensors are often less critical than those of the more sophisticated solution. In such a case, this particular bio-inspiration could be a false "good idea".

1.5. Process and pre-usage monitoring as a part of SHM

Sensors for Health Monitoring can be incorporated into the components during the manufacturing process of the composite. Thus, in a global approach including the processing stage, the sensors can be used first of all to monitor the processing parameters in order to optimize the initial properties of the material. The physical parameters of the material that can be monitored during the process are varied: refractive index, visco-elastic properties, conductivity, etc. A range of techniques is available allowing their on-line monitoring: electrical techniques [KRA 91, PIC 99], electro-mechanical impedance techniques using embedded piezo-patches [JAY 97, GIU 03], acousto-ultrasonics (or optical techniques using fiber-optic sensors [CHA 01, DEG 02a]). It could be interesting to mix such different sensors achieving a multidetection [CHA 00].

For temperature during the process and inside the composite, once again various optical fiber-based sensor systems are available. These are predominantly based on fluorescence decay measurements [LIU 00], fiber Bragg gratings [LIU 98, DEW 99] or modified extrinsic fiber Fabry–Perot sensors [DEG 02b].

There is an intermediate phase of the life of a structure that can need SHM too: between the end of the manufacturing process and the beginning of the functioning phase, for certain structures, a lot of handling and transportation operations take place. During this phase, which could be called the pre-usage phase of the structure, accidental loads, not known by the end-user, may occur and threaten the structure's reliability. A good illustration of such a risk is given in [GUN 99]. On January 17, 1997, the Delta II mission 241 failed when the rocket exploded after a flight of 12.5

seconds, with the consequence that the first of the new set of Global Positioning Satellites (GPS) was lost– see Figure 1.10. The occurrence of damage, caused by a handling overload while the rocket was being transported by road before firing, was strongly suspected. The remedy consisted of equipping the structure with an SHM system that registered the shocks occurring during the full pre-usage phase.

Figure 1.10. *Delta II mission 241 explosion, from [GUN 99], a catastrophic failure which could have been avoided by pre-usage health monitoring – a) road transportation: the rocket is inside the trailer, here detached; b) Delta II liftoff; c) the explosion, initiated from a crack in one of the graphite epoxy motors situated at the base of the rocket*

For this type of SHM, it is easier to detect the possible damaging events than the damage that is thought to have been caused. The sensors can be resistive strain gauges or strain-sensitive fiber-optic sensors for the quasi-static loads and acoustic emission sensors for impact type loads.

1.6. SHM as a part of system management

Health Management can be defined as the process of making appropriate decisions/recommendations about operation, mission and maintenance actions, based on the health assessment data gathered by Health Monitoring Systems [REN 02].

Figure 1.11 presents the general organization of SHM, and how it is included into a Health Management System. This general presentation is independent of the considered application domain. In the present section, more details are given on the different elements of Health Management and the information fluxes. Although this is done by taking the aircraft customer domain as an example – based on a keynote lecture by R. Ikegami from the Boeing Company [IKE 99] – it is representative of the way that SHM can be integrated in more general Health Management systems, whatever the application domain. Structural usage and damage parameters are registered by sensors and used by on-board data acquisition and signal processing equipment. The data from the sensors are transformed into information, related to the structural usage, the environmental history and the resulting damage, thanks to a usage and damage Monitoring Reasoner, which contains information processing algorithms. Predictive Diagnostic Models and Prognostics Models feed a Life Management Reasoner, which converts the information delivered by the Usage/Damage Monitoring Reasoner into knowledge about the structural health of the aircraft. This knowledge is then communicated to an Integrated Vehicle Health Management (IVHM) system, which disseminates the information to the flight crew, the operations and maintenance services, the Regulatory Agencies and the Original Equipment Manufacturer. Thus, a condition-based approach to aircraft inspection and maintenance is possible.

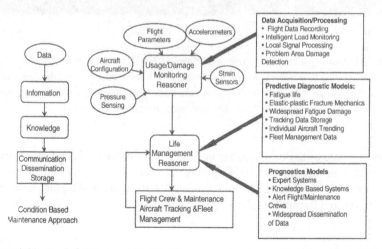

Figure 1.11. *Aircraft Structural Health Management system architecture, from [IKE 99]*

This approach has been refined in more recent papers. In particular, another person from the same company [GOG 03] gives a more comprehensive view of the interconnections between the various reasoners involved in the Structural Health Management architecture. In addition, a description of an IVHM system for air vehicles for the US Department of Defense (DoD) is given by Derriso [DER 03].

1.7. Passive and active SHM

SHM, like Non-Destructive Evaluation (NDE), can be passive or active. Figure 1.12 presents the possible situations in which both experimenter and examined structure are involved. The structure is equipped with sensors and interacts with the surrounding environment, in such a way that its state and its physical parameters are evolving.

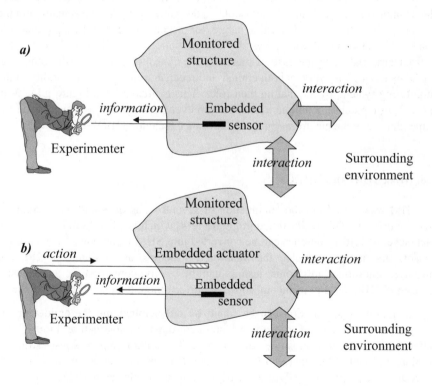

Figure 1.12. *The two possible attitudes of the experimenter defining:*
a) passive and b) active monitoring

If the experimenter is just monitoring this evolution thanks to the embedded sensors, we can call his action "passive monitoring". For SHM, this sort of situation is encountered with acoustic emission techniques detecting, for example, the progression of damage in a loaded structure or the occurrence of a damaging impact [DUP 99, STA 99].

If the experimenter has equipped the structure with both sensors and actuators, he or she can generate perturbations in the structure, thanks to actuators, and then, use sensors to monitor the response of the structure. In such a case, the action of the experimenter is "active monitoring". In the aforementioned example, the monitoring becomes active, by adding to the first piezoelectric patch, which is used as an acoustic emission detector, a second patch, which is used as an emitter of ultrasonic waves. The receiver, here, is registering signals, resulting from the interaction of these waves with a possible damage site, allowing its detection [WAN 99, LEM 00 b, PAG 02].

In classical NDE, the excitation is, generally, achieved using a device external to the examined structure, but the philosophy is the same. In SHM, the actuator and the sensor can be different or identical in nature, for instance, excitation by a piezoelectric patch and detection of the waves, by a fiber-optic sensor [LIN 02] or another piezoelectric patch. In the case of piezoelectric transducers, it is worth noting that the same device can work as both emitter and receiver, which gives flexibility to the monitoring system, by alternating their roles. This is illustrated in Figure 1.13. With piezoelectric patches, a unique transducer can even perform the two functions at the same time, as in the electromechanical impedance technique [BOI 02, GIU 03].

1.8. NDE, SHM and NDECS

SHM was born from the conjunction of several techniques and has a common basis with NDE. This is illustrated in Figure 1.14, which is taken from [CHA 99]. In fact, several NDE techniques can be converted into SHM techniques, by integrating sensors and actuators inside the monitored structure, as in Figure 1.12b). For instance, traditional ultrasonic testing can be easily converted in an acousto-ultrasonic SHM system, using embedded or surface-mounted piezoelectric patches.

An intermediate solution can be found by only embedding the emitter or the receiver, the other part of the system being kept outside the structure. Figure 1.15, taken from [WAL 99], illustrates this concept. The author calls it Non-Destructive Evaluation Ready Material (NDERM) concept. Perhaps a better denomination might be NDE Ready Structure (NDERS) or NDE Cooperative Structure (NDECS).

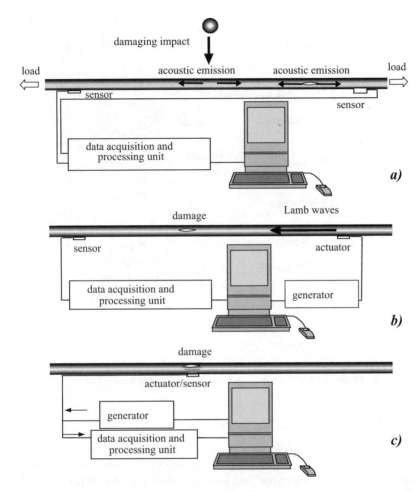

Figure 1.13. *Flexibility of monitoring techniques using piezoelectric patches: a) passive method: acoustic emission technique, b) active method: acousto-ultrasonic technique with generation of Lamb waves, c) active method: electromechanical impedance technique*

Such a solution is *a priori* interesting in two situations:

– when it is easy to position the emitter inside the structure, during the process, in a region where it is difficult, or impossible, to produce a stimulation from outside without demounting the structure;

– when it is possible to use for the detection a non-contact, full-field imaging system allowing rapid monitoring of a large part of the structure. This is possible, for instance, with techniques such as infrared thermography or shearography.

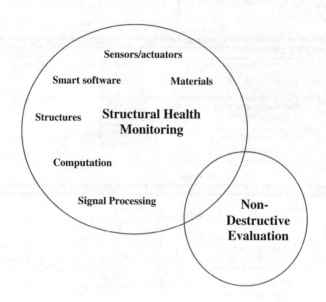

Figure 1.14. *The basic components of SHM, taken from [CHA 99]*

Several NDECS techniques are presently possible:

– by embedding in the material or bonding on the internal face of the structure a network of conductors, constituting a grid sensitive to the magnetic field, generated by an external electromagnetic antenna and crossing the structure made of conductive composite, such as carbon/epoxy composites. This technique has been proposed as an alternative to a fully integrated electromagnetic technique, such as the one called Hybrid Electromagnetic Layer Performing (HELP Layer) [LEM 00a];

– by using lock-in ultrasonic vibrothermography [ZWE 00] (a recent technique presently used in several laboratories) and generating ultrasounds, Lamb waves in particular [KRA 98], thanks to an embedded piezoelectric patch, the camera monitoring the surface thermal field produced by the interaction of the waves with the defect (delamination for instance);

– by using lock-in shearographic imaging of ultrasound, generated by an embedded piezoelectric patch, as demonstrated in [DUP 99, TAI 00].

Figure 1.16 presents images of Lamb waves, interacting with delaminations, obtained by the two existing NDECS techniques, mentioned above. Both images show the interaction of Lamb waves with a delamination caused by an impact in a carbon epoxy plate. The frequencies of the waves are 112 kHz for image 15a) and 68 kHz for image 15b). The waves are intensity modulated at a low frequency (0.033 Hz for thermography and 0.3 Hz for shearography). The full temperature scale of image 15a) is just 20 mK. Such sensitivity is possible thanks to the lock-in detection. Image 1.15b) shows the incident Lamb waves, generated by a circular piezoelectric patch, located near the upper right corner of the image and the diffraction, caused by the delamination, allowing a clear localization of the damage.

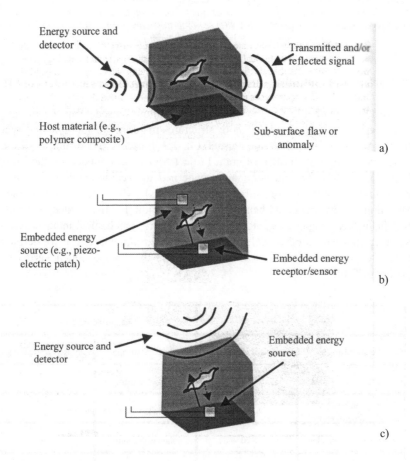

Figure 1.15. *NDE Cooperative Structures (NDECS), an intermediate solution between NDE and SHM, taken from [WAL 99]: a) conventional ultrasonic (surface contact) NDE; b) smart material with active and passive embedded sensors; c) NDECS: embedded elements improve the resolution and the depth of penetration of a conventional ultrasonic NDE system*

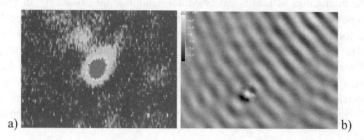

a)   b)

Figure 1.16. *Monitoring of an impact-generated delamination in a carbon-epoxy coupon using NDECS techniques: a) lock-in ultrasonic vibrothermography coupled with embedded piezo patch Lamb waves (from [KRA 98]); b) lock-in shearography coupled with embedded piezo patch Lamb waves (from [TAI 00])*

1.9. Variety and multidisciplinarity: the most remarkable characters of SHM

SHM is remarkable for the variety of techniques used. This is, in fact, the consequence of the diversity of both structures/materials to monitor and types of damage to detect. To illustrate this, Table 1.3 presents the results of a survey, carried out as part of the Brite-Euram Project MONITOR funded by the European Union [MON 95]. The questionnaire concerned the requirements of end-users in aeronautics (aircraft manufacturers and operators), who were asked to rank the inspection targets that could benefit from SHM solutions. If we extend the survey to other fields (civil engineering, atomic energy industry, mechanical industry, etc.) that use other materials at other scales and in other environments, and if we define more closely the possible damage to be detected, the diversity is much varied than it appears in Table 1.3.

Metallic structures		Composite structures	
Type of damage	End-users interested (%)	Type of damage	End-users interested (%)
Fatigue crack development	100	Impact damage	65
Corrosion bonding/ debonding of joints	82	Delamination	65
Stress corrosion cracking	70	Bonding-debonding	59
Impact damage	47		

Table 1.3. *Aeronautics end-users' needs that can be satisfied by SHM: percentage of respondents to the questionnaire who showed positive interest, from [BAR 97]*

To satisfy these needs, the variety of sensing techniques is tremendous; for a given type of damage, several techniques can be satisfactorily applied. The sensors which can be used are based on various physical phenomena and are made of very different materials. Although not considering the optical sensors, Table 1.4 shows the diversity of physical phenomena and sensor materials that can be used. All this explains why SHM, like all research in SMS, is eminently multidisciplinary.

Physical effect	Materials	
	Polymers	**Inorganics**
Passive sensors		
Piezoelectricity	Polyvinylidene fluoride Polyvinylidene fluoride trifluoroethylene Polyhydroxibutyrate Liquid crystalline polymers (flexoelectricity)	Piezoelectric zirconate titanate Zinc oxide Quartz
Pyroelectricity	Polyvinylidene fluoride Langmuir-Blodgett ferroelectric superlattices	Triglycine sulfate Lead-based lanthanum-doped zirconate titanate Lithium tantalate
Thermoelectricity (Seebeck effect)	Nitrile-based polymers Polyphthalocyanines	$Cu_{100}/Cu_{57}Ni_{43}$ Lead telluride Bismuth selenide
Photovoltaic	Polyacetylene/n-zinc sulfide Poly(N-vinyl carbazole)+merocyanine dyes Polyaniline	Silicon Gallium arsenide Indium antimonide
Electrokinetic	Polyelectrolyte gel ionic polymers	Sintered ionic glasses
Magnetostriction	Molecular ferromagnets	Nickel Nickel-iron alloys
Active sensors		
Piezoresistivity	Polyacetylene Pyrolized polyacrylonitrile Polyacequinones Polyaniline Polypyrrole Polythiophene	Metals Semiconductors
Thermoresistivity	Poly(p-phenylene vinylene)	Metals Metal oxides Titanate ceramics Semiconductors
Magnetoresistivity	Polyacetylene Pyrolized polyvinylacetate	Nickel-iron alloys Nickel-cobalt alloys
Chemioresistivity	Polypyrrole Polythiophene Ionic conducting polymers Charge transfer complexes	Palladium Metal oxides Titanates Zirconia
Photoconductivity	Copper phthalocyanines Polythiophene complexes	Intrinsic and extrinsic (doped) semiconductors

Table 1.4. *Materials for sensor design, from [DER 02]*

The type of sensors used to monitor structural health is heavily dependent on the types of structures that are to be monitored. Figure 1.17 presents the main types of sensors used in the two main fields of application: civil and aerospace engineering. These statistics are based on the communications given in the first two International Workshops on SHM held at the University of Stanford (1997 and 1999).

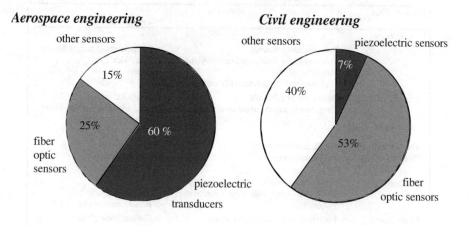

Figure 1.17. *The main types of sensors used for SHM, depending on the types of application: comparison between aerospace engineering and civil engineering, statistics based on the communications to the 1st and 2nd International Workshop on SHM at Stanford in 1997 and 1999, taken from [BAL 01a]*

For a narrower field defined by a specific type of structure and a specific type of damage (the SHM of composite structures with delaminations), the distribution of the various types of methods is given in Figure 1.18, for the period 1997–2003. In this case, the statistics are based on a wider bibliographical database of almost 1150 references. The distribution is similar to the one found for aerospace engineering (see Figure 1.17). As shown in Figure 1.18, in each family of sensors, there is a wide variety of specific sensors.

If we now consider the methods of monitoring, independently of the type of sensor used, again a very wide variety exists. This is shown in Figure 1.19, which presents the distribution of monitoring methods used in the same references.

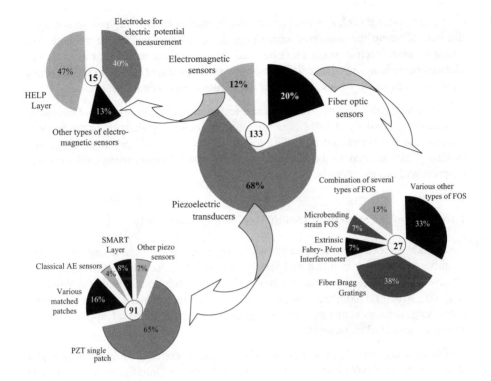

Figure 1.18. *Type of sensors used in the references relating to the monitoring of delaminations in composite structures. Statistics based on a general survey of SHM literature for the period 1997–2003 (1150 references)*

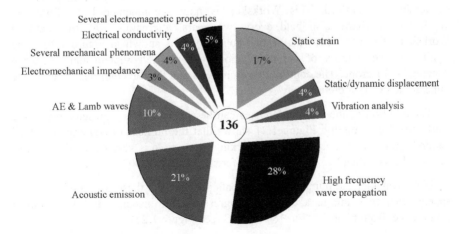

Figure 1.19. *Methods used for the monitoring of delaminations in composite structures*

In addition, for each specific sensor, several methods exist, varying according to the way of using the sensor or identifying the characteristics of the damage. For example, piezoelectric patches can be used for monitoring techniques as varied as electromechanical impedance, acoustic emission, propagation of high-frequency waves such as Lamb waves, analysis of random or modal vibrations, etc.

Considering this variety of sensors and techniques, the authors did not choose to present the state of art of SHM, according to the fields of application or the types of materials or structures, but according to the general types of sensors and their use instead. This seemed to be the best way to write homogeneous and coherent chapters without redundancy.

1.10. Birth of the Structural Health Monitoring Community

SHM is a recent field of research which appeared in the final decade of the last century. As seen above, it is a multidisciplinary domain. Researchers and engineers involved in it come from various more classical disciplines: structural vibration analysis, structural control, non-destructive evaluation, materials science, signal processing, sensors/actuators technology, etc. The main fields of application are aerospace and civil engineering.

Papers were initially presented at NDE, structural control conferences and later at conferences, devoted to the new topic of Smart or Intelligent Materials and Structures (for instance, the SPIE Conferences on Smart Structures and Materials).

The need for conferences totally devoted to SHM gave birth in 1997 to the International Workshop on SHM (IWSHM), created by Professor Fu-Kuo Chang. Since that date, the IWSHM has taken place every two years at the same place, the University of Stanford. This Workshop is now accompanied by a European Workshop on SHM, also held every two years, alternating with the Stanford Workshop. The usefulness of the two conferences is shown in Figure 1.20 – the origins of the authors of communications are really complementary between America and Europe. The contribution from Asia and Pacific countries is nearly the same at the two conferences.

The addition of the two conferences can be considered as representative of the world SHM community. Figure 1.21 shows that the number of countries in which research groups and/or industries are working in the field is very large: 37 countries sent speakers or attendees to these two conferences in 2002-2003.

If we consider the number of documents presented at these two conferences, it appears that they are in a constant progression, for instance, for the Stanford Conference, from 65 in 1997 to 184 in 2003 (see Figure 1.22).

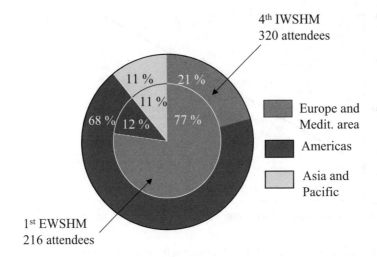

4th IWSHM
320 attendees

11 %   21 %

11 %

68 %   12 %   77 %

Europe and
Medit. area

Americas

Asia and
Pacific

1st EWSHM
216 attendees

Figure 1.20. *Respective attendance at the European Workshop on SHM (1st EWSHM, Cachan 2002) and at the International Workshop on SHM (4th IWSHM, Stanford, 2003)*

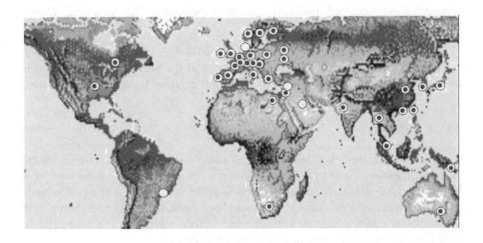

Figure 1.21. *World distribution of speakers and attendees at the fourth IWSHM (2003) and first EWSHM (2002) considered as a whole. White disks correspond to countries sending attendees and disks with a black center to countries sending speakers*

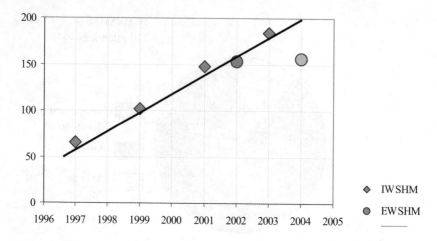

Figure 1.22. *Development of the SHM community: number of papers presented at the Workshops on SHM (IWSM and EWSHM). This evaluation is based on the published proceedings*

A parallel evolution has occurred for scientific journals. At the beginning, papers on SHM were mainly published in the two reviews, devoted to the more general topic of smart structures and materials: *Smart Materials and Structures* and *Intelligent Materials Systems and Structures*. For the last two years, however, a new journal entirely devoted to SHM has been published: the *Structural Health Monitoring International Journal*.

This evolution and the statistics, as mentioned above, prove the existence of an expanding community, illustrating the pertinence of the SHM philosophy. There is no doubt that the increase of activity on SHM will continue for some time to come.

1.11. Conclusion

In the context described in the previous paragraph, the need for courses on the subject is obvious. New researchers and engineers are attracted by the topic. For beginners, the field is rather difficult, because of the multidisplinarity of the subject. This is the reason for publishing the present book, which is based on a short course of continuing education, given at the Ecole Normale Supérieure of Cachan (Paris) by a group of specialists from several European countries.

The book is structured according to the types of monitoring techniques, independent of the fields of application. In each chapter, both sensing techniques and related data reduction techniques are described. Each chapter is conceived as an introduction to the monitoring technique, giving a large list of references, through which the reader can continue to explore the state of the art and the potential applications.

1.12. References

[BAL 01a] BALAGEAS D.L., "Le contrôle de santé de structure integer", in *Systèmes et microsystèmes pour la caractérisation*, Proceedings C2I Conference, vol. 2, Ed. F. Lepoutre, D. Placko and Y. Surrel, Hermes Sciences Publications, Paris 2001, pp. 14-43.

[BAL 01b] BALAGEAS D.L., "Structural Health Monitoring R & D at the European Research Establishments in Aerospace (EREA)", in *Structural Health Monitoring – The Demands and Challenges, Third International Workshop on Structural Health Monitoring*, September 12-14, 2001, Stanford, CA, ed. Fu-Kuo Chang, CRC Press, Boca Raton, London, New York, Washington D.C., 2001, pp.12-29.

[BAR 97] BARTELDS G., "Aircraft structural health monitoring, prospects for smart solutions from a European viewpoint", *Structural Health Monitoring – Current Status and Perspectives, Proceedings of the First International Workshop on Structural Health Monitoring*, Stanford, CA, September 18-20, 1997, Lancaster – Basel, Technomic Publishing Co, Inc, pp. 293-300.

[BER 03] BERAL B., SPECKMANN H., "Structure Health Monitoring (SHM) for aircraft structures: a challenge for system developers and aircraft manufacturers", Structural Health Monitoring 2003, from *Diagnostics & Prognostics to Structural Health Management, Proceedings of the 4th International Workshop on Structural Health Monitoring*, Stanford, CA, September 15-17, 2003, Lancaster, PA, DEStech Publications, Inc, pp. 12-29.

[BOI 02] BOIS C., HOCHARD C., "Measurement and modelling for the monitoring of damaged laminate composite structure", Structural Health Monitoring 2002, *Proceedings of the First European Workshop on Structural Health Monitoring*, Cachan, France, July 10-12, 2002, Lancaster, PA, DEStech Publications, Inc, pp. 425-432.

[CHA 00] CHAILLEUX E., SALVIA M., JAFFREZIC-RENAULT N., JAYET Y., MAZZOUZ A., SEYTRE G., "*In situ* multidetection cure monitoring of an epoxy-amine system", *Journal of Advanced Sci.*, 12, 3, 2000, pp. 291-297.

[CHA 01] CHAILLEUX E., SALVIA M., JAFFREZIC-RENAULT N., MATEJEC V., KASIC I., "*In situ* study of the epoxy cure process using a fiber optic sensor", *Smart Materials and Structures*, 10, 2001, pp. 1-9.

[CHA 99] CHANG F.-K., "Structural Health Monitoring: a summary report on the First International Workshop on Structural Health Monitoring, September 18-20, 1997", Structural Health Monitoring 2000, *Proceedings of the Second International Workshop on Structural Health Monitoring*, Stanford, CA, September 8-10, 1999, Lancaster – Basel, Technomic Publishing Co, Inc, pp. xix-xxiv.

[CHA 02] CHANG F.-K., "Ultra reliable and super safe structures for the new century", Structural Health Monitoring 2002, *Proceedings of the First European Workshop on*

Structural Health Monitoring, Cachan, France, July 10-12, 2002, Lancaster, PA, DEStech Publications, Inc, pp. 3-12.

[CHO 99] CHOI Y.K., SALVIA M., "Processing and modeling of adaptive glass-epoxy laminates with embedded shape memory alloys", Proceedings of the 10th International Conference on Adaptive Structures and Technologies, Paris, France, 1999, Lancaster-Basel, Technomic Publishing Co, Inc., pp. 221-228.

[DEG 02a] DEGAMBER B., FERNANDO G.F., "Process monitoring of fiber reinforced polymer composites", *Materials Research Bulletin*, Special issue on optical fiber sensors, March 2002.

[DEG 02b] DEGAMBER, B., DUMITRESCU O., FERNANDO G.F., "Microwave processing of thermosets: non-contact cure monitoring and fiber optic temperature sensors", Int. Conf. Fiber Reinforced Composites 2002, Newcastle, March 26-28, 2002, pp. 416-423.

[DER 02] DE ROSSI D., CARPI F., LORUSSI F., MAZZOLDI A., SCILINGO P., TOGNETTI A., "Electroactive polymer fibers and fabrics for distributed, conformable and interactive systems", Structural Health Monitoring 2002, *Proceedings of the First European Workshop on Structural Health Monitoring*, Cachan, France, July 10-12, 2002, Lancaster, PA, DEStech Publications, Inc, pp. 106-114.

[DER 03] DERRISO M.M., PRATT D.M., HOMAN D.B., SCHROEDER J.B., BORTNER R.A., "Integrated Vehicle Health Management: the key to future aerospace systems", Structural Health Monitoring 2003, from *Diagnostics & Prognostics to Structural Health Management, Proceedings of the 4th International Workshop on Structural Health Monitoring*, Stanford, CA, September 15-17, 2003, Lancaster, PA, DEStech Publications, Inc, pp. 3-11.

[DEW 99] DEWYNTER-MARTY V., FERDINAND P., BOCHERENS E., CARBONE R., BERENGER H., BOURASSEAU S., DUPONT M., BALAGEAS D., "Embedded fiber Bragg grating sensors for industrial composite cure process monitoring", *Journal of Intelligent Material Systems and Structures*, 9, 10, 1999, pp. 785-787.

[DRY 94] DRY C., "Timed release of chemicals in cementitious material after the material has hardened to repair cracks, rebond fibers, and increase flexural toughening", *Fracture Mechanics* 25th vol. ASTM. STP 1220, Philadelphia, 1994, pp. 123-127.

[DRY 96] DRY C., "Procedures developed for self repair of polymer matrix composites materials", *Composite Structures*, 35, 1996, pp. 263-269.

[DUP 99] DUPONT M., OSMONT D., GOUYON R., BALAGEAS D.L., "Permanent monitoring of damaging impacts by a piezoelectric sensor based integrated system", Structural Health Monitoring 2000, *Proceedings of the Second International Workshop on Structural Health Monitoring*, Stanford, CA, September 8-10, 1999, Lancaster-Basel, Technomic Publishing Co, Inc, pp. 561-570.

[FRA 03] FRANGOPOL D.M., "New directions and research needs in life-cycle performance and cost of civil infrastructures", Structural Health Monitoring 2003, from *Diagnostics & Prognostics to Structural Health Management, Proceedings of the 4th International*

Workshop on Structural Health Monitoring, Stanford, CA, September 15-17, 2003, Lancaster, PA, DEStech Publications, Inc, pp. 53-63.

[GAN 92] GANDHI M.V., THOMPSON B.S., *Smart Materials and Structures*, Chapman and Hall, 1992.

[GIU 03] GIURGIUTIU V., "Embedded Ultrasonics NDE with Piezoelectric Wafer Active Sensors", *Instrumentation, Mesure, Métrologie*, 3, 3-4, 2003, pp. 149-180.

[GOG 03] GOGGIN P., HUANG J., WHITE E., HAUGSE E., "Challenge for SHM transition to future aerospace systems", Structural Health Monitoring 2003, from *Diagnostics & Prognostics to Structural Health Management, Proceedings of the 4th International Workshop on Structural Health Monitoring*, Stanford, CA, September 15-17, 2003, Lancaster, PA, DEStech Publications, Inc, pp. 30-41.

[GOR 97] GORANSON U.G., "Jet transport structures performance monitoring", *Structural Health Monitoring – Current Status and Perspectives, Proceedings of the First International Workshop on Structural Health Monitoring*, Stanford, CA, September 18-20, 1997, Lancaster-Basel, Technomic Publishing Co, Inc, pp. 3-17.

[GUN 99] GUNN III L.C., "Operational Experience with Health Monitoring on the Delta II Program", Structural Health Monitoring 2000, *Proceedings of the Second International Workshop on Structural Health Monitoring*, Stanford, CA, September 8-10, 1999, Lancaster – Basel, Technomic Publishing Co, Inc, pp. 133-141.

[IKE 99] IKEGAMI R., "Structural Health Monitoring: assessment of aircraft customer needs", Structural Health Monitoring 2000, *Proceedings of the Second International Workshop on Structural Health Monitoring*, Stanford, CA, September 8-10, 1999, Lancaster – Basel, Technomic Publishing Co, Inc, pp. 12-23.

[JAY 97] JAYET Y., BABOUX J.C., "Monitoring the cycle of life of polymer based composites by an embedded piezoelectric element", *Proceedings of the 8th Int. Conference on Adapt. Structures*, Wakayama, Y. Murotsu *et al.* Eds, 1997, pp. 177-183.

[KRA 91] KRANBUEHL, D.E., "Continuous dielectric measurement of polymerizing systems", *Journal of Non-Crystalline Solids*, 131 part 2, 1991, pp. 930-934.

[KRA 98] KRAPEZ J.-C., TAILLADE F., GARDETTE G., BALAGEAS D., "La vibrothermographie par ondes de Lamb: vers une nouvelle méthode de CND?", (in French), *Journée Soc. Franç. des Thermiciens*, March 31, 1998, Châtillon.

[LEM 00a] LEMISTRE M., MARTINEZ D., BALAGEAS D.L., "Electromagnetic structural health monitoring for carbon-epoxy multilayer materials", *Proceedings of European COST F3 Conference*, Ed. J. A., Güemes, Madrid, Spain, 2000, pp. 687-695.

[LEM 00b] LEMISTRE M., OSMONT D., BALAGEAS D.L., "Active health monitoring system based on wavelet transform analysis of diffracted Lamb waves", *SPIE Proceedings*, vol. 4073, 2000, pp. 194-202.

[LIN 02] LIN M., POWERS W.T., QING X., KUMAR A., DEARD S.J., "Hybrid piezoelectric/fiber optic SMART layers for Structural Health Monitoring", Structural Health Monitoring 2002, *Proceedings of the First European Workshop on Structural*

Health Monitoring, Cachan, France, July 10-12, 2002, Lancaster, PA, DEStech Publications, Inc, pp. 641-648.

[LIU 98] LIU T., FERNANDO G.F., RAO Y.J., JACKSON D. A., ZHANG L., BENNION I., "Simultaneous strain and temperature measurements using a multiplexed fiber Bragg grating sensor and an extrinsic Fabry-Perot sensor", *Journal of Smart Structures and Materials*, 7, 1998, pp. 550-556.

[LIU 00] LIU T., FERNANDO G.F., ZHANG Z., GRATTAN K.T.V., "Simultaneous strain and temperature measurements in composites using an extrinsic Fabry-Perot sensor and a rare-earth doped fiber", *Sensors and Actuators-A Physical*, 80, 3, 2000, pp. 208-215.

[MON 95] MONITOR BRITE-EURAM Project No.: BE 95-1524, 1995.

[MON 74] MONTAGNA, W., RARAKKAL P.F., *The Structure and Function of Skin*, Academic Press, Inc., New York, 1974.

[MOT 99] MOTOKU M., VALDYA U.K., JANOWSKI G.M., "Parametric studies on self-repairing approaches for resin infused composites subjected to low velocity impact", *Smart Materials and Structures*, 8, 1999, pp. 623-638.

[OTT 88] OTT J., O'LONE R.G., "737 fuselage separation spurs review of safeguards", *Aviation Week and Space Technology*, 1988, May 9, pp. 92-95.

[PAG 02] PAGET C.A., GRONDEL S., LEVIN K., DELEBARRE C., "Damage detection in composite by a wavelet-coefficient technique", Structural Health Monitoring 2002, *Proceedings of the First European Workshop on Structural Health Monitoring*, Cachan, France, July 10-12, 2002, Lancaster, PA, DEStech Publications, Inc, pp. 313-320.

[PIC 99] PICHAUD S., DEUTEUTRE X., FIT A., STEPHAN F., MAAZOUZ A., PASCAULT J.P., "Chemorheological and dielecric study of epoxy-amine for processing control", *Polymer Intern.*, 48, 1999, pp. 1205-1218.

[REN 02] RENSON L., "Health Monitoring Systems for future reusable launchers", Structural Health Monitoring 2002, *Proceedings of the First European Workshop on Structural Health Monitoring*, Cachan, France, July 10-12, 2002, Lancaster, PA, DEStech Publications, Inc, pp. 65-75.

[ROG 93] ROGERS C.A., Intelligent Material Systems and Structures, Report from the Center for Intelligent Material Systems and Structures, Program Div. of Technomic Publishing Co, Inc, Lancaster, PA, 1993.

[STA 99] STASZEWSKI W.J., BIEMANS C., BOLLER C., TOMLINSON G.R., "Impact damage detection in composite structures – recent advances", Structural Health Monitoring 2000, *Proceedings of the Second International Workshop on Structural Health Monitoring*, Stanford, CA, September 8-10, 1999, Lancaster-Basel, Technomic Publishing Co, Inc, pp. 754-763.

[TAI 00] TAILLADE F., KRAPEZ J.-C., LEPOUTRE F., BALAGEAS D., "Shearographic visualization of Lamb waves in carbon epoxy plates interation with delaminations", *Eur. Phys. J.*, AP 9, 2000, pp. 69-73.

[TAK 00] TAKEDA N., "Development of structural health monitoring systems for smart composite structure systems", *Proceedings of the 11th International Conference on Adaptive Structures and Technologies*, Nagoya, Japan, 2000, pp. 269-276.

[WAL 99] WALSH S.M., "Practical issues in the development and deployment of intelligent systems and structures", Structural Health Monitoring 2000, *Proceedings of the Second International Workshop on Structural Health Monitoring*, Stanford, CA, September 8-10, 1999, Lancaster–Basel, Technomic Publishing Co, Inc, pp. 553-560.

[WAL 03] WALSH S.M., "A requirements-based approach to Structural Health Monitoring research, development, and application", Structural Health Monitoring 2003, from *Diagnostics & Prognostics to Structural Health Management, Proceedings of the 4th International Workshop on Structural Health Monitoring*, Stanford, CA, September 15-17, 2003, Lancaster, PA, DEStech Publications, Inc, pp. 79-87.

[WAN 99] WANG C.S., CHANG F.-K., "Built-in diagnostics for impact damage identification of composite structures", Structural Health Monitoring 2000, *Proceedings of the Second International Workshop on Structural Health Monitoring*, Stanford, CA, September 8-10, 1999, Lancaster–Basel, Technomic Publishing Co, Inc, pp. 612-621.

[WAN 97] WANG M.L., SATPATHI D., HEO G., "Damage detection of a model bridge using modal testing", *Structural Health Monitoring – Current Status and Perspectives, Proceedings of the First International Workshop on Structural Health Monitoring*, Stanford, CA, September 18-20, 1997, Lancaster–Basel, Technomic Publishing Co, Inc, pp. 589-600.

[YOS 96] YOSHIDA H., FUNAKI A., YANO S., "On the response and the responsive shape control of environmentally responsive composite with embedded Ti-Ni alloy as effectors", *Proceedings of the Third International Conference on Intelligent Materials*, Lyon, France, SPIE Proc. vol. 2779, 1996, Wilmington (PA), pp. 523-529.

[ZWE 00] ZWESCHPER T.H., DILLENZ A., BUSSE G., "Ultrasound lockin thermography – a NDT method for the inspection of aerospace structures", Proceedings of the 5th Conference on Quantitative Infrared Thermography (QIRT'2000), Reims (France), July 18-21, 2000, Lodz (Poland), Akademickie Centrum Graficzno-Marketingowe Lodart S.A., pp. 212-217.

Chapter 2

Vibration-Based Techniques for Structural Health Monitoring

2.1. Introduction

The safety of structures is one of the main issues in the consideration of aging bridges and buildings, aircraft or other aging structures. Structural Health Monitoring (SHM) methodology provides an interesting tool for the continuous monitoring of technical structures. The application of SHM methods provides many opportunities for prolonging the life of a structure by the early detection of damage. Besides the aspect of safety, the increased knowledge about the state of health of the structure can lead to an improved design of structures. During the 1994 Northridge earthquake in the USA, many damaged office buildings were supposed to be quake-proof. Cracks and breaks were found after a closer visual inspection. One of the consequences was the improvement of standards for earthquake-resistant design of structures. On the other hand, it also makes sense to develop methods that enable an immediate assessment of the safety of a structure after such a seismic event. In civil engineering, visual inspection is the most common method but it is labor-intensive and time-consuming. Disassembly of secondary parts is often required to gain access to the important load-bearing structural elements. Therefore, it is highly desirable to have a tool that will immediately answer questions of safety.

Other examples can be found in aeronautics industry. In one case, access to safety-critical structural elements during scheduled inspections, involving extensive disassembly, is an immense cost factor. Thus, alternative inspection methods are

---

Chapter written by Claus-Peter FRITZEN.

needed. In other cases, damage can only be detected at a rather late stage. Considerable repair and maintenance are the consequences. The possibility of monitoring structures by using integrated sensor networks, not only at the scheduled inspection intervals but on demand, improves the situation by discovering damage at a much earlier stage and continuously tracing further development of the damage. This clearly has a large economic impact, improves the safety of the structure and can lead to develop new principles in design. The latter is a new and very challenging idea, presented by Schmidt *et al.* [SCH 04]. When the damage state can be controlled, the expected weight reduction results from modifying today's damage tolerance concepts; in other words, less stringent damage scenarios may be assumed when SHM is globally applied [SCH 04].

The use of laminated high-performance reinforced plastics in aero-engineering, for example, introduces another factor of serious damage mechanism: the delamination of the plies by sudden impact, which causes tool drop, debris on the runway or bird strike. Delaminations weaken the structure and are undetectable by visual inspection; great care has to be taken to detect them through labor-intensive ultrasonic methods, which can only be carried out by persons with significant practical experience.

To contrast with these sudden overload situations, aging occurs in the structures, subjected to continuous vibrations at moderately low amplitudes. These vibrations can cause fatigue in the material and the initiation of fatigue cracks, especially at points of stress concentration, such as notches and material inhomogeneities like welded joints. The progressive character of crack growth in metals, under cyclic or stochastic loading, is clearly recognizable by fracture mechanics. For example, schedule-driven inspections, depending on the time of operation or fixed time spans, may lead to catastrophic failures, when the evolution of the damage between two inspections reaches a critical level. This situation is schematically presented in Figure 2.1.

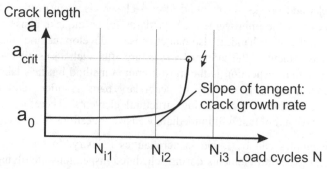

Figure 2.1. *Schematic picture of exponential crack growth under cyclic loading with constant vibration amplitude*

Under constant loading conditions, crack growth obeys an exponential law. At the beginning, a relatively small crack of size a_0 is tolerated, and after a first inspection at N_{i1} load cycles, it is discovered that the crack has only increased a bit; therefore, we apparently do not have to worry. Almost the same situation can be found after the second inspection. But then, the crack growth rate increases exponentially so that a critical crack length is reached, which causes unstable fracture and hence the failure of the structural part with possibly catastrophic consequences.

Usually, technical structures are designed for a certain maximum load level and many are designed for a limited lifetime. Both conditions may be violated during long times of operation, because the traffic load of a bridge increases over time or the operational load of a machine is increased due to production demands, for example.

Installation of large off-shore wind energy plants opens up a new field of application in SHM. There is an increased need for better remote monitoring because access is much more difficult compared to on-shore plants. Another challenge is the increased structural loading due to waves, strong wind and corrosion. For on-shore plants, the main focus of monitoring is on the machine part, especially the bearings and gears as well as the rotor blades. The off-shore technology requires additional monitoring of the tower and foundations as an integrated concept.

Over the last two decades, the interest in developing techniques for continuous long-term online monitoring of civil engineering structures, aircraft structures, railway systems, wind power stations or other machinery has greatly increased. Instead of inspecting the system at fixed intervals, we want to monitor its "health" in service on demand, by continuously extracting characteristic features which allow conclusions to be drawn on the structure's useful remaining lifetime, the time interval to the next maintenance/repair, or the need for immediate shutdown. In the abstract of [CHA 99], Chang has listed various research topics and applications in SHM as well as economic aspects. In the case of SHM, a structure with an integrated sensor and actuation system, and some kind of computer intelligence which processes the measured signals are needed. However, additional actuation systems usually mean additional costs. An interesting option is therefore to use the external power of the wind, traffic load, etc. The structure is then able to perform an automated self-diagnosis.

According to Rytter [RYT 93] (see Figure 2.2), we can proceed with a four-level damage assessment scale if we want to obtain more and more information on the damage. While level I only gives the information that the damage is present in the structure, level II provides the location(s) of the damage. At level III, the extent of the damage is evaluated. Therefore, a parametric model is needed to describe the damage (crack length, the size of a delamination, stiffness decrease or else a loss of mass). It should be mentioned that the determination of the type of damage is sometimes included as an extra step between level II and III [SOH 04]. The highest and most sophisticated level is the prognosis of the remaining lifetime. This requires the combination of the global structural model with local continuum damage models or fracture mechanics models, which are able to describe the evolution of damage or fatigue crack growth (see section 2.3). Approaches which deal with the reliability of systems, based on material damages (linear damage accumulation, fatigue damage, crack growth, etc.) and remaining lifetime considerations can be found in [CHE 01, CHE 03, WED 01, WAL 02, PEI 02, SOE 01, WOL 03], for example. Recently, Los Alamos National Laboratory has started an initiative on damage prognosis. The results of this workshop are described in [FAR 03].

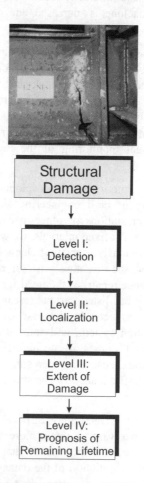

Figure 2.2. *The four diagnostic levels in structural damage assessment [RYT 93]*

There are still many unsolved problems at every level. For example, at the lowest level of damage detection, the challenges are to increase the sensitivity, detect small amounts of damage in an early state without getting false alarms, and separate the effects of the damage from the effects of the changes in environmental conditions.

Literature overviews can be found in [DOE 96, SOH 04, NAT 91, NAT 97, FRIS 97b, STA 04b, WOR 04, ZIM 92, AUW 03, MUE 03, CHAU 94, JAN 99]. Among others, the workshops on SHM [NAT 88, NAT 93, CHA 97, CHA 99, CHA 01, CHA 03, BAL 02, BOL 04], as well as the DAMAS and the SPIE conferences, provide a good insight into the current range of activities across the whole field. The publications [ISE 84, FRA 94, BAS 93, GER 98, GUS 00] discuss fault detection and isolation methods from the viewpoint of system theory and automation.

2.2. Basic vibration concepts for SHM

The principle which uses vibrations as characteristics of solid bodies to test their quality or consistency and distinguish good or bad conditions is widely used in daily life. For example, when buying drinking glasses, it is commonly accepted that the bright sound indicates a flawless glass, and a dull tone refers to something that is different about a piece. Traditionally, experienced traders tested the quality of large cheeses by knocking on them and listening to the sound. The degree of ripeness of melons can also be tested by a similar procedure. Many other examples can be found.

The same basic principle can also be applied to technical structures. The effects of material defects, like cracks or delaminations, on the dynamic behavior of a structure have been studied in [CAW 79, GUD 82, GUD 83, OST 90, LUO 00, OST 01], for example. The classical way to gain information on the measurement of an existing structure (with respect to its dynamic behavior) is to retrofit this structure with a set of sensors, excite it with an actuator or use the natural forces of the environment to process the data on computers (see the left-hand side of Figure 2.3). Motivated by concepts of smart structure, modern approaches to SHM include sensors, actuators and computational intelligence, which process and control the information from the early design phase to an integral part of the structure. That way, we try to mimic the biological system, composed of a nervous system (sensing), muscles (actuation) and a brain (processing, storing and recalling information). Communication between the smart structure and the environment can be achieved via data networks, such as the internet (see the right-hand side of Figure 2.3).

To see if something has changed, the measurement data have to be evaluated and some characteristic features, such as means, variances, maximum/minimum values, spectral information, etc., have to be extracted by pure signal analysis. We can compare the actual values with the reference values which derive from the undamaged state of the monitored system, and apply the time or frequency domain methods. Recently, combined time–frequency analysis, also called wavelet analysis, has become an important tool in data analysis [STA 04a]. Time–frequency methods can be used to analyze non-stationary events.

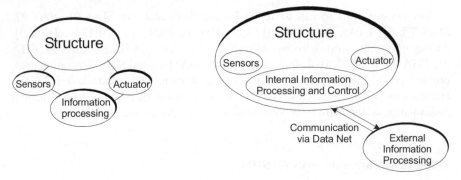

Figure 2.3. *Classical and modern SHM concepts; the modern concept has an intelligent structure with integrated sensors, actuators and processors*

The use of low-frequency excitation for SHM presents the advantage that the whole structure is excited to perform global motion. On one hand, the wavelength is usually larger than the local flaw, and the sensitivity for detection of incipient damage may be low. High-frequency excitation, on the other hand, yields higher sensitivity because the wavelength is smaller. However, vibration amplitudes rapidly decay when moving away from the excitation point. Thus, a much denser network of sensors is needed to cover a wider range of the structure. In SHM applications, piezoelectric elements are used for actuation and sensing because they can be surface-mounted or embedded into the material and are suitable for high-frequency excitation.

The impedance method is a very sensitive method for working in the higher frequency range (typically >30 kHz); changes in the impedance spectrum of the electromechanical system as a result of damage are compared. The impedance is determined from the voltage and the current of the piezoelectric actuator. The idea is that the damage in the structure changes its impedance and therefore the overall impedance of the coupled electromechanical system. A good overview is given by Park *et al.* [PAR 03] and further developments are described in [LOP 00, LOP 01, PAR 00a, PAR 00b, POH 01, ZAG 01].

Similar results can be found with the Stochastic Subspace Damage Detection Method [BAS 00, BAS 01] when operated in the higher frequency range [FRI 03b].

This method is described later in the chapter. Instead of spectral information, the correlation functions are calculated and a damage indicator is derived from these results.

However, it can be very difficult to link the physical interpretation of the results with the evaluation of the measurement data, if it is not sometimes impossible to gain such results. Thus in most cases, pure signal analysis only solve the detection problem (level I), but for the higher levels, additional information is needed. The use of quantitative or qualitative models to interpret the measurement data opens up the horizon for a much wider field of application in damage diagnosis. The group of quantitative models comprises all types of mathematical models. For example, structural models describe the static or dynamic behavior of the mechanical system, as well as the more generalized state space models known from system dynamics, which couple other engineering fields (hydraulics, magnetics, etc.). In control theory, the use of AR, ARX, ARMA models and other types within this family is widespread [GER 98, GUS 00, PIO 93, AND 97]. Independently from the type of model, it is generally admitted that the model is a duplicate of the real system, reproducing the same output, in the ideal case where model and real system are fed by the same input data. This provides an analytical redundancy and, as soon as the real system changes but the model does not, one can compare the two different outputs and draw conclusions from these so-called residuals. This will be worked out in more detail below. By using qualitative models, knowledge is cast into rules (if … then … rules), the measured data are processed, some characteristic features extracted, and the basis of the rules decided on the type of problem, which might have happened. This is the usual way to process information when problems occur, with a malfunctioning car for example.

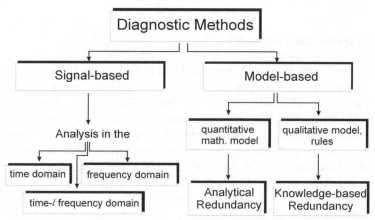

Figure 2.4. *Distinction between signal- and model-based methods*

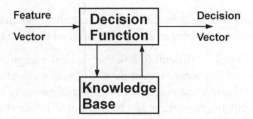

Figure 2.5. *Decision making supported by a knowledge base*

As suggested by Sohn *et al.* [SOH 01], Staczewski and Worden [STA 04a] and Worden [WOR 04], the damage diagnosis problem can also be considered in terms of pattern recognition. To solve this task, artificial neural networks are used. Neural networks can be trained in a supervised or unsupervised learning mode [BIS 95]. Supervised learning mode means that the network is trained by a set of target values from the network outputs. For each set of input patterns (such as a time series or characteristic features extracted from the raw data), a desired network output is specified (which is characteristic of a particular damage scenario). By adapting the network to the defined input–output relations (learning), the network is able to recognize certain damage scenarios when applied to measured data of the structure. Unsupervised learning does not involve the use of target data. Instead of learning an input–output mapping, the goal may be to model the probability distribution of the input data or to discover clusters in the data [BIS 95].

The methods which are used in the context of Outlier Analysis, Extreme Value Statistics and Sequential Probability Ratio Tests are described in [WOR 02b].

2.2.1. *Local and global methods*

Basically, we can distinguish between local and global methods. Local methods inspect the structure in a relatively small area, with ultrasonic waves, magnetic fields or by radiographic, eddy-current and thermal field methods. From calculations or experience with other structures, it is possible to guess where the "hot spots" of the concerned structure will be expected. In the cases where the sensors and actuators cannot be embedded in the structure, the area to be inspected has to be accessible. This may cause a problem in some applications. For an automated diagnosis based on mechanical waves, which are sent through the object, a relatively dense sensor network is required. Reflections of the waves, from boundaries or other discontinuities such as thickness jumps or holes, make it difficult to consider the effect of the damage. In some cases, damping and the spread of energy do not allow the use of a wider sensor mesh, when high frequency signals are applied. However, the local methods are very sensitive and are able to find small defects.

According to the classical methods of Non-Destructive Testing (NDT), the characteristics of ultrasound wave propagation in solid bodies can be used. In the context of SHM, these waves are generated by permanent installed actuators, which are either mounted on the surface or integrated into the material. The actuators are operated with a high-frequency excitation in the range (from higher kHz to lower MHz): see for example [IHM 03, IHM 04, STA 04a, LEM 01, KES 02, FRI 05, PAR 03, ZAG 01]. Especially in thin plate-like structures, waves propagate as guided waves, also called Lamb waves. The high-frequency excitation generates relatively short wavelengths (according to the relation $c = \lambda f$ where λ, f and c are the wavelength, the frequency and the phase velocity of the wave respectively), which represent the magnitude of the defect size in the structure. For example, the use of a 100 kHz signal produces a wave length of 20 mm for a longitudinal wave with a velocity of $c = 5000$ m/s. On the other hand, in the low-frequency range we work with structural modes whose characteristic size is similar to the dimensions of the complete structure or, for higher modes, to the size of the dimensions of a substructure.

However, the global methods use the fact that the local damage, which, for example, causes a reduction of a local stiffness, has an influence on the global behavior of the whole structure in terms of time and space. For example, the reduction of stiffness causes a decrease of eigenfrequencies, which are not local (spatial) but temporal quantities. Static deflections, due to a static load, characteristic vibration patterns from operational loads or natural vibrations, can be used for this purpose. These methods, based on low-frequency vibrations, monitor the whole system by looking at shifts of resonant frequencies, increases in damping or changes of vibration modes. These changes are used as features extracted from the raw data and they permit to distinguish between the undamaged and damaged states of the structure. The chosen features should be sensitive to the damage [NAT 97].

Usually, a coarser sensor network can be used even if sensors do not need to be close to the damaged area. However, global methods are less sensitive, and they usually have a lower spatial resolution compared to local methods. Nevertheless, sensitivity and spatial resolution can be improved by using a computational model to interpret the dynamic changes.

The global model-based damage identification can be considered as a special application of system identification methodology [CHA 99, NAT 97, FRI 98, DOE 96, FRI 01a, FRI 01b, FAR 04, LIN 91, LIN 99]. The damage causes characteristic local changes in stiffness, damping and/or mass, and, as a consequence of these changes, shifts of the dynamic characteristics, such as eigenfrequencies, modal damping and mode shapes, occur. The deviations between the actual dynamic properties and the undamaged state can be used to detect the damage and diagnose its location and extent. The combination of local and global methods is discussed in [SOH 04b, FRI 05].

2.2.2. Damage diagnosis as an inverse problem

As stated above, the goal of technical diagnosis is to determine whether a system (e.g. a structure) is damaged or not, and if it is the case, where and how severe this damage is. From the diagnostic result, it is decided what actions have to be initiated. To a certain extent, a parallel can be drawn between theses actions and a medical diagnosis. The physician checks some characteristic data, such as body temperature, blood pressure, heart rate, values from blood tests, etc. If there are any significant deviations from a normal state, he uses these symptoms to draw conclusions from his own experience, additional literature (which forms his knowledge base), or a heuristic model using some if–then rules, and diagnose the disease. This is the example of a typical inverse problem.

Figure 2.6 shows the basic structure of the direct problem. Given the input and the description of the process/system, we are interested in the response of the system. In that case, a typical task during the design phase of a system is to check the response due to different load cases.

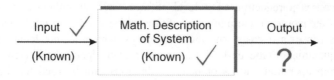

Figure 2.6. *Direct problem: given the input and a known model of the process, we search for the output*

Inverse problems, like the diagnostic task, can show some very basic difficulties. From a mathematical point of view, Hadamard has given the following classification to distinguish between the so-called "well-posed" and "ill-posed" inverse problems [BAU 87]. An inverse problem is called "ill-posed" if one of the following three conditions is not satisfied:

1) the inverse problem has a solution,

2) this solution is unique,

3) small disturbances of the measurement data only cause small deviations in the solution of the inverse problem.

If we look at our daily experience with inverse problems, many examples of ill-posed problems can be found. We observe but have no explanation from our experience ("knowledge-based model"). Maybe a disease shows symptoms, but as they have not yet been described in the literature, a diagnosis cannot be found. Then, the question of uniqueness arises. If the information about some symptoms is obtained after it has been subjected to tests, and if the same symptoms are characteristic to another disease, we cannot make a unique decision. It is sometimes due to a lack of complete information. In many cases, an increase of information after additional tests or measurements can solve the problem. Another question then arises. Consequently, when using the measurement method, the normal statistical variation of the measured information may lead to misinterpretation of the results. This makes our decision process highly unstable.

From a mathematical point of view, the commonly accepted ill-posed problem, which does not satisfy the condition in 3), is the mathematical operation of differentiation of noise-polluted data, in order to obtain the acceleration from measured displacement data by a double numerical differentiation for example. The statistical noise of the displacement is amplified by the differentiation, so that the acceleration data become very rough and the results may be useless. Moreover, it must be remembered that the integration of noisy data smoothes the results. However, if we do not have the information about the integration constants (the initial velocity and the initial displacement), the uniqueness of the solution can be debated.

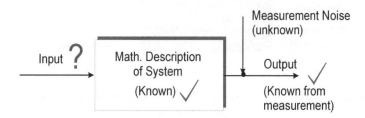

Figure 2.7. *Input problem: given the measured output of the system and the system description, the unknown input has to be determined*

The input problem in structural dynamics deals with determining the input loads of a structure on the basis of the measured responses, which can be strains, displacements, accelerations, etc. Practical examples are the determination of sudden impact loads on very sensitive high-performance composite materials, which may cause an internal delamination of the plies. This category of problems has been treated by [DOY 97, CHO 96, SEY 01a, SEY 01b, STA 04a, STA 04b]. Other similar problems are the determination of forces on off-shore structures (from the wave motion of the sea) or aircraft (wind gusts or special maneuvers). The problem here is twofold: the first question concerns the magnitude and the temporal course of the unknown force, and the second one deals with how these forces are spatially distributed over the structure.

The identification problem, as another category of inverse structural problems, deals with the question of how to determine the system properties from measured input and output signals, which can be altered by measurement noise (see Figures 2.8 and 2.9). Typically, the input signal is a test signal, which covers a certain frequency range of interest, such as a swept or stepped sine signal, a random or pseudo-random signal (see for example [NAT 92, EWI 00, HEY 98, ALL 93]). The most difficult task remains the "black box" problem. In this case, we only have the information about the input and the output, but both the structure of the model and the model parameters are unknown. This means that we do not know whether the system is linear or non-linear or what types of non-linearities describe the system. A check on non-linear behavior can be easily performed by testing the system and using equal test signals with different amplitudes. In the so-called category of ARMA models and other representatives of this group, the number of past time steps to be considered, is also unknown and has to be estimated [WAL 97, LJU 99].

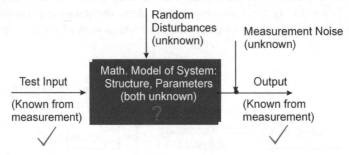

Figure 2.8. *Identification problem with unknown model structure and model parameters (black box model)*

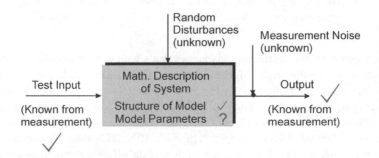

Figure 2.9. *Parameter identification from Input-Output measurements and known model structure*

If mathematical structure of the model is known, the simple problem of parameter identification only has to be taken into account. The estimation, in that case, implies that a defined statistical procedure, which deals with the random nature of the measurement data and maybe of the parameters, is to be defined. For this task, many strategies are adopted. Important elements can be found in the linear and non-linear parameter identification and the optimization literature [EYK 74, WAL 97, LJU 99], the statistical regression methods [WET 86, GLA 00] or the literature of inverse problems [BAU 87, LOU 89, HAN 94, HAN 98, TRU 97].

In many practical cases, we wish to get information about a system using an identification approach but the classical way of applying defined inputs, e.g. by means of a shaker, is not possible. On the other hand, immeasurable random forces from wind or traffic loads are present; see Figure 2.10. Based on the assumption of certain statistical properties of these random quantities, time-domain identification procedures such as the "Stochastic Subspace Identification (SSI)" method [PEE 99, PEE 00a, PEE 01a, PEE 01b] have been developed. Alternatively, a frequency domain approach developed by Brinker *et al.* [BRI 00], extracting for example modal data from an output-only measurement, can be used. These methods using the ambient vibration of the structure have received considerable attention as an intermediate step in damage diagnosis in all those cases where in-service structures have to be examined.

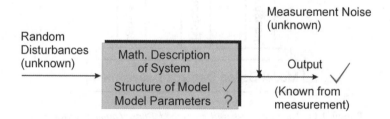

Figure 2.10. *Parameter identification from "output-only" measurements and a known model structure*

2.2.3. *Model-based damage assessment*

As illustrated in Figure 2.11, the model-based damage assessment is built on three major pillars. If one pillar represents a weak point, the whole building can collapse. These first two pillars are: the accurate measurement data including the pre-processing of the raw data, such as the extraction of modal data (eigenfrequencies, dampings and mode shapes); and the reliable reference model, which is obtained, for example, by updating an initial finite element model by means of the baseline measurement data of the undamaged structure. If it is possible, the model should be defined by the additional measurement data which were not used during the updating. This model represents our knowledge of the structure under consideration. The third pillar refers to the algorithm, which extracts the damage parameters using both the model and the measurement data. This algorithm strongly depends on the types of measurement data and computational model. Based on the updated reference model, the damage localization can be performed on the basis of the changes in the measured dynamic behavior. The model serves as a kind of interpreter, which translates changes of the measured data into changes in the structural model. The three-step procedure, consisting of updating the original model, localizing the areas of damage and quantifying the extent of damage by subsequently adapting the parameters using non-linear optimization procedures, was described in detail in earlier papers [FRI 91, FRI 99a, FRI 99b, FRI 99c, FRI 00].

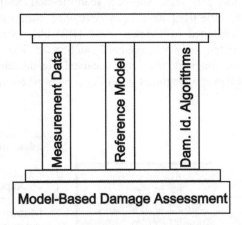

Figure 2.11. *The three basic pillars of model-based damage assessment*

Figure 2.12 shows that the rough scheme of an identification loop determines the damage location and extent. The system output and the model output are compared and the residuals are evaluated together with the model in order to localize and update the model parameters that have changed. Figure 2.14 shows a sketch of a system with integrated sensors and actuators coupled to a computer, to determine the location and extent of the damage via model-based assessment.

According to the control and automation literature, most methods are based on time series. The subject here is called "fault detection and isolation" [GER 98, GUS 00, BAS 93, LJU 99]. However, the view of a mechanical structure, as a general dynamic system, has transformed the boundaries into variable so that many of the methods can be used for the purpose of structural health monitoring. Typical models associated with the time domain are discrete time input–output models, which have an ARMAX structure (AR: Auto Regressive, MA: Moving Average, X: eXogenous input) or related structures (ARX, ARMA, etc.). The parameters of these models cannot usually be interpreted in a physical way. They do not contain stiffness or mass parameters, although they can be used to extract modal parameters and detect changes between the undamaged and damaged systems [SOH 01, GER 98, GUS 00, PIO 93, AND 97].

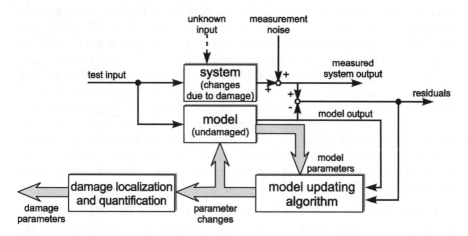

Figure 2.12. *Example of model-based diagnosis: comparison of measured, model output and model updating loop*

The methods which are more closely related to physical models are based on continuous or discrete time state-space models [JUA 85, JUA 94]. A mechanical model with second order time-derivatives can be easily transformed into a first-order state-space model. Filter-based techniques, for example, the Extended Kalman Filter, can also be used for parameter estimation purposes [JAZ 70, GEL 74, YOU 84, LEW 86], as used by [SEI 96, FRI 95] for the diagnosis of opening and closing cracks in rotating shafts.

The transformation of the input–output time signals into the frequency domain, using Fourier analyzers, creates auto- and cross-power spectral densities, which can determine the Frequency Response Functions (FRFs). FRFs are non-parametric characteristics of the behavior of a linear dynamic system, and they are band-limited. The limitation in the frequency range is given by the capabilities of the experimental devices (such as the frequency range of a shaker) and the chosen time sampling rate due to Shannon's theorem. The changes in the FRFs due to the damage can be used to detect, localize and quantify damage, as shown in [NAT 97,

FRI 98, FRI 99a, ZIM 95, NAT 92]. The use of changes of power spectral densities is described in [NAU 01].

A further step is to extract the modal parameters from the FRFs (eigenfrequencies, modal damping and real or complex mode shapes) [NAT 92, EWI 00, HEY 98, ALL 93]. The use of modal data can be considered as the classical approach in model updating and damage detection of structural systems (see [NAT 92, FRIS 95]). The modal parameters, as parametric models obtained from experimental data, represent the dynamics of the real system within a certain frequency range. Experimental Modal Analysis, in itself, is a huge field of scientific work and many different methods have been developed. They can be classified into frequency and time domain methods, single- or multi-dof (where dof means "degree of freedom") algorithms depending on whether only one single mode or multiple modes can be extracted in one step at a time. The sdof (single dof) methods are problematic as soon as two modes are closely spaced. To study systems with multiple eigenvalues, the so-called poly-reference methods have to be used in order to correctly extract the different modes associated with multiple poles. The inverse Fourier's transformation of the FRFs yields the Impulse Response Functions (IRFs). The data sets of the IRFs provide another possibility for extracting modal parameters in the time domain.

In most examples studied in this chapter, the change in the modal parameters, due to damage, is used for the task of damage assessment. Because the modal data provide the experimental data basis, it is fundamental that the modal analysis is carried out accurately.

In the case of ambient external forces resulting from stochastic wind loads, water waves or traffic, the input is not measurable. Modal analysis can be performed according to certain assumptions of the input signal (for example, stationary Gaussian white noise input). In order to identify modal parameters with ambient forces, time- and frequency-domain methods have been developed. For more details, see [VAN 93, PEE 99, PEE 00a, BAS 01, BRI 00].

[CAO 98, ZIM 99, SOH 99] propose that the "Ritz vectors" detect damages. The Ritz vectors are load dependent. Like the first Ritz vector, the static displacement vector due to a chosen load is taken. The second Ritz vector is constructed in such a way that it is orthogonal to the first vector and satisfies the boundary conditions using orthogonalization methods. In the following steps, the kth Ritz vector is constructed to be orthogonal to the previous vectors.

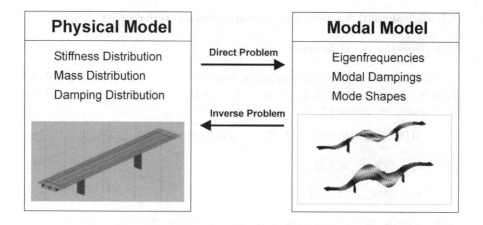

Figure 2.13. *Relationship between physical and modal models as direct and inverse problem*

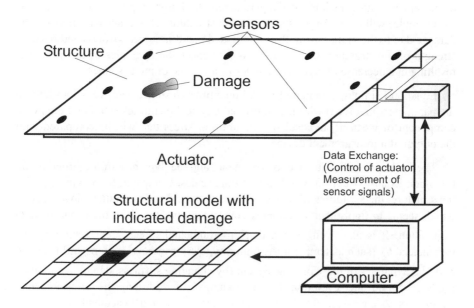

Figure 2.14. *Structure with integrated sensors and actuator*

]2.3. Mathematical description of structural systems with damage

2.3.1. *General dynamic behavior*

The dynamics of a general non-linear, time-varying damaged structure with the spatially discrete and coupled system of the non-linear equation of motion [1] and the evolution of damage [2] are described as follows:

$$M(\theta_d,\theta_e,x,t)\ddot{x}+g(x,\dot{x},\theta_d,\theta_e,t)=f(t,\theta_d,\theta_e) \qquad [1]$$

$$\dot{\theta}_d=\Gamma(\theta_d,\theta_e,x,\dot{x},t) \qquad [2]$$

$$y(t)=h(\theta_d,\theta_e,x,\dot{x},t) \qquad [3]$$

where M is the mass matrix, g the force vector of elastic forces, damping forces, etc. depending on the displacements x and the velocities $\dot{x}$, and t is the time. f is the vector of the external load acting on the system. The number of degrees of freedom (dof) is m. The non-linear function Γ describes the evolution of the damage parameters θ_d (crack length, play, loss of stiffness, loss of mass, etc.). The two differential equations [1] and [2] interact due to their coupling in the mechanical displacements and velocities, as the parameters do. Large amplitudes of vibration x, for example, will cause larger stresses in the structure. This again will increase the damage due to the vibrations in the system, and result in lower stiffness and residual strength of the structure. Lower stiffness influences the dynamics of the structure, resulting in a decrease of eigenfrequencies or a change of mode shapes.

Equation [3] links the internal model quantities (displacement, velocities and all the parameters involved) with the additional physical quantities $y(t)$. For example, a component of y can be an acceleration, strain or stress that will be compared with the output of a measurement device.

The evolution of the damage, on one hand, and the dynamics of the structure, on the other, usually occur on two different time scales. Compared to the vibrations of the structure, the evolution of damage is usually considered as a rather slow process, which seems to imply that θ_d remains constant during the short time span of data acquisition. It is also well known that environmental effects, represented by the parameters θ_e (such as temperature, humidity, etc.), can have a strong influence on both the dynamics of the vibrating system (by changing stiffness, damping and mass properties) and the evolution of the damage. In addition, the environmental parameters θ_e are assumed to be constant during vibration data acquisition.

2.3.1.1. *Linear systems*

Let us assume that the dynamics of the structure can be described by the linear equation of motion with m degrees of freedom:

$$M\ddot{x} + C\dot{x} + Kx = f(t) \qquad [4]$$

where M, C and K are the $m \times m$ mass, damping and stiffness matrices respectively. x is the displacement vector and f represents the external load of the system. To simplify, the assumption of proportional damping is often noted:

$$C = \alpha M + \beta K \qquad [5]$$

where only the two scalar constants α, β have to be determined.

Transforming this equation into the frequency domain delivers the complex algebraic equation:

$$\left(-\Omega^2 M + j\Omega C + K \right) X(\Omega) = F(\Omega) \qquad [6]$$

The expression:

$$Z(\Omega) = Z_\Omega = \left(-\Omega^2 M + j\Omega C + K \right) \qquad [7]$$

presents the complex dynamic stiffness. The product of dynamic stiffness and dynamic flexibility (represented by the FRF matrix $H(\Omega)$) resulting in the identity matrix I, we get an equation that can be used to correlate the M-C-K-model with the FRF measurement data:

$$\left(-\Omega^2 M + j\Omega C + K \right) H(\Omega) = I \qquad [8]$$

If the system is undamped or only lightly damped, the characteristic features of the system are the natural frequencies ω_i and the (real) normal modes φ_i, which can be calculated from the eigenvalue problem (with eigenvalue $\lambda_i = \omega_i^2$):

$$\left(K - \omega_i^2 M \right) \varphi_i = 0 \qquad [9]$$

The mode shapes are normalized to unit modal mass M_i for all further considerations:

$$\varphi_i^T M \varphi_i = M_i = 1 \qquad [10]$$

The mode shape vectors φ satisfy the orthogonality relation

$$\varphi_i^T M \varphi_k = 0 \quad for\ i \neq k \qquad [11]$$

2.3.1.2. *Relation between frequency response functions and modal data*

Modal transformation using the real eigenvalues and eigenvectors leads to representation of the FRF matrix for an excitation frequency Ω:

$$H(\Omega) = \sum_{i=1}^{m} \frac{\varphi_i\, \varphi_i^T}{\omega_i^2 - \Omega^2 + 2j\varsigma_i\omega_i\Omega} \qquad [12]$$

with $X(\Omega) = H(\Omega)F(\Omega)$. An element H_{kl} of the entire FRF matrix is:

$$H_{kl}(\Omega) = \sum_{i=1}^{m} \frac{\varphi_{ki}\, \varphi_{li}}{\omega_i^2 - \Omega^2 + 2j\varsigma_i\omega_i\Omega} \qquad [13]$$

describing the single output X_k due to a single dynamic input F_l. For the derivation of this relation and a much deeper discussion, refer to [HEY 98, NAT 92, FRIS 95].

The formulas show that FRF and modal data (including the dimensionless damping ratio ς_i for mode i) deliver the same information about the dynamic system, because the FRF matrix can be composed of the modal data.

The static flexibility H_{stat}, which describes the displacements resulting from a static load can be derived directly from the FRF:

$$H_{stat} = H(\Omega = 0) = \sum_{i=1}^{m} \frac{\varphi_i\varphi_i^T}{\omega_i^2} = \boldsymbol{\Phi}\boldsymbol{\Lambda}^{-1}\boldsymbol{\Phi}^T \qquad [14]$$

where:

$$\Lambda = diag(\lambda_1, \lambda_2, \cdots, \lambda_m) = diag(\omega_1^2, \omega_2^2, \cdots, \omega_m^2) \qquad [15]$$

and

$$\boldsymbol{\Phi} = \begin{bmatrix} \varphi_1 & \varphi_2 & \cdots & \varphi_m \end{bmatrix} \qquad [16]$$

For reasons of completeness, it should be mentioned that the more general case of non-proportional damping leads to complex eigenvalues λ_c, representing the frequency of vibration (imaginary part of the eigenvalue) and the damping (real part): $\lambda_{c,i} = \alpha_i + j\omega_i$. For stable systems, the real part is always negative. The eigenvectors also are complex eigenvectors: $\varphi_{c,i} = \varphi_{c,i,re} + j\varphi_{c,i,im}$. For weakly non-proportionally damped system, the eigenvalues and eigenvectors always appear as conjugate complex pairs: λ_i and $\lambda_{c,i}^* = \alpha_i - j\omega_i$; $\varphi_{c,i}$ and $\varphi_{c,i}^*$.

In contrast with real mode shapes, complex mode shapes, in general, do not have fixed nodes during one period of vibration, and the motion is simply not a perfect in-phase/out-of-phase motion with a phase shift of 0 or 180°.

For non-proportional viscous damping, the frequency response functions (FRFs) are:

$$H(\Omega) = \sum_{i=1}^{m} \frac{\varphi_{c,i}\varphi_{c,i}^{T}}{c_i\left(j\Omega - \lambda_{c,i}\right)} + \frac{\varphi_{c,i}^{*}\varphi_{c,i}^{*T}}{c_i^{*}\left(j\Omega - \lambda_{c,i}^{*}\right)} \qquad [17]$$

where c_i is a complex scalar normalization constant for the scaling of the complex eigenvectors. For more details, see [HEY 98]. This formula is one of the most important relations in experimental modal analysis, to extract eigenvalues and eigenvectors from measured frequency response functions, and subsequently determine eigenfrequencies and damping from the complex eigenvalue [ALL 93, HEY 98, EWI 00, NAT 92].

2.3.1.3. Incomplete modal data

It is not usually possible to measure all m modes, and a small subset of modes corresponding to the lower frequencies is only available: $m_{red} \ll m$. The reasons for this can be manifold, starting from the frequency bandwidth of the excitation signal, the frequency range of the sensors, the maximum sampling rate of the data acquisition device or high modal densities in the upper frequency range, which make it impossible to separate and extract these modes. Because of this, we can only take the sum over the known m_{red} modes. In order to take the unknown higher modes into account, the residual flexibility matrix H_{res} is introduced:

$$H(\Omega) = \sum_{i=1}^{m_{red}} \frac{\varphi_i\,\varphi_i^{T}}{\omega_i^2 - \Omega^2 + 2j\varsigma_i\omega_i\Omega} + H_{res} \qquad [18]$$

In the particular case of $\Omega = 0$, we get:

$$H_{stat} = \sum_{i=1}^{m_{red}} \frac{\varphi_i\,\varphi_i^{T}}{\omega_i^2} + H_{res} \qquad [19]$$

It is important to remember that the static flexibility matrix is the inverse of the stiffness matrix $H_{stat} = K^{-1}$, which allows determination of the residual flexibility, if the inverse of K is available.

2.3.2. State-space description of mechanical systems

The formulation of the diagnosis problem in the time domain, especially in state-space notation, is frequently used in control theory to identify faults in general technical systems. For this reason, the state-space description of mechanical systems in both continuous and discrete time formulations is presented here.

2.3.2.1. *Continuous-time state-space formulation of linear mechanical systems*

The vibration behavior of the mechanical structure can be described by the differential equation:

$$M\ddot{x} + C_d \dot{x} + K x = B_u u \qquad [20]$$

where again M, C_d and K are the $m \times m$ mass, damping[1] and stiffness matrix respectively, and x is the m-dimensional displacement vector. The $m \times n_u$ matrix B_u is the input matrix, $u(t)$ is the vector of external measurable inputs of length n_u, t is the (continuous) time. The input u can be any physical quantity, for example, a force, a ground motion or an input voltage, which is applied to a piezoelectric actuator. Together with the matrix B_u this yields a generalized force (or moment):

$$f(t) = B_u u(t) \qquad [21]$$

2.3.2.1.1. System equation

The state-space vector of a mechanical system is composed of the vectors of displacements and velocities:

$$z = \begin{pmatrix} z_1 \\ z_2 \end{pmatrix} = \begin{pmatrix} x \\ \dot{x} \end{pmatrix} \qquad [22]$$

and hence has the dimension $2m \times 1$. Premultiplication of equation [20] by M^{-1} gives:

$$\ddot{x} = -M^{-1}C_d \dot{x} - M^{-1}K x + M^{-1}B_u u \qquad [23]$$

and with $\dot{z}_1 = z_2 = \dot{x}$ we can write the equation of motion as a first order system:

$$\begin{pmatrix} x \\ \dot{x} \end{pmatrix}^{\bullet} = \begin{bmatrix} 0 & I \\ -M^{-1}K & -M^{-1}C_d \end{bmatrix} \begin{pmatrix} x \\ \dot{x} \end{pmatrix} + \begin{bmatrix} 0 \\ M^{-1}B_u \end{bmatrix} u(t) \qquad [24]$$

or, more briefly:

$$\dot{z} = A_c z + B_c u \qquad [25]$$

where the $2m \times 2m$ matrix A_c is called the system matrix:

$$A_c = \begin{bmatrix} 0 & I \\ -M^{-1}K & -M^{-1}C_d \end{bmatrix} \qquad [26]$$

---

[1] The damping matrix is indicated by the subscript d in order to avoid confusion with the measurement matrix C.

and the $2m \times n_u$ matrix B_c is the input matrix:

$$B_c = \begin{bmatrix} 0 \\ M^{-1}B_u \end{bmatrix}$$ [27]

The subscript c denotes that the state-space equation is formulated for continuous time t.

2.3.2.1.2. Measurement equation

The second part of the state-space equations relates the state space and the input vector to the n_m output quantities of the output vector $y(t)$. This is called the measurement equation:

$$y = C\,z + D\,u$$ [28]

Matrix C is called the output matrix and has size $n_m \times 2m$, D is the feed-through matrix of size $n_m \times n_u$. The values of y can be (absolute or relative) displacements or velocities, which are directly part of the state-space quantities of the system, but they can also be accelerations, strains, forces or electric voltages, etc. If adequate relations between x, u and y are found, the matrices C and D can be formulated. For displacements, velocities and acceleration, we can write:

Displacements:

$$y = C_x x = \begin{bmatrix} C_x; & 0 \end{bmatrix} z \; ; \; C = \begin{bmatrix} C_x; & 0 \end{bmatrix}, \text{ and } D = 0$$ [29]

Velocities:

$$y = C_v \dot{x} = \begin{bmatrix} 0; & C_v \end{bmatrix} z \; ; \; C = \begin{bmatrix} 0; & C_v \end{bmatrix}, \text{ and } D = 0$$ [30]

Accelerations:

$$y = C_a \ddot{x} = C_a \left[-M^{-1}C_d\, \dot{x} - M^{-1}K\, x + M^{-1}B_u\, u \right]$$ [31]

In the case of measured accelerations, the matrices C and D of the measurement equation are:

$$C = C_a \left[-M^{-1}K; \quad -M^{-1}C_d \right] \text{ and } D = C_a M^{-1} B_u$$ [32]

The matrices C_x, C_v, C_a aim to select the measured components of the full displacement, velocity or acceleration vector, respectively. In a similar way, it is also possible to express relative displacements, strains, etc.

2.3.2.2. *Discrete-time state-space model*

When working with real measurement data, the time series are usually sampled at equidistant intervals Δt using AD converters. The time axis is not continuous and the data are only available at discrete sampling instants t_k, $k = 0,1,2,3,...,n_t$ with $t_k = k \cdot \Delta t$. The sampling rate is $f_s=1/\Delta t$ and has to be chosen, so that the highest frequency component of the response can be observed.

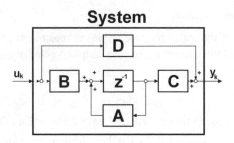

Figure 2.15. *Block scheme of a state space model*

In order to take care of Shannon's theorem and to avoid aliasing effects, analog low-pass filters have to be used. For more details concerning data-acquisition, see [NAT 92, EWI 00, HEY 98, ALL 93].

Given the initial state vector z_0 the solution of equation [25] (see [JUA 94]) at time t is:

$$z(t) = e^{A_c(t-t_0)}z_0 + \int_{t_0}^{t} e^{A_c(t-t_0)}B_c \, u(\tau)d\tau \qquad [33]$$

for $t > t_0$. The equation describes the evolution of the state $z(t)$, starting from z_0 under the action of the input $u(t)$. In particular, if we start with time $t_k = k\Delta t$ instead of t_0, and look for the next state at time $t_{k+1} = (k+1)\Delta t$, we get:

$$z(t_{k+1}) = e^{A_c(t_{k+1}-t_k)}z_k + \int_{t_k}^{t_{k+1}} e^{A_c(t_{k+1}-t_k)}B_c \, u(\tau)d\tau \qquad [34]$$

or with $t_{k+1} - t_k = \Delta t$:

$$z(t_{k+1}) = e^{A_c\Delta t}z_k + \int_{t_k}^{t_{k+1}} e^{A_c\Delta t}B_c \, u(\tau)d\tau \qquad [35]$$

The integration requires the continuous-time input u, which is not usually available, but can be interpolated between data points. The most common way is to assume that $u(\tau)$ is constant between two sampling times k and $k+1$:

$$u(\tau) = u(k\,\Delta t) \text{ for } k\Delta t \leq \tau < (k+1)\,\Delta t \qquad [36]$$

In this case we can take u out of the integral:

$$z(t_{k+1}) = e^{A_c \Delta t} z(t_k) + (\int_{t_k}^{t_{k+1}} e^{A_c \tau'} B_c \, d\tau') u(t_k) \qquad [37]$$

Now, with $z(t_k) = z(k\,\Delta t) = z_k$ and $u(t_k) = u(k\,\Delta t) = u_k$, we can write:

$$z_{k+1} = A\,z_k + B\,u_k \ ; \ k = 0, 1, 2, \cdots \qquad [38]$$

where we can easily identify the discrete-time system matrix A and the input matrix B as:

$$A = e^{A_c \Delta t} \qquad [39]$$

$$B = \int_0^{\Delta t} e^{A_c \tau'} d\tau' \, B_c \qquad [40]$$

The measurement equation of the continuous-time model did not contain any derivatives. Therefore, the measurement equations for the continuous- and the discrete-time state-space model are identical:

$$y_k = C\,z_k + D\,u_k \qquad [41]$$

2.3.2.3. Modal representation

The modal characteristics can be computed from the discrete-time state-space model system matrix A. Solving the eigenvalue problem:

$$(A - \delta_i I)\,\psi_i = 0 \qquad [42]$$

we get the complex eigenvalues $\Delta = diag\{\delta_1, \delta_2, ...\}$, arranged in the diagonal matrix Δ and the complex eigenvectors $\Psi = [\psi_1, \psi_2, ...]$: The diagonal matrix Δ contains the information about the damped natural frequencies and damping rates. The matrix $\Phi = C\,\Psi$ defines the mode shapes at the sensor points.

Using the modal information, the realization of a stochastic system (A, C) can be easily transformed into the realization $(\Delta, C\Psi)$: see, for example [BAS 00, BAS 01].

The desired physical eigenfrequencies and damping rates usually employed in vibration engineering are simply the imaginary and real parts of the diagonal matrix

$\mathbf{\Lambda}$, after transformation from the discrete-time domain to the continuous-time domain using the relation:

$$\lambda_i = \alpha_i + j\omega_i = \delta_{ci} = \frac{1}{\Delta t}\ln(\delta_i) \qquad [43]$$

or as a matrix:

$$\mathbf{\Lambda}_c = \ln(\mathbf{\Lambda})/\Delta t \qquad [44]$$

2.3.2.4. *Discrete-time state-space model with stochastic excitation and noise*

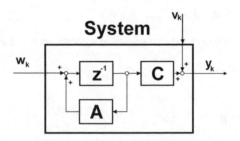

Figure 2.16. *Stochastic state space model*

If a deterministic measurable input u_k does not exist, but the system is excited by a stochastic process w_k, we get the discrete state space model:

$$z_{k+1} = A\, z_k + w_k \qquad [45]$$

$$y_k = C\, z_k + v_k \qquad [46]$$

In practice, w_k can result from stochastic wind or traffic loads on the structure under consideration. This means that the outputs and state variables are also stochastic quantities.

The noise processes w and v are both immeasurable signals. The most common assumption about w and v is that they have zero mean:

$$E\{w_k\} = 0 \qquad [47]$$

and:

$$E\{v_k\} = 0 \,\ldots \qquad [48]$$

where $E\{...\}$ denotes the expectation operator, and w and v are white-noise processes with co-variance matrices Q, R and S, describing the intensity of the system disturbances (Q), the measurement noise (R) and the cross coupling of the two (S):

$$E\left\{ w_k w_l^T \right\} = Q \, \delta_{kl} \qquad [49]$$

$$E\left\{ v_k v_l^T \right\} = R \, \delta_{kl} \qquad [50]$$

$$E\left\{ w_k v_l^T \right\} = S \, \delta_{kl} \qquad [51]$$

where the Kronecker δ_{kl} symbol is:

$$\delta_{kl} = \begin{cases} 1 & \text{if } k = l \\ 0 & \text{if } k \neq l \end{cases} \qquad [52]$$

and k and l are any arbitrary time instants [PEE 00a, BAS 00, JUA 94].

2.3.2.5. *Continuous-time state-space formulation of non-linear mechanical systems*

Starting with the equation of motion [1] and putting the force vector f on the right-hand side yields:

$$M(\theta_d, \theta_e)\ddot{x} = f(t, \theta_d, \theta_e) - g(x, \dot{x}, \theta_d, \theta_e, t) = f^*(x, \dot{x}, \theta_d, \theta_e, t) \qquad [53]$$

or, after premultiplying by the inverse mass matrix:

$$\ddot{x} = M^{-1}(\theta_d, \theta_e) f^*(x, \dot{x}, \theta_d, \theta_e, t) \qquad [54]$$

With the arrangement of the displacements and velocities as state-space vectors, equation [22], we get:

$$\begin{pmatrix} x \\ \dot{x} \end{pmatrix}^{\bullet} = \begin{pmatrix} \dot{x} \\ M^{-1}(\theta_d, \theta_e) f^*(x, \dot{x}, \theta_d, \theta_e, t) \end{pmatrix} \qquad [55]$$

from which continuous-time state-space formulation of a non-linear mechanical system follows the system equation [56] and the measurement equation [57]:

$$\dot{z} = \begin{pmatrix} z_1 \\ z_2 \end{pmatrix}^{\bullet} = \begin{pmatrix} z_2 \\ M^{-1}(\theta_d, \theta_e) f^*(z_1, z_2, \theta_d, \theta_e, t) \end{pmatrix} = \tilde{f}(z_1, z_2, \theta_d, \theta_e, t) \qquad [56]$$

$$y(t) = h(\theta_d, \theta_e, x, \dot{x}) \qquad [57]$$

The augmentation of the state-space vector with the damage parameters yields a coupled differential equation:

$$\begin{pmatrix} x \\ \dot{x} \\ \theta_d \end{pmatrix}^{\!\cdot} = \begin{pmatrix} \dot{x} \\ M^{-1}(\theta_d,\theta_e)f^*(x,\dot{x},\theta_d,\theta_e,t) \\ \Gamma(x,\dot{x},\theta_d,\theta_e,t) \end{pmatrix} \qquad [58]$$

with the augmented state-space vector:

$$z = \begin{pmatrix} z_1 \\ z_2 \\ z_3 \end{pmatrix} = \begin{pmatrix} x \\ \dot{x} \\ \theta_d \end{pmatrix} \qquad [59]$$

In [NAT 93], this type of dynamic equation, including the interaction of the dynamics of the structure with respect of the vibrations and the evolution of the damage, is called a holistic model. As pointed out earlier, the evolution of the damage (if no sudden breakdown occurs) happens on a very "slow" time scale, because the damage aggravates after days, months or years, while the oscillations of the mechanical structure take place on a much "faster" time scale with frequencies lying in a range from some Hz to some kHz or even much higher. This justifies the decoupling of the different dynamic equations again, assuming that the damage parameters are constant $\theta_d = const.$ or $\dot{\theta}_d = 0$ for the relatively short duration of only some milliseconds up to some minutes of the vibration data acquisition:

$$\begin{pmatrix} x \\ \dot{x} \\ \theta_d \end{pmatrix}^{\!\cdot} = \begin{pmatrix} \dot{x} \\ M^{-1}(\theta_d,\theta_e)f^*(x,\dot{x},\theta_d,\theta_e,t) \\ 0 \end{pmatrix} \qquad [60]$$

This representation allows using the Extended Kalman Filter (EKF) technique [JAZ 70, YOU 84, FRI 90, FRI 95, SEI 95a, SEI 95b] for simultaneous estimation of the state variables, displacements and velocities, together with the damage parameters as new state-space variables.

2.3.2.6. *Discrete-time state-space formulation of non-linear mechanical systems*

For discrete, sampled time $t_k = k\Delta t$, $k = 0,1,2,3,\ldots$, $t_{k+1} - t_k = \Delta t$, the solution of the state-space equations is obtained by integration, which can be done numerically:

$$z_{k+1} = z_k + \int_{t_k}^{t_{k+1}} \tilde{f}(z,\theta_e,\tau)\, d\tau . \qquad [61]$$

2.3.3. *Modeling of damaged structural elements*

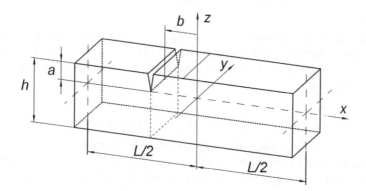

Figure 2.17. *Beam element with a transverse crack*

An overview of the modeling of damage in a structural member is given in [OST 01, FRIS 02]. However, different possibilities for considering the damage exist, and they will be explained in this section.

2.3.3.1. *Change of the element stiffness or damping*

This approach smears the effect of the damage over the whole element by reducing the element's stiffness. Regardless of the source of the real damage (cracks, corrosion, etc.), one view suggests that a loss of stiffness is to be studied instead of the stiffness of the original structural part. For cracks in concrete beams, this view has been used by Maeck [MAE 03c] and Eilbracht [EIL 95]. It can also be observed that the damage causes an increase of damping, due to plastic deformations or increased internal friction.

For example, we consider the stiffness matrix of a simple finite Euler–Bernoulli beam element with transverse nodal displacements and rotations in one plane:

$$k_{el} = \frac{EI}{l^3} \begin{bmatrix} 12 & 6l & -12 & 6l \\ & 4l^2 & -6l & 2l^2 \\ & & 12 & -6l \\ sym. & & & 4l^2 \end{bmatrix} \qquad [62]$$

The global change of stiffness results from the alteration in the bending stiffness $EI \rightarrow (EI + \Delta EI) = EI\,(1 + \Delta a)$. ΔEI is usually negative and we get a decrease of the stiffness and an increase of the flexibility of this finite element. If the beam deformation is bending and tensile, the stiffness can change into ΔEI (the bending stiffness) and ΔEA (the tensile stiffness). 3D-beams with shear (Timoshenko theory), plate or shell elements, or with any other type of element, can be studied the same way.

Other authors have determined the change in stiffness properties, according to a delamination, using a simple four-zone model [TRA 89, ZOU 96]. This approach requires the less detailed information about the damage and the smallest programming effort.

2.3.3.2. *Connecting local changes with global system matrices*

As a result of the damage, the system changes its characteristics, which may be expressed in the dynamic model, equation [4], by the stiffness matrix changes ΔK. If modal data is considered, the changes in the stiffness matrices alter the eigenvalues $\Delta \lambda_i = \Delta(\omega_i^2)$ and mode shapes $\Delta \varphi_i$ according to equation [9]:

$$\left(K_0 - \lambda_i\, M_0 \right) \varphi_i = 0 \;\rightarrow\; \left((K_0 + \Delta K) - (\lambda_i + \Delta \lambda_i)\, M_0 \right) (\varphi_i + \Delta \varphi_i) = 0 \qquad [63]$$

The change of the global behavior must be related to local parameters describing the damage. Thus, for each sub-region of the structure, such as finite element, a suitable correction parameter Δa_j is introduced to qualify the local change of stiffness:

$$\Delta K = \sum_{j=1}^{n} K_j \Delta a_j \qquad\qquad [64]$$

$$\Delta C = \sum_{j=1}^{n} C_j \Delta a_j \qquad\qquad [65]$$

$$\Delta M = \sum_{j=1}^{n} M_j \Delta a_j \qquad\qquad [66]$$

In this formulation, we have one parameter set Δa for all three matrices. If a certain parameter Δa_j only influences the damping matrix, we must set $M_j = K_j = 0$.

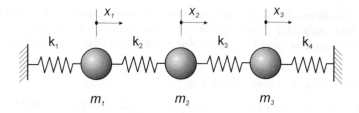

Figure 2.18. *Simple three-dof oscillator to demonstrate the meaning of dimensionless parameters Δa_j*

To symplify the approach, the three-dof oscillator, shown in Figure 2.18, is considered. The mass and stiffness matrices of this simple mechanical system are:

$$K = \begin{bmatrix} k_1 + k_2 & -k_2 & 0 \\ -k_2 & k_2 + k_3 & -k_3 \\ 0 & -k_3 & k_3 + k_4 \end{bmatrix} ; \ M = \begin{bmatrix} m_1 & 0 & 0 \\ 0 & m_2 & 0 \\ 0 & 0 & m_3 \end{bmatrix} \quad [67]$$

If we consider the spring stiffnesses as the parameters, which may be subject to change, the stiffness matrix can be expanded into:

$$K = \begin{bmatrix} k_1 & 0 & 0 \\ 0 & 0 & 0 \\ 0 & 0 & 0 \end{bmatrix} + \begin{bmatrix} k_2 & -k_2 & 0 \\ -k_2 & k_2 & 0 \\ 0 & 0 & 0 \end{bmatrix} + \begin{bmatrix} 0 & 0 & 0 \\ 0 & k_3 & -k_3 \\ 0 & -k_3 & k_3 \end{bmatrix} + \begin{bmatrix} 0 & 0 & 0 \\ 0 & 0 & 0 \\ 0 & 0 & k_4 \end{bmatrix} \quad [68]$$

Introducing the dimensionless parameter changes $\Delta a_j = \Delta k_j / k_j$, $j = 1, 2, 3, 4$:

$$\Delta K = \begin{bmatrix} k_1 & 0 & 0 \\ 0 & 0 & 0 \\ 0 & 0 & 0 \end{bmatrix} \Delta a_1 + \begin{bmatrix} k_2 & -k_2 & 0 \\ -k_2 & k_2 & 0 \\ 0 & 0 & 0 \end{bmatrix} \Delta a_2 ...$$

$$+ \begin{bmatrix} 0 & 0 & 0 \\ 0 & k_3 & -k_3 \\ 0 & -k_3 & k_3 \end{bmatrix} \Delta a_3 + \begin{bmatrix} 0 & 0 & 0 \\ 0 & 0 & 0 \\ 0 & 0 & k_4 \end{bmatrix} \Delta a_4$$

or:

$$\Delta K = K_1 \Delta a_1 + K_2 \Delta a_2 + K_3 \Delta a_3 + K_4 \Delta a_4 \quad [69]$$

where the substructure matrices K_j can be clearly identified. In the case of more complex structures where the finite element method is used, the substructure matrices K_j can be derived from the element stiffness matrices or groups of several element stiffness matrices. For an automated use of this method, the FE program has to provide access to the element matrices.

The alterations, due to damage and environmental changes, must be separated because their magnitude must be different. Environmental effects are excluded in this article. Temperature effects are treated in [PEE 00b, PEE 01b, FRI 03b].

2.3.3.3. *Detailed finite element modeling*

Computationally, the damage is considered as a detailed finite element modeling, e.g. modeling a cracked component using solid elements and 2D or 3D mesh refinement around the crack tips. To open and close the crack, material contact of the crack surfaces has to be taken into account to describe the problem properly. In [DIR 88], the stiffness behavior of a transverse crack in a rotating shaft was investigated for various crack depths and during one full revolution of the shaft, where the crack can be fully open, fully closed or in any partially closed intermediate state. While this approach yields the most accurate results, it is generally too costly for practical use within damage diagnosis.

2.3.3.4. *Calculation of stiffness properties using local fracture-mechanics concepts*

This more advanced concept makes it possible to combine local fracture-mechanics concepts of stress singularities at the crack tip with the global stiffness/flexibility properties of the structural element. The aim is to get a direct relationship between physical crack size, location in a structural element and the loss of stiffness due to this crack. The concept is applicable to various crack shapes, geometries of the structure and load cases. It has also been applied to rotating shafts, where the crack is closed, fully open or partially open during one revolution. We can only present a rough idea of this concept here and refer to the literature [OST 90, KRA 92a, KRA 92b, OST 01, RIZ 90, THE 90, GOU 88] for a more detailed presentation of the idea and for its practical application.

First, *Castigliano*'s famous theorem is used to derive an important relationship between the elastic strain energy U and the flexibility of a structural element under consideration. The flexibility coefficient h_{ij} can be calculated from the second derivative with respect to the loads (which are the forces and moments at the nodes of the finite elements):

$$h_{ij} = \frac{\partial^2 U}{\partial F_i \partial F_j}$$

The energy release rate G (defined under linear-elastic fracture mechanics conditions):

$$G = -\frac{\partial \Pi}{\partial A_F} = \frac{\partial U}{\partial A_F}$$

presents the reduction of the (internal and external) potential and the increase of strain energy as a function of the increase of the fracture surface A_F. Furthermore, the energy release rate is related to the stress intensity factors K_I, K_{II} and K_{III}:

$$G = \beta \frac{K_I^2}{E} + \beta \frac{K_{II}^2}{E} + (1+v) \frac{K_{III}^2}{E}$$

with:

$$\beta = \begin{cases} 1 & \text{for plane stress} \\ (1-v^2) & \text{for plane strain} \end{cases}$$

The stress intensity factors (SIF) have been provided in the fracture mechanics literature for various geometric and loading conditions (E is Young's modulus, v is the Poisson's ratio). The SIFs can also be found in handbooks [TAD 73].

With increasing crack length and fracture surface A_F, the structure becomes more flexible and the compliance increases. The increase of compliance can be expressed in terms of the energy release rate and the stress intensity factors.

Introducing the energy release rate and the stress intensity factors into the strain energy expression yields:

$$h_{ij} = h_{ij0} + \frac{\partial^2}{\partial F_i \partial F_j} \int_{A_F} G \, dA_F$$

$$= h_{ij0} + \frac{\partial^2}{\partial F_i \partial F_j} \int_{A_F} \left(\beta \frac{K_I^2}{E} + \beta \frac{K_{II}^2}{E} + (1+v) \frac{K_{III}^2}{E} \right) dA_F$$

[70]

The integration is carried out over the crack surface A_F. The coefficient h_{ij0} is the integration constant representing the original flexibility. The stress intensity factors linearly depend on the load quantities, so that the calculation of the partial derivative with respect to F_i and F_j is not difficult. By means of the flexibility coefficients h_{ij}, which are dependent on the size and shape of the crack surface A_F, the stiffness matrix of the structural element can be calculated. The results show the fact that at the early phase when cracks are small, the increase of flexibility is very small as well.

2.4. Linking experimental and analytical data

2.4.1. Modal Assurance Criterion (MAC) for mode pairing

In the case where modes from experiments and analysis are compared, the practical problem of how to determine which analytical mode i fits best to an experimental mode k arises. This is extremely important for the calculation of the mode shape and eigenfrequency residuals. We can evaluate the similarity between two modes using the Modal Assurance Criterion (MAC), which delivers values between 0 and 1. A value of 1 shows a perfect correlation between the two modes, while 0 means no similarity between these two modes. The MAC value [HEY 98, ALL 93]:

$$MAC_{ik} = \frac{(\varphi_{i,A}^T \, \varphi_{k,B})^2}{(\varphi_{i,A}^T \, \varphi_{i,A})(\varphi_{k,B}^T \, \varphi_{k,B})} \qquad [71]$$

is defined by comparing a data set of mode shapes A with another set B, where, for example, set A can be the analytical set and B the measured set. The MAC works independently from the individual scaling of measured and analytical mode shape vectors. An alternative version of the MAC can be defined by considering the mass distribution:

$$MAC_{ik}^* = \frac{(\varphi_{i,A}^T \, M \, \varphi_{k,B})^2}{(\varphi_{i,A}^T M \, \varphi_{i,A})(\varphi_{k,B}^T M \, \varphi_{k,B})} \qquad [72]$$

However, care should be taken if the number of sensors does not allow a sufficient number of mode shape components to adequately represent the mode shapes. The extreme case where the eigenvector consists of one mode shape component always leads to a MAC value of 1 regardless of how well or badly the modes really correlate.

2.4.2. Modal Scaling Factor (MSF)

Assumingly, a corresponding mode pair from sets A and B using the MAC is found. However, mode shapes can be normalized in an arbitrary way. As mentioned earlier, the mode shapes are often mass normalized so that the generalized mass is equal to 1; see equation [10]. Even if the magnitude is the same, two corresponding mode shapes may have a different sign (phase shift of 180°). Thus, to compare analytical and experimental mode shapes, a consistent scaling is to be used. The measured mode shape may be scaled to the analytical mode shape by multiplying it by the scalar MSF [FRIS 95]:

$$\varphi_{i,B} = MSF \cdot \varphi_{i,A} \qquad [73]$$

The MSF can easily be determined by pre-multiplication of the last equation with the transposed measured mode shape vector:

$$MSF = \frac{(\varphi_{i,B}^T \varphi_{i,A})}{(\varphi_{i,B}^T \varphi_{i,B})} = \frac{(\varphi_{i,A}^T \varphi_{i,B})}{(\varphi_{i,B}^T \varphi_{i,B})}$$ [74]

Alternatively, a weighted MSF can be introduced using the mass matrix M:

$$MSF^* = \frac{(\varphi_{i,B}^T M \varphi_{i,A})}{(\varphi_{i,B}^T M \varphi_{i,B})} = \frac{(\varphi_{i,A}^T M \varphi_{i,B})}{(\varphi_{i,B}^T M \varphi_{i,B})}$$ [75]

2.4.3. Co-ordinate Modal Assurance Criterion (COMAC)

While the MAC compares two modal data sets A and B mode-wise, summing over all degrees of freedom in the scalar product, the COMAC investigates each individual degree of freedom by summing over all modes k. The COMAC value for a certain dof i is (see for example [HEY 98]):

$$COMAC_k = \frac{(\sum_i \varphi_{ki,A} \varphi_{ki,B})^2}{(\sum_i \varphi_{ki,A} \varphi_{ki,A})(\sum_i \varphi_{ki,B} \varphi_{ki,B})}$$ [76]

A value of 1 indicates good correlation. In [HEY 98], it is pointed out that consistent scaling has to be used because the COMAC is very sensitive to the way the modes are scaled.

2.4.4. Damping

In most cases, the results of eigenvalue analysis in FE codes are based on the conservative problem using the mass and stiffness matrix only, which leads to real eigenvalues and eigenvectors. If the real structure is only lightly damped (or proportionally damped), we can directly compare the natural frequencies and mode shapes of the analytical model with the measurement results. However, if damping of the structure leads to complex eigenvectors, we must either extract the real eigenvectors from the complex experimental modal analysis results or include damping in the model:

$$\left(\lambda_c^2 M + \lambda_c C + K\right) \varphi_c = 0$$ [77]

and compare the complex results directly.

2.4.5. *Expansion and reduction*

One problem that arises in the context of linking analysis and test results with some of the approaches discussed later is that the dimensions ($m \times n$) of the system matrices K_0 and M_0 are usually much larger than the number of the measured dofs: $m \gg m_{meas}$. This means either to reduce the system matrices to the number of the measured dofs or conversely expand the measured modal vectors to the full size of the FE model matrices [OCA 89, GYS 90, FRIS 95, LIN 91, ZIM 01].

Next, it is assumed that each measurement node geometrically coincides with a corresponding node in the model. The goal is to expand a given vector of measurements x_m with length m_{meas} to a full set, which is compatible with the model size m. The vector of the unknown displacements is given x_u by means of x_m and the transformation rules following from the numerical model. The expanded vector can then be written as:

$$x_{m,expd} = \begin{bmatrix} x_m \\ x_u \end{bmatrix} = T\,x_m,$$ [78]

where T is a linear transformation matrix. The differences in the expansion methods depend on the approach to constructing the transformation matrix T. As shown later, the relationship between the expanded eigenvector $\varphi_{i,expd}$ and the measured components $\varphi_{i,m}$ can be expressed by the same transformation matrix T:

$$\varphi_{i,expd} = T\,\varphi_{i,m}$$ [79]

2.4.5.1. *Static and dynamic expansion*

The commonly used method of determination of the transformation matrix T, called dynamic expansion, which includes the special case of static expansion, will be discussed in this section. [GUY 65] has introduced this method for the purposes of model reduction, but the same method for expansion of the displacement vectors can be used. The application of these methods requires that the forces $f_u = 0$ at the dofs correspond to x_u. Starting with the definition of the dynamic stiffness, equation [7]:

$$Z(\Omega) = -\Omega^2 M + j\Omega C + K$$ [80]

and subdividing the vector of displacements into the known components x_m and the unknown components x which we need to determine to get the full set:

$$x_{m,expd} = \begin{bmatrix} x_m \\ x_u \end{bmatrix}$$ [81]

This partitioning leads to:

$$\begin{bmatrix} Z_{mm}(\Omega) & Z_{mu}(\Omega) \\ Z_{um}(\Omega) & Z_{uu}(\Omega) \end{bmatrix} \begin{pmatrix} x_m \\ x_u \end{pmatrix} = \begin{pmatrix} f_m \\ f_u \end{pmatrix}$$ [82]

where the first and the second rows lead to:

$$Z_{mm} x_m + Z_{mu} x_u = f_m$$ [83]

$$Z_{um} x_m + Z_{uu} x_u = f_u$$ [84]

From the second equation, it follows that:

$$x_u = Z_{uu}^{-1} \left(f_u - Z_{um} x_m \right)$$ [85]

With $f_u = 0$ we obtain $x_u = -Z_{uu}^{-1} Z_{um} x_m$ so that the frequency-dependent transformation matrix T_Ω finally becomes:

$$x_{m,expd} = \begin{pmatrix} x_m \\ x_u \end{pmatrix} = \begin{bmatrix} I \\ -Z_{uu}^{-1}(\Omega) Z_{um}(\Omega) \end{bmatrix} x_m = T_\Omega x_m$$ [86]

If we use the equation of motion, for instance for an undamped linear system, the transformation matrix T_Ω can be written as in the mass and stiffness matrix:

$$T_\Omega = \begin{bmatrix} I \\ -\left(K_{uu} - \Omega^2 M_{uu} \right)^{-1} \left(K_{um} - \Omega^2 M_{um} \right) \end{bmatrix}$$ [87]

For the special case where we want to expand eigenvectors as response data, we can write $f_m = 0$ and $\Omega = \omega_i$ and directly use the expansion equations derived before the expansion of the eigenvectors:

$$\left(\begin{bmatrix} K_{mm} & K_{mu} \\ K_{um} & K_{uu} \end{bmatrix} - \omega_i^2 \begin{bmatrix} M_{mm} & M_{mu} \\ M_{um} & M_{uu} \end{bmatrix} \right) \begin{pmatrix} \varphi_{i,m} \\ \varphi_{i,u} \end{pmatrix} = \begin{pmatrix} 0 \\ 0 \end{pmatrix},$$ [88]

so that we get the same transformation matrix:

$$T_{\omega,i} = \begin{bmatrix} I \\ -\left(K_{uu} - \omega_i^2 M_{uu} \right)^{-1} \left(K_{um} - \omega_i^2 M_{um} \right) \end{bmatrix}$$ [89]

The expanded eigenvector is:

$$\varphi_{m,expd,i} = T_{\omega,i}\, \varphi_{m,i}$$ [90]

If the corresponding eigenfrequency is chosen as the frequency of the dynamic expansion, $\Omega = \omega_i$, the result of the expansion process is exactly $T_{\omega,i}$ provided that there are no additional influences of measurement errors or modeling errors. Choosing frequencies close to the eigenfrequencies will give good approximations. The disadvantage is that the transformation matrix has to be calculated for each frequency. Alternatively, the static expansion matrix can be used, setting $\Omega = 0$ as a special case of the dynamic expansion. The transformation matrix T then reduces to the simpler expression:

$$T_{stat} = \begin{bmatrix} I \\ -K_{uu}^{-1} K_{um} \end{bmatrix}.$$ [91]

2.4.5.2. Reduction

The expansion process can be inverted to reduce the model to the measurement dofs. Eliminating the displacements x_u by setting $x = T_\Omega x_m$ and premultiplying the left-hand side by T_Ω^T yields to a viscously damped system:

$$\left(-\Omega^2 T_\Omega^T M T_\Omega + j\Omega T_\Omega^T C T_\Omega + T_\Omega^T K T_\Omega \right) x_m = T_\Omega^T f_m.$$ [92]

The reduced system matrices of size $m_{meas} \times m_{meas}$ can be written as:

$$M_{0,red} = T_\Omega^T M_0 T_\Omega$$ [93]

$$C_{0,red} = T_\Omega^T C_0 T_\Omega$$ [94]

$$K_{0,red} = T_\Omega^T K_0 T_\Omega$$ [95]

2.4.5.3. SEREP expansion

This expansion method was introduced by O'Callahan and Avitable [OCA 89]. The acronym SEREP stands for System Equivalent Reduction-Expansion Process. Instead of using the system matrices, mass normalized real mode shapes determined by finite element models are used to build up the transformation matrix T. Again, the co-ordinates are subdivided into those that can be measured and those that are unknown and will be determined by the expansion process. A modal transformation approach is used to determine the physical co-ordinates in terms of modal co-ordinates:

$$\begin{pmatrix} x_m \\ x_u \end{pmatrix} = \Phi q = \begin{pmatrix} \Phi_L & \Phi_H \end{pmatrix} \begin{pmatrix} q_L \\ q_H \end{pmatrix}$$ [96]

The set of modes is partitioned into a set of lower modes Φ_L (which are known from the numerical calculation) and set of higher modes Φ_H (which are not known).

The number of modes to be considered in $\boldsymbol{\Phi}_L$ should be at least equal to the number of the measured dofs $(\dim\{\boldsymbol{x}_m\})$, as shown in later sections. The equation system is further partitioned so that we obtain:

$$\begin{pmatrix} \boldsymbol{x}_m \\ \boldsymbol{x}_u \end{pmatrix} = \begin{pmatrix} \boldsymbol{\Phi}_{mL} & \boldsymbol{\Phi}_{mH} \\ \boldsymbol{\Phi}_{uL} & \boldsymbol{\Phi}_{uH} \end{pmatrix} \begin{pmatrix} \boldsymbol{q}_L \\ \boldsymbol{q}_H \end{pmatrix} \qquad [97]$$

It is then assumed that the modal co-ordinates of the higher modes have such a small influence on the frequency range of interest that $\boldsymbol{q}_H = \boldsymbol{0}$ can be set. This means that the upper and lower rows of the last equation are reduced to:

$$\boldsymbol{x}_m = \boldsymbol{\Phi}_{mL}\, \boldsymbol{q}_L \qquad [98]$$

$$\boldsymbol{x}_u = \boldsymbol{\Phi}_{uL}\, \boldsymbol{q}_L \qquad [99]$$

The modal dofs are eliminated by multiplying the upper row by the inverse $\boldsymbol{\Phi}_{mL}^{-1}$ (provided that the inverse exists) from the left-hand side:

$$\boldsymbol{q}_L = \boldsymbol{\Phi}_{mL}^{-1}\, \boldsymbol{x}_m\,. \qquad [100]$$

Putting this into the lower equation for the "u" co-ordinates, the final result for the expanded displacement vector is:

$$\boldsymbol{x}_{m,\text{exp}} = \begin{pmatrix} \boldsymbol{x}_m \\ \boldsymbol{x}_u \end{pmatrix} = \begin{pmatrix} \boldsymbol{I} \\ \boldsymbol{\Phi}_{uL}\, \boldsymbol{\Phi}_{mL}^{-1} \end{pmatrix} \boldsymbol{x}_m \qquad [101]$$

This yields the desired *SEREP* transformation matrix:

$$\boldsymbol{T}_{SERP} = \begin{pmatrix} \boldsymbol{I} \\ \boldsymbol{\Phi}_{uL}\, \boldsymbol{\Phi}_{mL}^{-1} \end{pmatrix}. \qquad [102]$$

If the number of modes is not equal to the number of measured dofs, the matrix $\boldsymbol{\Phi}_{mL}$ is no longer a square matrix and the inverse cannot be calculated. Then, the pseudo-inverse $\boldsymbol{\Phi}_{mL}^{+}$ is used instead of the inverse $\boldsymbol{\Phi}_{mL}^{-1}$:

$$\boldsymbol{T}_{SEREP} = \begin{pmatrix} \boldsymbol{I} \\ \boldsymbol{\Phi}_{uL}\, \boldsymbol{\Phi}_{mL}^{+} \end{pmatrix} \qquad [103]$$

In the case where fewer mode shapes are used for the expansion than we have measured dofs, the transformation matrix will be rank deficient.

2.4.6. *Updating of the initial model*

After generating the initial model of the structure and comparing the results with measurement data, the problem may arise that the quality of the predicted model is not as good as it was expected. The reasons for the deviation between measurement and model prediction can be manifold:

– measurement errors (deterministic/stochastic),

– modeling errors (idealized modeling of geometry and boundary conditions, inadequate material laws, inaccurate parameter values for material properties, wrong type of finite elements for the purpose, etc.).

Concentrating on the modeling aspect here, the so-called model updating methods can be used. Model updating deals with the methodology, which localizes and corrects erroneous complex computational models, which are generated by finite element (FE) codes or Multibody System (MBS) codes. Methods for the primary step of localization of the area where the model has been modeled inadequately have been proposed by [LAD 89, EWI 00, LAL 88, LAL 89, LIN 91, FRI 92, FRI 98, FRI 99b, FRIS 97a]. In most cases, automated model correction is restricted to parameter identification. The other error sources are corrected by engineering experience. Generally, all methods for model correction originate in mathematical optimization [NAT 92, LIN 97b, LIN 99]. Finding the right parameter values by trial and error is a very difficult task, especially for a larger number of uncertain parameters. We can formulate this task by minimization of an objective function f (also called cost function or loss function) depending on the set of uncertain parameters θ, which can be material constants, spring stiffnesses, damping constants, friction parameters, etc:

$$\underset{\theta}{Min}\{ f(\theta) \mid h(\theta) = 0, g(\theta) \le 0\} \qquad [104]$$

The functions h and g are equality and inequality constraints which have to be satisfied by the parameters θ. The scalar function f describes the "distance" between the measurement data and the corresponding model predictions. By minimization of f, the model is adaptable to the measurement data, so that the distance becomes smaller. We define residuals as:

$$r_j(\theta) = y_j^{meas} - y_j^{model}(\theta) \qquad [105]$$

where the y values are any two corresponding quantities of measurement and model.

Putting the single residual term into vector form yields:

$$
\begin{pmatrix} r_1(\boldsymbol{\theta}) \\ r_2(\boldsymbol{\theta}) \\ \vdots \\ r_m(\boldsymbol{\theta}) \end{pmatrix} = \begin{pmatrix} y_1 \\ y_2 \\ \vdots \\ y_m \end{pmatrix}^{meas} - \begin{pmatrix} y_1(\boldsymbol{\theta}) \\ y_2(\boldsymbol{\theta}) \\ \vdots \\ y_m(\boldsymbol{\theta}) \end{pmatrix}^{model} \quad \text{or } \boldsymbol{r}(\boldsymbol{\theta}) = \boldsymbol{y}^{meas} - \boldsymbol{y}^{model}(\boldsymbol{\theta})
$$

[106]

which describes the difference between a characteristic measured value and the corresponding calculated value. The y values can be time-response data, eigenfrequencies, components of mode shape vectors, frequency-response data (split into real and imaginary parts if complex), forces or anything else which it makes sense to compare.

The simplest form of residual minimization is based on the minimization of an ordinary unweighted Least Squares (LS) function:

$$
f(\boldsymbol{\theta}) = \sum_{j=1}^{j_{max}} \left[y_j^{meas} - y_j^{model}(\boldsymbol{\theta}) \right]^2 = \left\| \boldsymbol{y}^{meas} - \boldsymbol{y}^{model}(\boldsymbol{\theta}) \right\|^2
$$
$$
= \left(\boldsymbol{y}^{meas} - \boldsymbol{y}^{model}(\boldsymbol{\theta}) \right)^T \left(\boldsymbol{y}^{meas} - \boldsymbol{y}^{model}(\boldsymbol{\theta}) \right)
$$

[107]

where $\|\cdots\|$ denotes the Euclidian vector norm.

Introducing individual weighting of the residuals in order to take the variances of different sensors, etc., into account leads us to the Weighted Least-Squares function (WLS):

$$
f(\boldsymbol{\theta}) = \sum_{j=1}^{j_{max}} w_j \left[y_j^{meas} - y_j^{model}(\boldsymbol{\theta}) \right]^2
$$
$$
= \left(\boldsymbol{y}^{meas} - \boldsymbol{y}^{model}(\boldsymbol{\theta}) \right)^T \boldsymbol{W} \left(\boldsymbol{y}^{meas} - \boldsymbol{y}^{model}(\boldsymbol{\theta}) \right)
$$

[108]

where w_j is a scalar positive weighting coefficient and $\boldsymbol{W} = \text{diag}(w_1, w_2, \ldots, w_{jmax})$ is a diagonal weighting matrix. The more general case of a non-diagonal weighting matrix coupling the different off-diagonal elements is also possible, but the proper determination of the weighting coefficients is much more difficult.

The special case where the weighting matrix is chosen to be the inverse covariance matrix of the residuals:

$$
\boldsymbol{W} = \boldsymbol{\Sigma}^{-1}
$$

[109]

leads us to the Gauss–Markov method. This special use of the weighting matrix yields estimates of the parameters with minimum variance on the assumption that the covariance matrix is known and the measurement noise has a Gaussian distribution:

$$f(\theta) = \left(y^{meas} - y^{model}(\theta) \right)^{T} W \left(y^{meas} - y^{model}(\theta) \right) \dots$$
$$+ \left(\theta - \theta_0 \right)^{T} W_{\theta} \left(\theta - \theta_0 \right)$$

[110]

Compared to the objective function of equation [108], a second quadratic term is added. This term increases if the parameters θ move away from their original values θ_0. W_{θ} is a symmetric, positive definite weighting matrix. It introduces additional information about the parameters. W_{θ} relates to the confidence in the initial parameters θ_0. Low values in W_{θ} mean low confidence, high values meaning high confidence. High values in the weighting matrix W_{θ} lead to large values of f so that the optimizer prefers to keep the parameters close to the original values θ_0. Low values in W_{θ} means: θ can move away from θ_0 without a significant increase of f. It is important to keep a good balance between the two weighting matrices W and W_{θ}.

It is accepted that the whole minimization process in a more general framework of Bayes estimation or Maximum Likelihood estimation maximizing expressions of probability. Because this goes far beyond the scope of this chapter, refer to [NAT 92, FRIS 95, EYK 74, WAL 97, TEU 03]. In [MUE 02, MUE 03]; multi-objective cost functions are used for the optimization instead of one scalar function.

A further comment should be made on the regularizing effect of the second term in equation [110]. To recap, the identification problem is an inverse problem, which becomes ill-posed if information is missing. Unfortunately, the measurement data are usually incomplete in respect of the measured degrees of freedom (the number of measurement locations is much smaller than the number of model degrees of freedom in any practical case) and in respect of the frequency range that is bounded by the experimental devices (sampling rates, storage capacities, frequency characteristics of the sensors, actuators and amplifiers, etc.).

In other words, if information on the measurement alone is partially missing, the result of the objective function with a multitude of parameter solutions (non-unique) will be minimum. The dilemma is solved by adding information directly on the parameters (see the second term in equation [110]).

If the number of parameters is larger than the number of measurement data, the system which can be solved by the regularization term in equation [110] is undetermined.

2.4.6.1. *Updating using modal data*

With modal data as measurement information, the following approach can be used:

$$f(\boldsymbol{\theta}) = \sum_i w_{\lambda,i} \left(\frac{\lambda_{i,meas} - \lambda_i(\boldsymbol{\theta})}{\lambda_{i,meas}} \right)^2 + \sum_i w_{\varphi,i} \left\| \boldsymbol{\varphi}_{i,meas} - MSF_{ii} \, \boldsymbol{\varphi}_i(\boldsymbol{\theta}) \right\|^2 \qquad [111]$$

While the first term minimizes the "distance" between the model and measured eigenfrequencies (eigenvalues), the second term minimizes the difference between the measured and model eigenvectors. Each term can be individually weighted by the weighting coefficients w. Different scaling of measured and calculated eigenvectors must be compensated, e.g. by the modal scale factor MSF. Application of the MAC has to be carried out first, to ensure that the correct mode pairs are compared in the summation.

For most of our updating applications we have used the following alternative approach, which turned out to be well suited:

$$f(\boldsymbol{\theta}) = \sum_{i=1}^{m} w_{\lambda,i} \left(\frac{\lambda_{i,meas} - \lambda_i(\boldsymbol{\theta})}{\lambda_{i,meas}} \right)^2 + w_{\varphi,i} \left(1 - MAC_{ii} \right)^2 \qquad [112]$$

While the first term minimizes the "distance" between the model and measured eigenfrequencies (eigenvalues), the second term maximizes the MAC values of corresponding mode pairs, and thus minimizes $1 - MAC$ ($0 \leq MAC \leq 1$). This brings the eigenvectors into optimum agreement. The MAC does not compare the eigenvector components by directly calculating the difference between the components. It maximizes the correlation via the MAC value. An advantage of this formulation is that different scaling of measured and calculated eigenvectors is automatically compensated by the MAC calculation.

2.4.6.2. Updating using frequency response functions

To compare the complex frequency response function (FRFs), the following cost function is used:

$$f(\boldsymbol{\theta}) = \sum_{k,l,\mu} w_{FRF,kl}^{Re} \left(\mathrm{Re}\{H_{kl,meas}(\Omega_\mu) - H_{kl}(\Omega_\mu, \boldsymbol{\theta})\} \right)^2 \cdots$$
$$+ \quad w_{FRF,kl}^{Im} \left(\mathrm{Im}\{H_{kl,meas}(\Omega_\mu) - H_{kl}(\Omega_\mu, \boldsymbol{\theta})\} \right)^2 \qquad [113]$$

For weakly damped structures, the magnitudes of the FRFs close to the resonances can be several orders higher than in the non-resonant regions. This can cause severe convergent problems with the minimization algorithms because small changes in the parameters create only small shifts in the eigenfrequencies but very large changes in the FRFs or in the differences of the FRFs. The application of exponential windows, which adds artificial damping and which is easy to compensate in the model, may reduce this unexpected effect.

2.4.6.3. *Gradient and gradient-free methods*

If gradient-based methods such as the conjugate gradient method or the Newton and Quasi-Newton (DFP, BFGS, etc.) methods [HIM 72, WAL 97] are used, the necessary but not sufficient condition for the minimum is that the partial derivatives of f with respect to all parameters equals zero:

$$\frac{\partial f}{\partial \theta_j} = 0 \ , j = 1, 2, \ldots, n \tag{114}$$

Partial derivatives of the objective function with respect to the parameters are needed:

$$\frac{\partial f}{\partial \theta_j} = -2 \left(y^{meas} - y^{model}(\theta) \right)^T W \frac{\partial}{\partial \theta_j} \left(y^{model}(\theta) \right) \tag{115}$$

the parameters are sometimes calculated numerically. In derivative-free optimization methods such as genetic algorithms, these derivatives do not have to be determined [BOH 05, GOL 89]. The genetic programming algorithms revealed to be efficient at the early phase of the search for the minimum, while at the later phase the gradient-based methods are more efficient [FRI 04a].

The aspect of derivatives of the natural frequencies, mode shapes or forced vibration responses in the context of damage identification will be discussed again later in this chapter.

The vector functions h and g respectively introduce equality and inequality constraints in constrained optimization which can also take additional information into account. The variation of the parameter j is then restricted to lower and upper bonds:

$$\theta_j^{low} \leq \theta_j \leq \theta_j^{up} \tag{116}$$

This makes sense in the cases where a material parameter such as Young's modulus or the density of a material is precisely or at least known to vary only in a certain small range. In other cases, a parameter can only take a positive number.

2.5. Damage localization and quantification

2.5.1. *Change of the flexibility matrix*

2.5.1.1. *Method of Pandey and Biswas*

Pandey and Biswas [PAN 94] suggested to use the static flexibility matrix to display the damaged areas. As shown in equation [14], the flexibility matrix can be written as a mode superposition:

$$H_{stat,0} \approx \sum_{i=1}^{m_{red}} \frac{\varphi_{0,i}\, \varphi_{0,i}^{T}}{\omega_{0,i}^{2}} \qquad\qquad [117]$$

$$H_{stat,d} \approx \sum_{i=1}^{m_{red}} \frac{\varphi_{d,i}\, \varphi_{d,i}^{T}}{\omega_{d,i}^{2}} \qquad\qquad [118]$$

$$\Delta H_{stat} = H_{stat,0} - H_{stat,d} \qquad\qquad [119]$$

It should be noted that the residual flexibility terms representing the higher modes are missing. However, higher modes less contribute to the flexibility matrix because of larger eigenfrequency values in the denominator. Pandey and Biswas [PAN 94] propose the following procedure: for each measurement dof j they determine the maximum absolute values of the jth column of ΔH_{stat}:

$$\overline{\delta}_{j} = \max_{i} \left| \Delta h_{ij,stat} \right| \qquad\qquad [120]$$

In order to detect and locate damage in the structure, they use the quantity $\overline{\delta}_{j}$ as a measure of change in flexibility for each measurement location.

2.5.1.2. Damage Locating Vectors (DLV)

Based on the change of the flexibility matrix, Bernal [BER 00, BER 02] proposes a method that uses Damage Locating Vectors (DLV). The main idea is to compute a set of load vectors L, called DLVs, which have the property of inducing stress fields whose magnitude is either zero in the damaged elements or small in the presence of truncations and approximations. The DLVs are associated with sensor co-ordinates and are systematically computed in the null space[2] of the change of the measured flexibility matrix:

$$(H_{stat,0} - H_{stat,d})\, L = \Delta H_{stat} L = 0 \qquad\qquad [121]$$

This means that the load vector L produces identical deformations when applied to the undamaged and the damaged structure:

$$x_{stat} = H_{stat,0}\, L = H_{stat,d}\, L \qquad\qquad [122]$$

Although the two systems are locally different because of the presence of damage, identical displacements for the two cases by the special choice of the load vector are obtained. This result can only be achieved if these different structural parts are stress-free or their strain energy is zero – which is the same. This is used to

---

[2] Given a rank-deficient matrix A, all solution vectors $x \neq 0$ satisfying $Ax = 0$ lye in the null space of A.

localize the damaged area, see [BER 00]. To satisfy equation [121], there is a second possibility:

$$H_{stat,0} \; L = H_{stat,d} \; L = 0$$

This possibility occurs if the matrices $H_{stat,0}, H_{stat,d}$ are themselves rank deficient and their null spaces have a non-zero intersection [BER 00]. In this case the load vectors L are not DLVs and the damage cannot be located. A sufficient condition for rank deficiency is that the number of identified modes is smaller than the number of sensors. The modal data are determined by means of the ERA approach, see [JUA 94]. For further details, see [BER 00, BER 02].

2.5.2. Change of the stiffness matrix

Ewins and Sidhu [EWI 00] have derived a method for the updating of an analytical model using measured mode shapes. This method can be applied to display changes from the stiffness matrix:

$$\Delta K = K_d - K_0 \tag{123}$$

to the dynamics of the structure:

$$\Delta K = K_{0,red} \; (K_0^{-1} - K_d^{-1}) \; K_{0,red} = K_{0,red} \; (H_{stat,0} - H_{stat,d}) \; K_{0,red} \tag{124}$$

The matrix $K_{0,red}$ is the condensed stiffness matrix, reduced from the original analytical stiffness matrix K_0 to the measured degrees of freedom, according to:

$$K_{0,red} = K_{0,mm} - K_{0,mu} K_{0,uu}^{-1} K_{0,um} \tag{125}$$

The flexibility matrices (equation [124]) are calculated using the incomplete set of measured modes from the undamaged and damaged states.

Another group of methods determines the stiffness matrix K_d by minimizing the change of the stiffness matrix [BAR 78, BAR 82]:

$$J = \frac{1}{2} \left\| N^{-1} (K_d - K_0) \; N^{-1} \right\| = Min. \tag{126}$$

with the weighting matrix $N = M_0^{1/2}$ under the constraints of the eigenvalue problem and the symmetry of the stiffness matrix:

$$(K_d - \omega_{d,i}^2 M)\varphi_{d,i} = 0 , \; i = 1,...,r \le m \; \text{ and } \; K_d = K_d^T \tag{127}$$

A similar procedure is presented by [BER 83] with an updated mass matrix as weighting matrix instead of the original analytical mass matrix. The results of these methods and some variations are compiled by Friswell and Mottershead [FRIS 95].

2.5.3. Strain-energy-based indicator methods and curvature modes

2.5.3.1. The strain energy method

This method, developed by Stubbs *et al.* [STU 95], has been used for further studies [WOR 01, MAE 03c]. The derivation of the method is tailored to an Euler–Bernoulli beam which aim is to find changes in bending stiffness. The basic idea is that the change in the curvature of the vibration modes is a good indicator for local damage [PAN 91]. For an Euler–Bernoulli beam the strain ε is:

$$\varepsilon = \kappa z \qquad [128]$$

where κ is the curvature of the beam axis and z is the co-ordinate perpendicular to the beam axis, starting from the neutral axis. For small deformations, it is commonly accepted that the curvature is approximately equal to the second derivative of the beam deflection w:

$$\kappa \approx w'' \qquad [129]$$

We follow the basic ideas presented by [STU 95], although the notation has been adapted to the symbols used in this chapter. The total strain energy of the whole beam can be obtained by integration over the full beam length L:

$$U_{tot} = \frac{1}{2} \int_0^L EI[w''(x)]^2 \, dx \qquad [130]$$

Note that the shear deformation is neglected here. Basically, the bending stiffness EI can vary along the beam axis. Considering only a small section of the beam, the portion of strain energy for this beam section of length Δl is:

$$U_{el,j} = \frac{1}{2} \int_{\Delta l} EI[w''(x)]^2 \, dx \qquad [131]$$

The modal strain energy can be obtained by replacing the general displacement w by the continuous mode shapes $\varphi_i(x)$. Thus, the modal strain energy for the whole beam or a part of it from the expressions is written as:

$$U_{tot,i} = \frac{1}{2} \int_L EI \, [\varphi''(x_i)]^2 \, dx \qquad [132]$$

$$U_{el,j} = \frac{1}{2} \int_{\Delta l} EI[\varphi''(x_i)]^2 \, dx \qquad\qquad\qquad [133]$$

Introducing the ratio of the element energy according to the total energy for the undamaged (0) and the damaged (d) states:

$$F_{i,el}^d = \frac{U_{el,i}^d}{U_{tot,i}^d} \qquad\qquad\qquad [134]$$

$$F_{i,el}^0 = \frac{U_{el,i}^0}{U_{tot,i}^0} \qquad\qquad\qquad [135]$$

Obviously, the sum over all elements is $\sum_{el} F_{i,el}^d = \sum_{el} F_{i,el}^0 = 1$ and it is assumed that the $F_{i,el}^d \ll 1$ and $F_{i,el}^0 \ll 1$. This implies that it can be approximately set that:

$$1 + F_{i,el}^d \cong 1 + F_{i,el}^0 \qquad\qquad\qquad [136]$$

Substituting the expressions [134] and [135] into the last equation yields:

$$1 = \frac{1 + (U_{el,i}^d / U_{tot,i}^d)}{1 + (U_{el,i}^0 / U_{tot,i}^0)} = \frac{(U_{el,i}^d + U_{tot,i}^d)}{(U_{el,i}^0 + U_{tot,i}^0)} \frac{U_{tot,i}^0}{U_{tot,i}^d} \qquad\qquad\qquad [137]$$

By making some further assumptions, Stubbs, Kim and Farrar [STU 95] developed the following formula:

$$\beta_{el,i} = \frac{EI_{el}^0}{EI_{el}^d} = \frac{\int_{\Delta l}[\varphi_i''^d(x)]^2 \, dx + \int_L[\varphi_i''^d(x)]^2 \, dx}{\int_{\Delta l}[\varphi_i''^0(x)]^2 \, dx + \int_L[\varphi_i''^0(x)]^2 \, dx} \frac{\int_L[\varphi_i''^0(x)]^2 \, dx}{\int_L[\varphi_i''^d(x)]^2 \, dx} = \frac{Num_{el,i}}{Den_{el,i}} \qquad [138]$$

To account for all available modes, a single indicator is derived in [STU 95]:

$$\beta_{el} = \frac{\sum_i Num_{el,i}}{\sum_i Den_{el,i}} \qquad\qquad\qquad [139]$$

Furthermore, the indicator is normalized by:

$$Z_{el} = \frac{\beta_{el} - \mu_\beta}{\sigma_\beta} \qquad\qquad\qquad [140]$$

It is commonly recognized that the numerical computation of the second-order derivative causes problems when data are noisy. In his PhD thesis, Maeck [MAE 03c] used Lagrange interpolation polynomials to calculate the second-order derivatives. To avoid rough curvatures, he used smoothing by means of regularization methods.

Farrar *et al.* use a slightly different derivation in their report [FAR 00]. Instead of equation [137], they set:

$$F_{i,el}^d \cong F_{i,el}^0 \qquad [141]$$

on the basis of the argument that the damage is primarily located at one single sub-region and, therefore, the fractional energy will remain approximately constant in the undamaged sub-regions. Putting the fractional energies [132] to [135] into equation [141] leads to:

$$\beta_{el,i}^* = \frac{EI_{el}^0}{EI_{el}^d} = \frac{\int_{\Delta l}[\varphi_i^{\prime\prime d}(x)]^2\,dx}{\int_{\Delta l}[\varphi_i^{\prime\prime 0}(x)]^2\,dx}\frac{U_{tot,i}^0}{U_{tot,i}^d} = \frac{Num_{el,i}^*}{Den_{el,i}^*} \qquad [142]$$

As before, the summation over all modes yields:

$$\beta_{el}^* = \frac{\sum_i Num_{el,i}^*}{\sum_i Den_{el,i}^*} \qquad [143]$$

This damage indicator gives less accurate results if strain energies of a considered element are small. In this case, two small values must be divided. In the case of noisy data this may lead to accurate problems. The method has also been extended to plate structures, see for example [COR 99].

2.5.3.2. *Adaptation of the damage indicator method to finite element formulation*

In finite beam elements, the beam deflection is usually expressed by a polynomial approximation as a function of the nodal displacements and rotations. Thus, the second derivative of the curvatures can also be expressed in terms of nodal displacements (including rotations). From the derivation of element formulation using energy principles, it is accepted that the strain energy in the whole structure and at element level can be expressed by:

$$U_{tot} = \frac{1}{2}u^T K\,u \qquad [144]$$

$$U_{el,j} = \frac{1}{2}u^T K_{el,j}u \qquad [145]$$

The modal strain energy is:

$$U_i = \frac{1}{2} \varphi_i^T K \, \varphi_i \qquad\qquad [146]$$

$$U_{el,i} = \frac{1}{2} \varphi_i^T K_{el} \, \varphi_i \qquad\qquad [147]$$

This formulation avoids calculating the curvatures. Providing the mode shapes which are mass normalized $\varphi_i^T M \varphi_i = 1$, we know that the energy expressions:

$$\varphi_i^T K \varphi_i = \lambda_i = \omega_i^2 \qquad\qquad [148]$$

are related to the eigenfrequencies, which are known from the measurements in the undamaged and damaged states. Now, the damage indicator (equation [142]) can be reformulated:

$$\beta_{el,i}^* = \frac{EI_{el}^0}{EI_{el}^d} = \frac{\frac{1}{2}EI_{el}^0 \int\limits_{\Delta l} [\varphi_i''^{\,d}(x)]^2 \, dx}{\frac{1}{2}EI_{el}^0 \int\limits_{\Delta l} [\varphi_i''^{\,0}(x)]^2 \, dx} \left(\frac{\lambda_{i,0}}{\lambda_{i,d}} \right) = \frac{Num_{el,i}^*}{Den_{el,i}^*} \qquad\qquad [149]$$

where the strain energies of the structure are replaced by the eigenvalues. The element strain energy expression in the numerator is calculated with the (known) bending stiffness of the undamaged element and the (known, measured) mode shapes of the damaged element:

$$U_{el,i}^{0d} = \frac{1}{2}EI_{el}^0 \int\limits_{\Delta l} [\varphi_i''^{\,d}]^2 \, dx = \frac{1}{2} \varphi_{i,d}^T K_{el,0} \, \varphi_{i,d} \qquad\qquad [150]$$

By analogy, the expression:

$$U_{el,i}^0 = \frac{1}{2}EI_{el}^0 \int\limits_{\Delta l} [\varphi_i''^{\,0}]^2 \, dx = \frac{1}{2} \varphi_{i,0}^T K_{el,0} \, \varphi_{i,0} \qquad\qquad [151]$$

is calculated with the original element stiffness and the mode shapes of the undamaged system. The resulting expression for the damage indicator becomes very simple and no numerical calculation of curvatures is required:

$$\beta_{el,i}^* = \frac{EI_{el}^0}{EI_{el}^d} = \frac{U_{el,i}^{0d}}{U_{el,i}^0} \left(\frac{\lambda_{i,0}}{\lambda_{i,d}} \right) = \frac{Num_{el,i}^*}{Den_{el,i}^*} \qquad\qquad [152]$$

Again, the summation over all available modes yields:

$$\beta_{el}^* = \frac{\sum\limits_i Num_{el,i}^*}{\sum\limits_i Den_{el,i}^*} \qquad\qquad [153]$$

The change of the bending stiffness according to the damage parameters introduced in equation [152] is:

$$\beta_{el}^* = \frac{EI_{el}^0}{EI_{el}^d} = \frac{1}{1+\Delta a_{el}}$$ [154]

which yields the relative reduction of stiffness:

$$\Delta a_{el} = \frac{1}{\beta_{el}^*} - 1$$ [155]

In a similar way, the damage indicator (equation [139]) can be adapted to the FE formulation.

2.5.4. MECE error localization technique

MECE is an acronym for Minimization of Errors in the Constitutive Equations. A formulation of this indicator method based on a continuum mechanical description was given in [LAD 89] and different FE based formulations in matrix notation can be found in [PAS 99]. The error indicator for the jth substructure is basically derived from a normalized strain energy expression using the ith mode:

$$e_{j,i} = \frac{(u_i - v_i)^T K_{el,j}(u_i - v_i)}{\frac{1}{2} u_i^T K_0 u_i + \frac{1}{2} v_i^T K_0 v_i}$$ [156]

Here, the displacement vector is v as the expanded mode shape:

$$v_i = \varphi_{i,\exp d}$$ [157]

resulting from a dynamic expansion of the ith mode shape of the damaged structure. Once v_i is known, u_i can be calculated from:

$$K_0 u_i = \omega_{i,meas}^2 M_0 v_i$$ [158]

which shows the mismatch between the eigenvector and eigenvalue of the damaged system and the matrices corresponding to the undamaged system. If there is no mismatch, then u and v are identical and all error indicators are zero. The higher the error indicator e, the larger is the modeling error at sub-structure j. To take all modes into account, the damage indicator can be calculated by the following summation:

$$e_j = \frac{\sum_i (u_i - v_i)^T K_{el,j}(u_i - v_i)}{\frac{1}{2} \sum_i u_i^T K_0 u_i + v_i^T K_0 v_i}$$ [159]

As is shown in [PAS 99], the strain energy can also spread throughout the structure in the ideal no-noise, no-expansion case. Another indicator-based method was presented in [STU 90].

2.5.5. *Static displacement method*

Before treating dynamic methods, static displacements are briefly considered. The elastic properties of a linear mechanical structure are characterized by the stiffness matrix K_0 only. Applying a static load f_i to this system (load case i), the displacement vector follows from the solution of the linear equation system:

$$K_0 x_{0i} = f_i \qquad [160]$$

If the structure is damaged, the stiffness matrix locally or globally changes, so that $K = K_0 + \Delta K$. The same force is applied with the difference that, in the damaged case, the new displacement vector x_d, which differs from the previous one according to the location, type and extent of the damage, is obtained:

$$(K_0 + \Delta K) x_{di} = f_i \qquad [161]$$

Combining the last two equations gives:

$$(K_0 + \Delta K) x_{di} = K_0 x_{0i} \qquad [162]$$

or:

$$-\Delta K \, x_{di} = K_0 (x_{di} - x_{0i}) \qquad [163]$$

Replacing the matrix ΔK using equation [64]:

$$-\sum_j (K_j x_{di}) \, \Delta a_j = K_0 (x_{di} - x_{0i}) \qquad [164]$$

Premultiplying with the inverse stiffness matrix yields:

$$-\sum_j (K_0^{-1} K_j x_{di}) \, \Delta a_j = (x_{di} - x_{0i}) \text{ or } \sum_j s_{i,j} \, \Delta a_j = r_i \qquad [165]$$

or as an equation system:

$$S_i \, \Delta a = r_i \qquad [166]$$

Changes in the static displacements are used here to establish a relation with the stiffness parameters. Taking several load cases into account, an over-determined equation system can be built up, from which the Δa_j can be calculated.

In practice, the static displacements can be generated, for example, by a heavy test vehicle moving slowly along a bridge, while the static displacement due to this loading is measured.

Ben Haim [BEN 92] has shown in his method of "Selective Sensitivity" that one can make special parameters sensitive while other parameters are totally insensitive in the case where special force configurations are chosen. Later, the theory was expanded to the dynamic case using modal data [COG 94].

2.5.6. *Inverse eigensensitivity method*

This well-established method is based on the fact that the change of the eigenvalues and eigenvectors (due to the damage) can be approximated by a linear Taylor expansion depending on the correction parameters Δa_j:

$$\Delta \bar{\lambda}_i = \frac{\Delta \lambda_i}{\lambda_i} \approx \sum_{j=1}^{n} \frac{1}{\lambda_i} \frac{\partial \lambda_i}{\partial a_j} \Delta a_j \qquad [167]$$

and:

$$\Delta \varphi_i \approx \sum_{j=1}^{n} \frac{\partial \varphi_i}{\partial a_j} \Delta a_j . \qquad [168]$$

The partial derivatives of the modal quantities with respect to the correction parameters can be calculated according to either [FOX 68] or [NEL 76]. Practical experience has shown that relative changes of the eigenvalues can be used instead of absolute changes. Especially for the higher modes, the changes of the eigenfrequency/eigenvalue are larger than for the lower modes. The derivative of equation [9] with respect to a parameter a_j yields:

$$(K - \lambda_i M)\frac{\partial \varphi_i}{\partial a_j} = -\left[\frac{\partial K}{\partial a_j} - \lambda_i \frac{\partial M}{\partial a_j} - \frac{\partial \lambda_i}{\partial a_j} M \right] \varphi_i \qquad [169]$$

Premultiplication of the transposed eigenvector φ_i^T and the assumption that K and M are symmetric matrices ($M = M^T$ and $K = K^T$) yield:

$$0 = -\varphi_i^T \left[\frac{\partial K}{\partial a_j} - \lambda_i \frac{\partial M}{\partial a_j} - \frac{\partial \lambda_i}{\partial a_j} M \right] \varphi_i \qquad [170]$$

By rearranging the last equation, also taking into account that the normalization is determined by $\varphi_i^T M_0 \varphi_i = 1$, one gets the eigenvalue sensitivity:

$$\frac{\partial \lambda_i}{\partial a_j} = \frac{\partial (\omega_i^2)}{\partial a_j} = \boldsymbol{\varphi}_i^T \left(\frac{\partial \boldsymbol{K}}{\partial a_j} - \lambda_i \frac{\partial \boldsymbol{M}}{\partial a_j} \right) \boldsymbol{\varphi}_i \; ; \; \lambda_i = \omega_i^2 \tag{171}$$

or:

$$\frac{\partial \lambda_i}{\partial a_j} = \frac{\partial (\omega_i^2)}{\partial a_j} = \boldsymbol{\varphi}_i^T (\boldsymbol{K}_j - \lambda_i \boldsymbol{M}_j) \, \boldsymbol{\varphi}_i \tag{172}$$

The original approach by Fox and Kapoor [FOX 68] of calculating the eigenvector derivatives starts from the fact that any m-dimensional vector (here the vector of the eigenvector derivatives) is determined by the basis of m linearl independent vectors, which are in this case the m mode shape vectors:

$$\frac{\partial \boldsymbol{\varphi}_i}{\partial a_j} = \sum_{\mu=1}^{m} a_{ji\mu} \, \boldsymbol{\varphi}_\mu \tag{173}$$

with unknown coefficients $a_{ji\mu}$ (j and i indicate the parameter index and the current mode shape index, respectively, while μ is a summation index). Replacing the derivative in equation [169] by [173] and making use of the orthogonality condition [11] yields:

$$a_{ji\mu} = \frac{\boldsymbol{\varphi}_\mu^T (\boldsymbol{K}_j - \lambda_i \boldsymbol{M}_j) \boldsymbol{\varphi}_i}{\lambda_i - \lambda_\mu} \quad \text{for} \; i \neq \mu \tag{174}$$

The coefficient for $i = \mu$ can be derived by differentiation of the orthogonality relation [11] with respect to the jth parameter. Straightforward calculation leads to:

$$a_{jii} = \frac{1}{2} \boldsymbol{\varphi}_i^T \boldsymbol{M}_j \, \boldsymbol{\varphi}_i \tag{175}$$

Using this procedure, the calculation of the eigenvector derivative shows the disadvantage that all n eigenvectors are required to calculate the complete sum. However, for large models, the calculation of all eigenvectors is impractical. An alternative was proposed by Nelson [NEL 76]. Nelson's algorithm only requires the ith eigenvector for the ith derivative. This method takes more computational time but the benefit is much more accurate than the Fox/Kapoor method when the summation is incomplete. In condensed form the derivatives are also presented by Friswell and Mottershead in [FRIS 95].

The residuals of the measured eigenvalues and eigenvectors, as well as the model based partial derivatives, are arranged in the following way [FRI 99b]:

$$\boldsymbol{S}_{modal} \, \Delta \boldsymbol{a} = \boldsymbol{r}_{modal} \tag{176}$$

where:

$$S_{modal} = \begin{bmatrix} \dfrac{1}{\lambda_1}\dfrac{\partial \lambda_1}{\partial a_1} & \dfrac{1}{\lambda_1}\dfrac{\partial \lambda_1}{\partial a_2} & \cdots & \dfrac{1}{\lambda_1}\dfrac{\partial \lambda_1}{\partial a_n} \\[2mm] \dfrac{1}{\lambda_2}\dfrac{\partial \lambda_2}{\partial a_1} & \dfrac{1}{\lambda_2}\dfrac{\partial \lambda_2}{\partial a_2} & \cdots & \dfrac{1}{\lambda_2}\dfrac{\partial \lambda_2}{\partial a_n} \\[2mm] \vdots & \vdots & & \vdots \\[2mm] \dfrac{\partial \{\varphi_1\}}{\partial a_1} & \dfrac{\partial \{\varphi_1\}}{\partial a_2} & \cdots & \dfrac{\partial \{\varphi_1\}}{\partial a_n} \\[2mm] \dfrac{\partial \{\varphi_2\}}{\partial a_1} & \dfrac{\partial \{\varphi_2\}}{\partial a_2} & \cdots & \dfrac{\partial \{\varphi_2\}}{\partial a_n} \\[2mm] \vdots & \vdots & & \vdots \end{bmatrix} ; \ r_{modal} = \begin{bmatrix} \Delta \bar{\lambda}_1 \\ \Delta \bar{\lambda}_2 \\ \vdots \\ \{\Delta \varphi_1\} \\ \{\Delta \varphi_2\} \\ \vdots \end{bmatrix}. \qquad [177]$$

with which the measured residuals between the undamaged 0 and the damaged d state:

$$r_{modal} = \begin{bmatrix} r_\lambda \\ r_\varphi \end{bmatrix}, \qquad\qquad\qquad [178]$$

with:

$$\begin{aligned} r_\lambda &= \begin{bmatrix} \dfrac{\lambda_{1,d} - \lambda_{1,0}}{\lambda_{1,0}}, & \dfrac{\lambda_{2,d} - \lambda_{2,0}}{\lambda_{2,0}}, & \cdots \end{bmatrix}^T \\[3mm] &= \begin{bmatrix} \dfrac{\omega_{1,d}^2 - \omega_{1,0}^2}{\omega_{1,0}^2}, & \dfrac{\omega_{2,d}^2 - \omega_{2,0}^2}{\omega_{2,0}^2}, & \cdots \end{bmatrix}^T \end{aligned} \qquad [179]$$

and:

$$r_\varphi = \begin{bmatrix} \varphi_{1,d} - \varphi_{1,0} \\ \varphi_{2,d} - \varphi_{2,0} \\ \vdots \end{bmatrix} \qquad\qquad [180]$$

can be explained. As discussed in section 2.4.1, the problem of proper mode pairing of the damaged and undamaged structures can be solved using the MAC criterion. Many authors, for example Cawley and Adams [CAW 79], only used the eigenfrequencies for damage identification. Usually, eigenfrequencies can be more accurately measured than the eigenvectors. However, the eigenvectors include important spatial information.

2.5.7. *Modal force residual method*

2.5.7.1. *The basic concept*

If the measured eigenfrequencies and expanded mode shapes of the damaged system are introduced into the eigenvalue problem [9], we get a mismatch resulting in a modal force residual:

$$\Delta f_{i,0} = \left(K_0 - \omega_{i,m}^2 M_0 \right) T_\omega \varphi_{i,m} \qquad [181]$$

Now, a matrix ΔK which keeps the norm of the residual vector Δf small for all modes I is to be found:

$$\Delta f_i = \left(K_0 + \Delta K - \omega_{i,m}^2 M_0 \right) T_\omega \varphi_{i,m} \overset{!}{=} 0 \qquad [182]$$

Rearranging the last equation and replacing $T_\omega \varphi_{i,m} = \varphi_{i,m,expd}$ leads to:

$$-\Delta K \varphi_{i,m,expd} = (K_0 - \omega_{i,m}^2 M_0)\varphi_{i,m,expd} \qquad [183]$$

Introducing the correction parameters Δa according to $\Delta K = \sum_j K_j \Delta a_j$ the equation is obtained by:

$$-\sum_j (K_j \varphi_{i,m,expd}) \Delta a_j = (K_0 - \omega_{i,m}^2 M_0)\varphi_{i,m,expd} \qquad [184]$$

which after considering all modes i and arranging them blockwise yields $S \Delta a = r$ with:

$$s_{i,j} = -K_j \varphi_{i,m,expd} \qquad [185]$$

$$r_i = \Delta f_{i,0} = \left(K_0 - \lambda_{i,m} M_0 \right) \varphi_{i,m,expd} \qquad [186]$$

which are put together as:

$$S = \begin{bmatrix} s_{1,1} & s_{1,2} & \cdots & \cdots s_{1,n_p} \\ \vdots & \vdots & & \vdots \\ s_{i,1} & s_{i,2} & s_{i,j} & \cdots s_{i,n_p} \\ \vdots & & & \vdots \end{bmatrix} \qquad [187]$$

$$r = \begin{bmatrix} r_1 \\ r_2 \\ r_i \\ \vdots \end{bmatrix} \qquad [188]$$

The solution of this equation system, which is generally overdetermined, makes it possible to determine the unknown correction parameters Δa describing the damage.

2.5.7.2. Formulation of the modal force residual method with condensed system matrices

One approach is to work with the expanded vector of dimension m (and matrices M_0 and K_0 of size $m \times m$) as shown earlier in this chapter. The other possibility (computer memory saving) is to reduce the order of the number of measurement dofs $m_{meas} \ll m$. By premultiplying equation [182] by T_ω^T we get:

$$\Delta f_{i,red} = T_\omega^T \left(K_0 + \Delta K - \omega_{i,meas}^2 M_0 \right) T_\omega \varphi_{i,meas} \qquad [189]$$

Introducing the dynamically condensed mass, stiffness and substructure matrices:

$$M_{0,red} = T_\omega^T M_0 T_\omega; \quad K_{0,red} = T_\omega^T K_0 T_\omega;$$
$$K_{j,red} = T_\omega^T K_j T_\omega \qquad [190]$$

and using the parameters according to equation [64], the modal force residual is obtained by:

$$\varepsilon_i = \Delta f_{i,red} = \left(K_{0,red} - \omega_{i,m}^2 M_{0,red} \right) \varphi_{i,m}$$
$$= -\sum_{j=1}^{n} (K_{j,red} \, \varphi_{i,m}) \Delta a_j \qquad [191]$$

If this equation, for all modes i according to equations [185] and [188], is arranged in one equation system, the solution for Δa can be obtained in the same manner, solving [176] and using parameter subset selection as shown in [FRI 99a, FRI 99b].

In contrast to the linearization of the last section, this method is linear in the parameters and, from this point of view, it is exact. However, the expansion/reduction is a source of possible inaccuracy. The baseline measurement data are not used explicitly here.

2.5.7.3. Weighted modal force residual method

An alternative formulation of the MFR method is now presented. It is accepted that the result of input residual methods such as the MFR method is very sensitive to errors. These errors are the expansion errors or stochastic noise in the data:

$$\Delta f_{i,0} = \left(K_0 - \lambda_{i,m} M_0 \right) \varphi_{i,m,expd} \qquad [192]$$

To compensate the residual error, a change in the stiffness matrix ΔK (no mass changes) is introduced by:

$$0 = \left(K_0 + \Delta K - \lambda_{i,m} M_0 \right) \varphi_{i,m,expd} \qquad [193]$$

On the other hand, the mismatch between the actual data and the 0-reference model can be expressed by:

$$K_0 \psi_i = \lambda_{i,m} M_0 \varphi_{i,m,expd} \qquad [194]$$

Combining the last two equations yields:

$$-\Delta K \varphi_{i,m,expd} = K_0 (\varphi_{i,m,expd} - \psi_i) \qquad [195]$$

Premultiplying this equation by the inverse stiffness matrix $K_0^{-1} = H_0$, which is the static flexibility matrix, yields:

$$-H_0 \Delta K \varphi_{i,m,expd} = (\varphi_{i,m,expd} - \psi_i) \qquad [196]$$

This step acts as the residual filter or weighting matrix of the residuals. Finally, replacing the global change ΔK by the correction parameters yields:

$$-\sum_j H_0 K_j \varphi_{i,m,expd} \, \Delta a_j = (\varphi_{i,m,expd} - \psi_i) \qquad [197]$$

where the sensitivity expressions for a mode i and a parameter j are:

$$s_{i,j} = -H_0 K_j \varphi_{i,m,expd} \qquad [198]$$

and the corresponding residual is:

$$r_i = \varphi_{i,m,expd} - \psi_i \qquad [199]$$

which enables us to formulate an equation system of the type: $S\Delta a = r$.

2.5.7.4. *Modification by Hu* et al.

A method proposed by Hu *et al.* [HU 01], which is also based on the force residual, can be directly derived from the previous steps by premultiplying equation [189] by the transposed kth eigenvectors of the undamaged system:

$$\varphi_{k,0}^T \Delta f_{i,red} = \varphi_{k,0}^T \left(K_{0,red} - \omega_{i,d}^2 M_{0,red} \right) \varphi_{i,d} = -\sum_{j=1}^{n} (\varphi_{k,0}^T K_{j,red} \, \varphi_{i,d}) \Delta a_j \qquad [200]$$

The vectors $\varphi_{k,0}^T$ and $\varphi_{i,d}$ both are measured vectors. The next aim is the elimination of the stiffness matrix, which is considered to be the most erroneous part of equation [200]. Because:

$$\boldsymbol{\varphi}_{k,0}^{T} \boldsymbol{K}_{0,red} = \omega_{i,0}^{2} \ \boldsymbol{\varphi}_{k,0}^{T} \boldsymbol{M}_{0,red} \tag{201}$$

the stiffness matrix can be replaced by the mass matrix in equation [200], which leads to:

$$\left(\omega_{k,0}^{2} - \omega_{i,d}^{2} \right) \left[\boldsymbol{\varphi}_{k,0}^{T} \ \boldsymbol{M}_{0,red} \ \boldsymbol{\varphi}_{i,d} \right] = -\sum_{j=1}^{n} (\boldsymbol{\varphi}_{k,0}^{T} \boldsymbol{K}_{j,red} \ \boldsymbol{\varphi}_{i,d}) \Delta a_{j} \tag{202}$$

This gives as many equations as there are combinations k, i. Only the mass matrix and the sub-stiffness matrices are retained. As measurement information the eigenfrequencies and mode shapes of both the damaged and the undamaged structures are used. Two mechanisms are at work here: the deviation of the eigenfrequencies and the mistuning of the orthogonality condition by $\boldsymbol{\varphi}_{i,d}$ in square brackets. The $\boldsymbol{\varphi}_{0}$ and $\boldsymbol{\varphi}_{d}$ are normalized such that $\boldsymbol{\varphi}_{0,i}^{T} \boldsymbol{M}_{0,red} \boldsymbol{\varphi}_{0,i} = 1$ and $\boldsymbol{\varphi}_{d,k}^{T} \boldsymbol{M}_{0,red} \boldsymbol{\varphi}_{d,k} = 1$ for all possible i, k.

2.5.7.5. Minimum-rank perturbation technique

Zimmermann introduced his Minimum-Rank Perturbation Technique (MRPT) to determine the minimum-rank perturbation of the stiffness matrix such as the measured and the analytical modal properties are in agreement [ZIM 99, ZIM 94]. The mass matrix is assumed to be correct. The mode shape vectors have size which is compatible with the system matrices. We start with the eigenvalue problem and with:

$$-\Delta \boldsymbol{K} \boldsymbol{\varphi}_{i,m,expd} = (\boldsymbol{K}_{0} - \omega_{i,m}^{2} \boldsymbol{M}_{0}) \boldsymbol{\varphi}_{i,m,expd} = \boldsymbol{r}_{i} \tag{203}$$

Putting the measured eigenvectors column by column into the matrix $\boldsymbol{\Phi}$ and arranging the eigenfrequencies in a diagonal matrix $\Lambda = diag(\omega_{1}^{2}, \omega_{2}^{2}, \ldots, \omega_{m_{red}}^{2})$ yields:

$$\boldsymbol{K}_{0} \boldsymbol{\Phi}_{m,expd} - \boldsymbol{M}_{0} \boldsymbol{\Phi}_{m,expd} \Lambda = -\Delta \boldsymbol{K} \boldsymbol{\Phi}_{m,expd} = \boldsymbol{R} \tag{204}$$

with:

$$\boldsymbol{R} = [\boldsymbol{r}_{1}, \boldsymbol{r}_{2}, \ldots] \tag{205}$$

Now, according to [ZIM 99], the solution for the MRPT problem is:

$$\Delta \boldsymbol{K} = -\boldsymbol{R} (\boldsymbol{R}^{T} \boldsymbol{\Phi}_{m,expd})^{-1} \boldsymbol{R}^{T} \tag{206}$$

Detailed derivations can be found in [ZIM 94]. This method has also been applied for FRFs and for Ritz-vectors by Zimmerman et al. (see [ZIM 99]).

2.5.8. *Kinetic and strain energy-based sensitivity methods*

2.5.8.1. *Modal kinetic and strain energy*

Another residual for damage detection based on the changes of the modal kinetic energy of the elements is proposed in the following section. This method was introduced by [FRI 04a, BOH 05]. For a conservative, undamaged 0-system which performs free vibrations in the *i*th mode:

$$u(t) = \varphi_i \sin(\omega_i t) \qquad\qquad [207]$$

The velocities follow from the time derivative of *u*:

$$\dot{u}(t) = \omega_i \varphi_i \cos(\omega_i t) \qquad\qquad [208]$$

The strain energy of the structure is:

$$U_i(t) = \frac{1}{2} u^T K\, u \ = \frac{1}{2} \varphi_i^T K \varphi_i\, \sin^2(\omega_i t) \qquad\qquad [209]$$

and kinetic energy is:

$$T_i(t) = \frac{1}{2} \dot{u}^T M\, \dot{u} \ = \frac{1}{2} \omega_i^2 \varphi_i^T M \varphi_i\, \cos^2(\omega_i t) \qquad\qquad [210]$$

which are both positive for any time instant *t*:

$$U_i(t) + T_i(t) = \frac{1}{2} \varphi_i^T K \varphi_i \sin^2(\omega_i t) + \frac{1}{2} \omega_i^2 \varphi_i^T M \varphi_i \cos^2(\omega_i t) = const. \qquad\qquad [211]$$

In particular for the two phase angles, $\omega_i t = 0$ and $\pi/2$ where sin and cos take the maximum value of 1, we get:

$$U_{i,\max} = \frac{1}{2} \varphi_i^T K \varphi_i = T_{i,\max} = \frac{1}{2} \omega_i^2 \varphi_i^T M \varphi_i \qquad\qquad [212]$$

the equality between the maximum modal kinetic energy $T_{i,\max}$ and the maximum modal strain energy $U_{i,\max}$ for a certain mode *i*. Omitting the subscript "max" and introducing 0 for the reference state, we can write:

$$T_{i,0} = U_{i,0} \qquad\qquad [213]$$

2.5.8.2. *Modal kinetic energy-based sensitivity method*

This method was recently introduced by [BOH 03, FRI 04]. The total modal kinetic energy is distributed along all substructures or finite elements *el* so that the total energy can be split into the contributions of each substructure/element:

$$T_{i,0} = \sum_{el=1}^{n} T_{i,0,el} = \frac{1}{2}\omega_{i,0}^2 \sum_{el=1}^{n} \boldsymbol{\varphi}_{i,0}^T \boldsymbol{M}_{0,el} \boldsymbol{\varphi}_{i,0} \qquad [214]$$

As explain earlier, damage, like a crack, will alter the eigenfrequencies and the mode shapes and hence the energy distribution and the total energy. Because the law of energy conservation is also valid for the damaged d system, we get:

$$T_{i,d} = U_{i,d} \qquad [215]$$

It has to be kept in mind that the mode shapes are mass-normalized:

$$\boldsymbol{\varphi}_{i,0}^T \boldsymbol{M}_0 \boldsymbol{\varphi}_{i,0} = \boldsymbol{\varphi}_{i,d}^T \boldsymbol{M}_d \boldsymbol{\varphi}_{i,d} = 1 \qquad [216]$$

with $\boldsymbol{M}_0 = \boldsymbol{M}_d$.

The change of the maximum kinetic energy in an element el between the undamaged 0-state and the damaged d-state becomes:

$$T_{i,el}^* = \left(T_{i,d,el} - T_{i,0,el}\right) = \frac{1}{2}\left(\omega_{i,d}^2 \boldsymbol{\varphi}_{i,d}^T \boldsymbol{M}_{0,el} \boldsymbol{\varphi}_{i,d} - \omega_{i,0}^2 \boldsymbol{\varphi}_{i,0}^T \boldsymbol{M}_{0,el} \boldsymbol{\varphi}_{i,0}\right) \qquad [217]$$

The kinetic energy T is approximated by a linear Taylor series expansion at point 0 so that the changes of the modal kinetic energy T^* in an element el due to a stiffness change of parameter j can be expressed by:

$$T_{i,el}^* \approx \sum_j \frac{\partial}{\partial a_j}\left(T_{i,0,el}\right) \Delta a_j =$$

$$= \sum_j \left(\frac{\partial \omega_{i,0}}{\partial a_j} \omega_{i,0} \boldsymbol{\varphi}_{i,0}^T \boldsymbol{M}_{0,el} \boldsymbol{\varphi}_{i,0} + \omega_{i,0}^2 \frac{\partial \boldsymbol{\varphi}_{i,0}^T}{\partial a_j} \boldsymbol{M}_{0,el} \boldsymbol{\varphi}_{i,0}\right) \Delta a_j, \qquad [218]$$

provided the element mass matrix $\boldsymbol{M}_{0,el}$ is symmetric. Using equation [217] and [218], a quadratic sensitivity matrix can be built up for each mode i, which contains the changes in the modal kinetic energy of the elements according to a damage occurrence parameterized by Δa_j:

$$S_{i,T_{el,j}^*} = \frac{\partial}{\partial a_j}\left(T_{i,0,el}\right) = \frac{\partial \omega_{i,0}}{\partial a_j} \omega_{i,0} \boldsymbol{\varphi}_{i,0}^T \boldsymbol{M}_{0,el} \boldsymbol{\varphi}_{i,0} + \omega_{i,0}^2 \frac{\partial \boldsymbol{\varphi}_{i,0}^T}{\partial a_j} \boldsymbol{M}_{0,el} \boldsymbol{\varphi}_{i,0} \qquad [219]$$

The residual of the change in the modal kinetic energy of the elements can be calculated using the measured eigenfrequencies $\omega_{i,meas}$ and eigenvectors $\boldsymbol{\varphi}_{i,meas}$ of the undamaged and the damaged states and the mass matrix, which is unaffected by the damage:

$$r_{i,T_{el}^*} = \frac{1}{2}\left[\omega_{i,d,meas}^2 \varphi_{i,d,meas}^T M_{0,el}\varphi_{i,d,meas} - \omega_{i,0,meas}^2 \varphi_{i,0,meas}^T M_{0,el}\varphi_{i,0,meas}\right] \qquad [220]$$

Arranging the sensitivity expressions and the residuals for all modes i and for all elements el, an equation system is then:

$$S_{MKE}\Delta a = r_{MKE} \qquad [221]$$

where the index MKE denotes Modal Kinetic Energy. For large complex structures, the number of measured dofs does not usually match the number of model dofs, so expansion of the mode shapes or reduction of the mass matrices is necessary. Using the MAC criterion, the appropriate pairing of corresponding modes is found. Compared to methods using the strain energy concept, the T^* method presents the advantage that changes in the structure, due to a damage occurrence, are taken into account with the changes of the measured modal properties in combination with the mass matrices of the unchanged elements, which are usually less uncertain than the stiffness matrices. Only when expansion or reduction techniques are necessary does the stiffness matrix which is mysteriously affected by the damage have to be implicitly used. This would actually be the case where using the other approach as well. From this point of view, the T^* method is expected to yield more accurate damage identification results compared to a strain energy concept.

2.5.8.3. Modal strain energy-based sensitivity method

In the same way as for the modal kinetic energy, similar expressions can be developed for the modal strain energy distributed along all substructures or finite elements el for mode i:

$$U_{i,0} = \sum_{el=1}^{n} U_{i,0,el} = \frac{1}{2}\sum_{el=1}^{n} \varphi_{i,0}^T K_{0,el}\varphi_{i,0} \qquad [222]$$

As studied earlier, a damage occurrence will change the eigenfrequencies and the mode shapes, and hence modify the energy distribution and the total energy. For the damaged structure, one can write:

$$U_{i,d} = \sum_{el=1}^{n} U_{i,d,el} = \frac{1}{2}\sum_{el=1}^{n} \varphi_{i,d}^T K_{d,el}\varphi_{i,d} = \dots$$

$$= \frac{1}{2}\sum_{el=1}^{n} \varphi_{i,d}^T K_{0,el}\varphi_{i,d} + \frac{1}{2}\sum_{el=1}^{n} \varphi_{i,d}^T \Delta K_{el}\varphi_{i,d} \qquad [223]$$

It has to be kept in mind that the mode shapes are always mass-normalized according to the unchanged mass matrix $\varphi_{i,0}^T M_0\varphi_{i,0} = \varphi_{i,d}^T M_0\varphi_{i,d} = 1$.

The change of the maximum kinetic energy in an element *el* between the undamaged 0-state and the damaged *d*-state becomes:

$$U^*_{i,el} = \left(U_{i,d,el} - U_{i,0,el}\right) = \frac{1}{2}\left(\boldsymbol{\varphi}^T_{i,d}\boldsymbol{K}_{d,el}\boldsymbol{\varphi}_{i,d} - \boldsymbol{\varphi}^T_{i,d}\boldsymbol{K}_{0,el}\boldsymbol{\varphi}_{i,d}\right) + \frac{1}{2}\boldsymbol{\varphi}^T_{i,d}\boldsymbol{\Delta K}_{el}\boldsymbol{\varphi}_{i,d} \quad [224]$$

The third term on the right-hand side is unknown because the changes in the element stiffnesses are not known beforehand. Now, the strain energy is approximated by a linear Taylor series expansion (where the reference point is the 0-system) so that the changes in the modal strain energy in the element *el* due to a stiffness change of parameter *j* can be expressed by:

$$U^*_{i,el} \approx \sum_j \frac{\partial}{\partial a_j}\left(U_{i,el}\right)\Delta a_j = \sum_j \left(\frac{\partial \boldsymbol{\varphi}^T_{i,0}}{\partial a_j}\boldsymbol{K}_{0,el}\boldsymbol{\varphi}_{i,0} + \frac{1}{2}\boldsymbol{\varphi}^T_{i,0}\frac{\partial \boldsymbol{K}_{0,el}}{\partial a_j}\boldsymbol{\varphi}_{i,0}\right)\Delta a_j \quad [225]$$

provided the element mass matrix $\boldsymbol{K}_{0,el}$ is symmetric. The second term into brackets equals zero if the parameter *j* does not influence the element stiffness matrix *el*. This means that, in the case where each element is connected with one parameter *j*, the second term will only appear on the diagonal *el* = *j*. Equation [224] still has the undetermined part $\frac{1}{2}\boldsymbol{\varphi}^T_{i,d}\boldsymbol{\Delta K}_{el}\boldsymbol{\varphi}_{i,d}$ on its right-hand side. However, whether this term can be expressed by $\boldsymbol{\Delta K}_{el} = \sum_j \frac{\partial \boldsymbol{K}_{el}}{\partial a_j}\Delta a_j$, it can be brought to the left-hand side.

Now, a quadratic sensitivity matrix can be created for each mode *i*, which contains the changes in the modal strain energy of the elements according to a damage occurrence parameterized by Δa_j :

$$\boldsymbol{S}_{i,U^*_{el,j}} = \frac{\partial \boldsymbol{\varphi}^T_{i,0}}{\partial a_j}\boldsymbol{K}_{0,el}\boldsymbol{\varphi}_{i,0} + \frac{1}{2}\boldsymbol{\varphi}^T_{i,0}\frac{\partial \boldsymbol{K}_{0,el}}{\partial a_j}\boldsymbol{\varphi}_{i,0} - \frac{1}{2}\boldsymbol{\varphi}^T_{i,d}\frac{\partial \boldsymbol{K}_{0,el}}{\partial a_j}\boldsymbol{\varphi}_{i,d} \quad [226]$$

The residual of the change in the modal strain energy of the elements (with the third term eliminated from the right-hand side) can be calculated using the measured eigenvectors $\boldsymbol{\varphi}_{i,0,m}$ and $\boldsymbol{\varphi}_{i,d,m}$ of the undamaged and the damaged states, respectively, and the element stiffness matrices of the undamaged system:

$$r_{i,U^*_{el}} = \frac{1}{2}\left[\boldsymbol{\varphi}^T_{i,d,m}\boldsymbol{K}_{0,el}\boldsymbol{\varphi}_{i,d,m} - \boldsymbol{\varphi}^T_{i,0,m}\boldsymbol{K}_{0,el}\boldsymbol{\varphi}_{i,0,m}\right] \quad [227]$$

Arranging the sensitivity expressions and the residuals for all modes *i* and for all elements *el*, an equation system can be:

$$\boldsymbol{S}_{MSE}\Delta\boldsymbol{a} = \boldsymbol{r}_{MSE} \quad [228]$$

where the index *MSE* denotes Modal Strain Energy. What was said about the application of expansion methods in the context of the kinetic energy method is basically valid in that case. One difference remains, which concerns the involved vector components. The primary co-ordinates used to calculate the kinetic energies are the translational degrees of freedom (which are at least more dominant for the lower modes). Because, in practice, translations are usually the quantities being measured, the expansion errors are smaller for the translational dofs. On the other hand, the rotations have to be fully expanded. If bending is the predominant deformation, the strain energies will be highly determined by the rotational dofs so that the conclusion can be drawn that the strain energy expressions are less accurate than the kinetic energy expressions as a result of expansion errors in the rotational dofs.

2.5.9. *Forced vibrations and frequency response functions*

Before going into more explanations, the following example (see [FRI 97]), which shows a base plate and five storeys, connected by elastic strips, is taken into account. The intention is then to demonstrate the change of the dynamic behavior following a "loss" of stiffness between the two storeys 3–4 and 1–2.

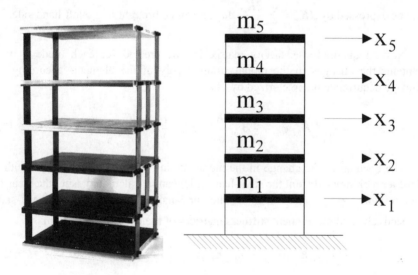

Figure 2.19. *Five-dof test structure and mechanical model*

Figure 2.20 shows the Frequency Response Functions (FRFs) before and after damage. Both the resonance peaks and the antiresonances are shifted to lower frequencies. The mathematical model is represented by the dotted and dashed lines. Both resonance peaks and antiresonances are shifted to lower frequencies due to damage. The figure also shows that the model (dashed line) matches the experimental curve (solid line) almost perfectly. Noise is distorting the measured signal around the first antiresonance, which is characterized by a low signal-to-noise ratio, above 90 Hz. The system is excited by an impact hammer and the response was measured by an accelerometer. This example is studied in [FRI 97, JEN 99].

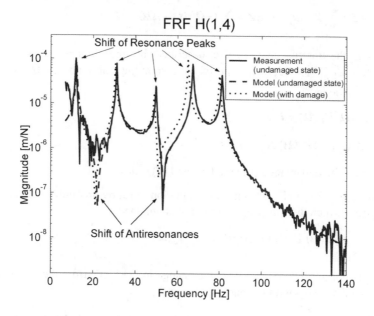

Figure 2.20. *Frequency Response Function H_{14} with and without damage between storeys no. 3 and 4*

2.5.9.1. *Exact formulation of parameter identification problem*

As shown earlier, the equation of motion of the linear mechanical system can be written in the frequency domain as:

$$\left(-\Omega^2 M + j\Omega C + K\right) X(\Omega) = F(\Omega) \qquad [229]$$

or:

$$\left[Z(\Omega)\right] X(\Omega) = F(\Omega) \qquad [230]$$

The output $X(\Omega)$ follows from simple multiplication of the input $F(\Omega)$ with the FRF matrix H: $X(\Omega) = H(\Omega)F(\Omega)$. It is immediately apparent that the FRF matrix H is the inverse of the dynamic stiffness matrix Z: $Z(\Omega) = H(\Omega)^{-1}$. Starting from a reference model (subscript 0) and representing the undamaged system by:

$$Z_0(\Omega) = (-\Omega^2 M_0 + j\Omega C_0 + K_0) = H_0(\Omega)^{-1}$$ [231]

the changes $\Delta Z = Z_d - Z_0$ due to the damage by the famous linear substructure approach are written as:

$$\Delta Z = \sum Z_j \Delta a_j \quad \text{or} \quad \Delta Z = \sum_j (-\Omega^2 M_j + i\Omega C_j + K_j) \Delta a_j$$ [232]

Again, the matrix changes are described by the dimensionless correction parameters Δa_j, $j = 1, 2, \ldots, n_p$ and the corresponding sub-matrices M_j, C_j and K_j. The two relations:

$$Z_0(\Omega)H_0(\Omega) = I \quad \text{and}$$ [233]

$$[Z_0(\Omega) + \Delta Z(\Omega)] H_d(\Omega) = I$$ [234]

are valid for the undamaged (0) and damaged (d) states.

Equations [233] and [234] introduce equation [232] by premultiplying both sides by H_0, from which the following expression is obtained:

$$-\sum_j [H_0(\Omega) Z_j(\Omega) H_d(\Omega)] \Delta a_j = H_d(\Omega) - H_0(\Omega)$$ [235]

This important relationship links the change in the FRFs from the original (0) to the damaged state (d) with the parameter changes Δa_j.

In the case where only one column (no. l) of the FRF matrix has been measured, it is immediately found that:

$$-\sum_j [H_0(\Omega) Z_j(\Omega) H_{l,d}(\Omega)] \Delta a_j = H_{l,d}(\Omega) - H_{l,0}(\Omega)$$ [236]

Another question arises: how can this equation be completely modified when some components of the model have been measured? Using a measurement matrix C_{meas}, which only consists of zeros and ones, in order to delete those positions where no measurement information is available, one can write:

$$-\sum_j C_{meas} [H_0(\Omega) Z_j(\Omega) H_{l,d}(\Omega)] \Delta a_j = C_{meas}[H_{l,d}(\Omega) - H_{l,0}(\Omega)]$$ [237]

Both sides of the equation are premultiplied by the matrix C_{meas}. Arranging this information for frequencies ω_v and splitting the complex equation into real and imaginary part yields the Output Error formulation for the FRFs:

$$S_{OE,FRF} \, \Delta a = r_{OE,FRF} \qquad\qquad [238]$$

In the ideal case of noise-free and full measurement information, this method yields the correct parameters in one step.

2.5.9.2. A linear Taylor expansion as an alternative

An alternative is to develop an equation system based on a direct linear Taylor expansion. Neglecting higher order terms, one obtains:

$$\sum_j \frac{\partial H(\Omega)}{\partial a_j} \Delta a_j = H_d(\Omega) - H_0(\Omega). \qquad\qquad [239]$$

With the identity $Z\,H = I$ (equation [8]), applying the differentiation to this product yields:

$$Z\frac{\partial H}{\partial a_j} + \frac{\partial Z}{\partial a_j} H = 0 \qquad\qquad [240]$$

Straightforward calculation leads to:

$$\frac{\partial H}{\partial a_j} = -Z^{-1}\frac{\partial Z}{\partial a_j} H = -H\frac{\partial Z}{\partial a_j} H \qquad\qquad [241]$$

and in the case where the undamaged state (0) is the point at which the Taylor series is developed:

$$H_j(\Omega) = \frac{\partial H(\Omega)}{\partial a_j} = -H_0(\omega)\frac{\partial Z(\Omega)}{\partial a_j} H_0(\Omega) = -H_0(\Omega)\,Z_j(\Omega)\,H_0(\Omega) \qquad [242]$$

The last expression is almost identical to the term in brackets on the left-hand side of equation [235]. The only difference is that H_0 is present instead of H_d on the right-hand side of H_j. At first glance, this small difference does not involve important consequences. However, the assumption that small changes in the parameter cause small changes in the FRFs is critical when using a linear Taylor approximation close to the poles (the resonances). The different positions of the poles for the 0 and the d systems yield large differences in the FRFs and hence in the expression for H_j (equation [242]). Especially for slightly damped systems, the shift of the poles/eigenfrequencies causes large deviations of the FRFs because the peaks of the FRFs reach very large values.

The conclusion is that H_d, rather than H_0, should be used for the calculation of the sensitivities H_j. However, the analytically correct approach usually involves expanding the columns of the measured H_d. Experience shows that expansion errors are smaller than the linearization errors in this case.

2.5.9.3. Equation error (input residual method) using FRFs

Starting again from the identity $Z\,H = I$, equations [233] and [234] are used:

$$[Z_0(\omega) + \Delta Z(\omega)]\,H_d(\omega) = I$$

It immediately follows that:

$$[\Delta Z(\omega)]\,H_d(\omega) = I - Z_0(\omega)H_d(\omega) \qquad [243]$$

Expressing the change of the dynamic stiffness by the parameter changes Δa_j:

$$\Delta Z = \sum_j (-\omega^2 M_j + j\omega C_j + K_j)\,\Delta a_j \qquad [244]$$

one gets:

$$\sum_j [(-\omega^2 M_j + j\omega C_j + K_j)\,H_d(\omega)]\,\Delta a_j = I - Z_0(\omega)H_d(\omega) \qquad [245]$$

While the FRFs $H_{kl,0}$ can be calculated from the model, the $H_{kl,d}$ are uncertain measured quantities because of superimposed noise. Missing measurement values in $\{H_{l,d}\}$ must be reconstructed by expansion techniques. This means that equation [245] is only satisfied by approximation. If equation [245] is arranged for different frequencies ω_v and the complex equation is split up into its real and imaginary parts, the system of real linear equations based on Input Error formulation for FRFs is:

$$S_{IE,FRF}\,\Delta a = r_{IE,FRF} \qquad [246]$$

It is a common error to believe that more parameters could be determined because many more frequencies ω_v are available. There is a close relationship between the vibration modes and the FRFs, as shown in equation [12]. The information in the FRFs depends on the number of modes included in the frequency band. If the summation is not complete, which is usually the case in practical applications, information from the higher frequency range (higher modes) is lacking.

It is not recommended to use an identification formulation which includes the direct inversion of the FRF matrix; starting from:

$$[Z_0(\omega) + \Delta Z(\omega)] = H_d^{-1}(\omega) \qquad [247]$$

and solving for the parameters in ΔZ:

$$[\Delta Z(\omega)] = H_d^{-1}(\omega) - Z_0(\omega) \qquad [248]$$

The inversion of a matrix containing noise-corrupted values leads to very inaccurate results.

2.5.9.4. *The projected input residual method*

The problem of incompatibility of the large size of the system matrices ($m \times m$) on one hand and on the other hand the fact that only a reduced measured set is available ($m_{red} \ll m$) has been studied by Oeljeklaus [OEL 98, OEL 99, OEL 00] by using a different approach. The input residual for the forced vibrations in the frequency domain is defined by:

$$r_{IE} = F(\Omega) - Z_\Omega \, x(\Omega) \text{ with } Z_\Omega = -\Omega^2 M + j\Omega C + K$$

At this stage, the further evaluation of the residual requires the full-sized displacement vector x. In earlier sections, the expansion of the measured vector was presented: $x_{m,expd} = T_\Omega x_m$, where T was a frequency dependent on the expansion matrix. Instead, Oeljeklaus uses a projection which enables him to suppress the unmeasured degrees of freedom. The fact that not all components are measured is expressed by the measurement matrix C_m, which has dimension $m_{red} \times m$ and consists of zeros (eliminating the unmeasured components) and ones (maintaining the measured components):

$$x_m = C_m x \qquad [249]$$

The vector:

$$\tilde{x} = C_m^T C_m x \qquad [250]$$

again has the full size m, but it is filled with zeros at those positions where measurement information is missing. Now, a projection matrix P is constructed in such a way that the unmeasured components are masked out and hence their actual value does not matter:

$$P = I - (Z_\Omega \bar{C}_m^T)(Z_\Omega \bar{C}_m^T)^+ \qquad [251]$$

The matrix $\bar{C}_m$ is the complement of C_m, mapping the full displacement vector onto the unmeasured components. I again is the identity matrix and the $(..)^+$ denotes the pseudo-inverse. For the full derivation of the projection matrix, see [OEL 98].

The projected input residual vector now becomes:

$$r_{PIR} = P[F_m(\Omega) - Z_\Omega C_m^T C_m x(\Omega)] \qquad [252]$$

whose norm has to be minimized. Because the projection itself depends on the parameter, the problem is a non-linear minimization task. This method is also sensitive to noisy data, as basically are all input residual formulations. The method has been applied by [HUT 02, LOE 04].

2.5.9.5. *Antiresonances*

Lallement *et al.* [LAL 92] and He *et al.* [HE 94] have pointed out that the zeros (or antiresonances, where $H_{ik}(\Omega_{AR}) = 0$) introduce useful additional information about a dynamic system. While the resonance peaks are defined by the system poles, which appear in the denominator of each FRF, the zeros/antiresonances are different for each individual FRF. Mottershead [MOT 98] has investigated the sensitivities of the zeros according to parameter changes. He has shown that the sensitivities of the antiresonances can be expressed by sensitivities of eigenvalues and mode shapes. Also, the modes with eigenfrequencies closest to the zero most significantly contribute to the sensitivities. This information has not yet been widely used for updating and damage identification.

2.5.9.6. *Transmissibilities*

The complex transmissibility ratio (TR) is defined as the ratio of two Fourier transformed output signals Y_i and Y_j. i and j denote different sensor positions/directions; see Figure 2.21:

$$T_{ij}(\Omega) = \frac{Y_i(\Omega)}{Y_j(\Omega)} \tag{253}$$

The ratio can also be used in the case of ambient vibration with output-only measurements if the different channels are measured simultaneously. In the single input–multiple output case (SIMO), the structural response can be expressed by the FRFs: $Y_i(\Omega) = H_{ik}(\Omega)F_k(\Omega)$. If the transmissibility ratio is calculated, the force F_k cancels out and T_{ij} becomes:

$$T_{ij(k)} = \frac{H_{ik}(\Omega)}{H_{jk}(\Omega)} \tag{254}$$

It becomes clear that the excitation point (k) is important. As shown by Johnson and Adams [JOH 02, JOH 04], the representation of the response by its zeros and poles is very useful:

$$H_{ik}(\Omega) = g_{ik} \frac{(j\Omega - z_{ik1})(j\Omega - z_{ik2})\cdots(j\Omega - z_{ikn})}{(j\Omega - \lambda_1)(j\Omega - \lambda_2)\cdots(j\Omega - \lambda_n)} \tag{255}$$

where g is a gain factor, the zs are the zeros and the λs are the poles, which are equal to the eigenvalues of the system. Because the poles (eigenvalues of the system) appear in the denominator of all responses, they are also influenced by all parameters of the system. If the transmissibility ratio is calculated, the system poles are eliminated by the division of the two responses and only the zeros (antiresonances) remain in the equation. As shown in [JOH 02, JOH 04], the T_{ij} possess some good localization properties which are used to localize damage, based on measurement data; see also [SAM 00]. Comparing the TRs for the undamaged and the damaged cases, a damage indicator can be defined for each index combination i, j of channels, for example by:

$$ DI_{T_{ij}} = \sqrt{ \sum_v \left(1 - \frac{\left| T_{ij}^d (\Omega_v) \right|}{\left| T_{ij}^0 (\Omega_v) \right|} \right)^2 } \qquad [256] $$

Manson et al. [MAN 02] have used the transmissibility ratios to train neural networks for pattern recognition.

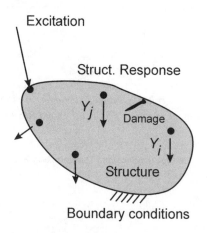

Figure 2.21. *Determination of transmissibility ratios from measured output spectra Y(Ω)*

For the five-dof structure shown in Figure 2.23, an example of transmissibility ratios which are computed from a pair of FRFs, $H_{4,4}$ and $H_{3,4}$ is shown; see equation [254]. The dashed curves are calculated from the FRFs of the undamaged state and the solid curves show the damaged state. The damage represent a reduction of stiffness of about 20% between dofs 3 and 4. The different transmissibility ratio curves can be seen in the middle part of Figure 2.22. From these curves, the relative changes $1 - \left| T_{ij}^d (\Omega_v) \right| / \left| T_{ij}^0 (\Omega_v) \right|$ are determined; see the lower part of this figure. The summation along the frequency axis according to equation [256] yields the damage indicator $DI_{T4,3}$.

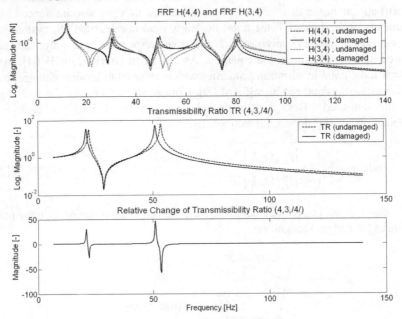

Figure 2.22. *Calculation of the relative change of the transmissibility ratio due to damage from a set of two FRFs H(4,4) and H(3,4)*

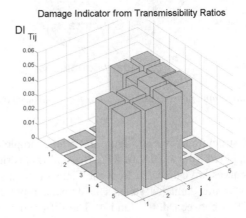

Figure 2.23. *Change of transmissibility ratios for the five-storey oscillator*

The whole matrix for all combinations i, j is shown in Figure 2.23. Besides the fact that the diagonal elements are always zero, the blocks [1:3, 1:3] and [4:5, 4:5] show very small values. This means that almost no change occurs in the transmissibility ratios between these dofs. The stiffness reduction was located between dofs 3 and 4. For such a chain-like oscillator, if the damage is located between two degrees of freedom, the transmissibility changes are large and the damage indicator shows significant increase. Otherwise, the indicator is small; see also [JOH 02]. This allows damage localization from measured data alone.

2.5.9.7. Impedance method

A good overview is given by Park *et al.* [PAR 03], and further developments of the impedance method are described in [LOP 00, LOP 01, PAR 00a, PAR 00b, POH 01, PAR 03, ZAG 01, GIU 03]. The impedance method is a very sensitive method, working in the higher frequency range (typically > 30 kHz) comparing changes in the impedance spectrum of the electromechanical (EM) system due to damage. The basic requirement is one piezoelectric element, which works as actuator and sensor. Practically, the impedance of the EM-system is determined by the input voltage and the output current of the piezoelectric actuator; see Figure 2.24. The idea is that the damage in the structure changes the structural impedance and hence the resulting impedance spectrum of the coupled EM-system. The complex impedance $Z_{struct}(\Omega)$ of the mechanical structure (which is a ratio of velocity to force at the position of the PZT[3] actuator/sensor) can be derived from the continuous or discrete mass, damping and stiffness properties:

$$\frac{V(\Omega)}{I(\Omega)} = Z_{EM}(\Omega) = \frac{1}{j\Omega C}\left(1 - \kappa^2 \frac{Z_{struct}(\Omega)}{Z_{PZT}(\Omega) + Z_{struct}(\Omega)}\right)^{-1} \qquad [257]$$

where C is the zero load capacity and κ is an electromechanical cross-coupling coefficient of the piezoelectric transducer; see [GIU 03]. The real part of the actual spectrum is compared to the stored reference spectrum of the undamaged system and a level I-detection can be performed. The impedance method is a qualitative method which possesses a local character. At high frequencies, the structural resonances are localized and highly sensitive to local damage. Furthermore, actuator energy is dispersed into the structure so that effects from the damage can be only seen close to the actuator position. In general, the sensing range is closely related to the material properties of the host structure, its geometry, the frequency ranges being used and the properties of the PZT materials. However, frequencies higher than 500 kHz are not recommended because the sensing region becomes too small. It is estimated that a sensing radius of 0.4 m can be reached in composite structures and up to 2 m in a metal beam [PAR 03].

---

[3] PZT denotes the material lead (Pb) zirconate titanate.

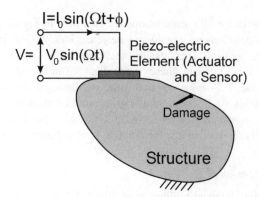

Figure 2.24. *Measurement of the impedance of the coupled electromechanical system from voltage and current*

2.6. Solution of the equation system

Depending on the number of data sets and the number of parameters, the equation:

$$S \, \Delta a = r \qquad\qquad [258]$$

may be determined, over- or under-determined. Considering the ordinary case of an ill-posed (over-determined) equation system, the remaining error has to be minimized. Therefore, an error vector is introduced by ε and is presented:

$$S \, \Delta a = r + \varepsilon \qquad\qquad [259]$$

The correlation of the columns of the sensitivity matrix and the residuals r can be calculated by the following equation:

$$\kappa_j = \frac{r^T s_j}{\left\| s_j^T s_j \right\| \left\| r^T r \right\|} \quad -1 \leq \kappa_j \leq 1; \ \cos\phi_j = \kappa_j \qquad [260]$$

The weighted scalar product of s_j and r expresses the coincidence of the jth column of the sensitivity matrix and the residual vector. If the jth column is orthogonal to r then $\kappa_j = 0$. In practice, there are many parameters leading to very small numbers, for example $\kappa_j = 10^{-1} \cdots 10^{-2}$. It is reasonable to skip these parameters. The other extreme, $|\kappa_j| = 1$, means that the jth parameter alone is able to represent r. A threshold value is defined according to the current problem to render a parameter unnecessary if:

$$\left| \kappa_j \right| \le \kappa_{th} \, .$$

This allows a certain preselection of parameters.

Minimizing the weighted sum of the components of ε (using the positive definite weighting matrices W_ε):

$$J = \varepsilon^T W_\varepsilon \varepsilon \quad \rightarrow \min . \tag{261}$$

leads to the following equation system:

$$(S^T W_\varepsilon S) \varDelta a = S^T W_\varepsilon r \tag{262}$$

which must be solved by suitable numerical methods. The direct solution of equations [258] or [262] usually yields to bad results if the data are polluted by measurement noise or other errors and the condition number of S is large. Due to the nature of the ill-posed problem, small disturbances are amplified so that the calculated solution often strongly differs from the original solution. The problem is that changes in two neighboring elements of a finite element model may nearly affect the dynamics of the structure so that almost linearly dependent columns of the sensitivity matrix are obtained. In regression, this problem is known as multicollinearity. The finer the discretisation is, the more strongly this effect is pronounced. Therefore, it is important to use regularization methods to solve equations [258] and [262]; see [NAT 97, BAU 87, LOU 89, HAN 94, HAN 98, TRU 97] concerning stabilization of the solution. In the context of damage detection, it is important to keep the number of unknowns as small as possible. This means that large number of possible damage candidates must be reduced to a minimum in order to ensure good results. These two options are discussed in the following sections.

2.6.1. Regularization

By minimizing the weighted sum of the components ε and constraining the norm of the parameter vector $\varDelta a$ to "small" values, the extended weighted least-squares (EWLS) functional [NAT 97, NAT 92] is:

$$J = \varepsilon^T W_\varepsilon \varepsilon + \varDelta a^T W_a \varDelta a \quad \rightarrow \min . \tag{263}$$

where W_ε and W_a are weighting matrices. The influence of the penalty term, which increases if the parameter changes become large, depends on the balance between the two weighting matrices. Minimizing J according to $\varDelta a$ yields:

$$(S^T W_\varepsilon S + W_a) \varDelta a = S^T W_\varepsilon r \tag{264}$$

For $W_a = \gamma^2 I$, the Tykhonov–Phillips regularization is obtained [NAT 97, LOU 89, FRI 99a, FRI 99b, FRI 99c]:

$$(S^T W_\varepsilon S + \gamma^2 I)\varDelta a = S^T W_\varepsilon r \qquad [265]$$

The proper determination of the unknown regularization parameter γ is extensively studied in the documents on ill-posed problems [BAU 87, LOU 89, HAN 94, HAN 98, TRU 97]. Typical approaches are the L-curve method or the cross-validation method.

2.6.2. Parameter subset selection

If we deal with complex structures and assume that each substructure/finite element is a potential location for damage, we get a large set of unknown parameters in $\varDelta a$. Thus, special attention has to be paid to the problem of the high dimensionality of the parameter space, which has to be reduced as far as possible. The reduced subset of the dominant parameters must be able to describe the damage scenario and should finally concentrate on those parameters corresponding to the damaged sub-region(s) of the structure. A method of determining the set of the most dominant parameters was proposed by Lallement and Piranda [LAL 90], leading to a suboptimal solution. Defining the error vector:

$$\varepsilon = r - S\,\varDelta a \qquad [266]$$

this intuitive strategy starts with one single parameter, which leads to the largest reduction of the error norm of:

$$e = \|\varepsilon\|^2 = \varepsilon^T \varepsilon = (r - S\,\varDelta a)^T (r - S\,\varDelta a) = \|(r - S\,\varDelta a)\|^2 \qquad [267]$$

With e_0 as the error norm with all parameter changes equal to zero $\varDelta a = 0$:

$$e_0 = r^T r = \|r\|^2 \qquad [268]$$

a relative error measure can be introduced:

$$e_{rel} = \frac{e(\varDelta a)}{e_0} = \frac{\|r - S\,\varDelta a\|^2}{\|r\|^2} ;\ \ 0 \le e_{rel} \le 1 \qquad [269]$$

$$\varDelta a_k = \varPi \begin{pmatrix} \varDelta a_{red,k} \\ 0 \end{pmatrix} \qquad [270]$$

which starts with 1 for all changes equal to zero $\varDelta a = 0$ and drops to 0 in the ideal case where the parameters are able to describe the residual vector r entirely. The matrix Π describes the permutation of the parameters so that the elements of the reduced parameter set are at their original positions.

In the first step, the parameter Δa_ρ is chosen (while keeping all other parameters at zero) so that we obtain a maximum reduction of the error measure e_{rel}. This means that n_p solutions with one parameter have to be tried out if Δa has n_p components. The next parameter yields the largest error reduction while maintaining the parameter Δa_ρ from the first step. This means $(n_p - 1)$ solution steps with a two-parameter vector. The procedure is carried out with three parameters and so on. Finally, it is observed that taking more parameters into account does not improve the solution any further and the relative error cannot be reduced significantly. This is the point at which the procedure stops.

A similar strategy based on QR-decomposition was proposed in [JEN 99, FRI 98]; see the following sections. Applications of subset selection to mechanical structures are described in [AHM 97, FRIS 97a].

2.6.2.1. Forward/backward selection

Finding a model with a minimum number of parameters which contribute significantly to the solution of the problem is extensively explained in the literature of multivariate regression analysis [MIL 90, WET 86, GLA 00]. Here "significantly" is expressed in terms of statistical quantities such as the F-value. This criterion has a direct statistical foundation and is less heuristic than the method described above. In regression analysis, the notions of *forward* and *backward* selection are distinguished.

The forward parameter selection process basically works in the same way as the procedure described above, starting with no parameter and subsequently increasing the number by one until no further significant improvement can be achieved. The backward selection, however, works the other way round, beginning with all parameters and subsequently eliminating those components which do not significantly worsen the solution when they are omitted from the solution vector. Backward selection is computationally much more laborious because the selection process begins with many steps, each solving a regression problem with a large number of unknowns. Forward and backward selection can be combined in order to use the specific strengths of both so as to select a certain number of parameters, then eliminate some of them by backward selection, then use forward selection again, and so on. One can observe that the selected parameters, which have made a significant contribution to the solution at the beginning, are again eliminated later, because it turns out that the combination of some newly chosen parameters makes other first selected parameters obsolete. Forward/backward selection is described for example in [WET 86, GLA 00, MIL 90]. The criterion based on the F-statistic during forward selection is:

$$F = \frac{MS_{k|1,2,...,k-1}}{MS_{res|1,2,...,k}} \qquad [271]$$

First, let us define $SS_{k|1,2,...,k-1}$ as the incremental sum of squares associated with the adding of the kth parameter, given that the first significant parameters $1, ..., k-1$ are already in the regression equation. There is one degree of freedom associated with the incremental sum of squares insofar as $MS_{k|1,2,...,k-1} = SS_{k|1,2,...,k-1}/1$. The incremental sum of squares (without data weighting) is:

$$SS_{k|1,2,...,k-1} = \|\varepsilon_{k-1}\|^2 - \|\varepsilon_k\|^2 \qquad [272]$$

where the error vectors are computed with $k-1$ and k parameters, respectively. Thus, for k parameters, the error is:

$$\varepsilon_k = r - S_{red,k}\Delta a_{red,k} \qquad [273]$$

$MS_{res|1,2,...,k}$ is the mean square residual for the regression equation containing the parameters $1,..., k$:

$$MS_{res|1,2,...,k} = \frac{SS_{res|1,2,...,k}}{n-k} = \frac{\|\varepsilon_k\|^2}{n-k} \qquad [274]$$

where n is the number of regression equations. Note that the bias value is sometimes also taken into the regression analysis. In this case, the denominator is $n-k-1$ [GLA 00]. Finally it can be written as:

$$F_k = \frac{\|\varepsilon_{k-1}\|^2 - \|\varepsilon_k\|^2}{\|\varepsilon_k\|^2/(n-k)}, \qquad [275]$$

When the F-value has been tested with all parameters which have not yet been taken into account in the regression equation, the parameter is finally chosen so that it delivers the maximum value for F. In addition, F must satisfy the condition for entry:

$$F \geq F_{in} \qquad [276]$$

The backward selection step follows more or less similar rules as described above. The F-value can be calculated by equation [275].

The parameter with the minimum increase of the incremental sum of squares is chosen and the kth parameter can be deleted if:

$$F \leq F_{out} \qquad [277]$$

In the stepwise regression forward and backward selection are combined.

The choice of F_{in} and F_{out} is defined as the stopping rule for entering and deleting variables from the active solution vector. For further details on how these values are chosen, see [MIL 90, WET 86, GLA 00].

2.6.2.2. QR-decomposition with forward selection

The regularization effect using orthogonal decomposition techniques such as the QR algorithm is obtained by reduction of the original parameter space to a smaller subspace so that the problem [259] will be numerically much more stable and at the same time yield a sufficiently accurate solution. The key question is what this subspace must look like.

The sensitivity matrix S is decomposed into a product of an orthogonal matrix Q and an upper triangular matrix R, while a column interchange is performed for S, which is stored in the permutation matrix Π:

$$S\,\Pi = QR \qquad \qquad [278]$$

New instrumental co-ordinates β_i are introduced that describe the contribution of the ith normalized orthogonal vector q_i and are calculated by means of Gram–Schmidt orthogonalization [GOL 96]:

$$Q\,R\underbrace{\left\{\Pi^T \Delta a\right\}}_{\beta} = r \qquad \qquad [279]$$

or:

$$\sum_{j=1}^{n_p} q_j\,\beta_j = r \text{ with } \left\|q_j\right\| = 1 \qquad \qquad [280]$$

Due to the orthogonality of the q_j, the factors β_j can be determined directly from the scalar product:

$$\beta_j = q_j^T r \qquad \qquad [281]$$

So far, the order of the column permutation can be freely chosen to get the vectors q_j. The standard approach is sensitivity based and the permutation is performed in such a way that the magnitudes of the diagonal elements of the matrix R have descending order [GOL 96]. The algorithm was applied in [FRI 98, JEN 99] and is abbreviated as QRD for later use.

To get a first impression of how much each parameter can contribute to the right-hand side (rhs) of equation [259], a correlation coefficient κ_j, $j = 1,...,n_p$ is calculated (equation [260]). The scalar product of s_j and r is the projection of the change of the residuals due to a change of the jth parameter in the residuals, expressing the similarity of the jth column of the sensitivity matrix and the residual vector:

$$\kappa_i = \frac{s_j^T r}{\|s_j\| \ \|r\|} ; \quad -1 \le \kappa_j \le 1 \tag{282}$$

The norm $\|s_j\|$ is a measure of sensitivity. $|\kappa_j| = 1$ means that the jth parameter alone is sufficient to represent the rhs, while $\kappa_j = 0$ means that there is no correlation between this parameter and the residuals. Now, the column permutation in S is performed in such a way as to use a column s_j, which – after an orthogonalization step – contributes most to the reduction of the norm of the rhs $\|r\|$. The procedure is started with that parameter a_j whose corresponding column s_j yields $\kappa_{max} = \max\left\{ |\kappa_j|, j = 1, 2, ..., n_p \right\}$ such that $q_1 = \left\{ s_i \right\} / \|s_i\|$.

Finally, evaluation of the error using [280] and [281] leads to:

$$\|\varepsilon\|^2 = \left\| r - \sum_{j=1}^{n_p} q_j \beta_j \right\|^2 = \|r\|^2 - \sum_{j=1}^{n_p} \beta_j^2 = \|r\|^2 \left(1 - \sum_{j=1}^{n_p} \overline{\beta}_j^2 \right) \tag{283}$$

with:

$$\overline{\beta}_j = \frac{q_j^T r}{\|r\|} \tag{284}$$

From equation [281], it is deduced that q_j yields no contribution to the error reduction, if it is orthogonal to the right-hand side, because then $\beta_j = 0$. Small β_j yield small contributions for the error reduction. Hence, it makes sense only to choose columns having large β_j. The column interchange during orthogonalization is performed in such a way that $|\beta_1| \ge |\beta_2| \ge |\beta_3| \ge ... \ge |\beta_{np}|$ where the first $k \le n_p$ columns with the largest β_j are only used so that $|\beta_j| / \|r\| = |\overline{\beta}_j| \ge \tau$ where τ is a user-defined threshold. With this reduced set the equation system is solved:

$$R\left\{ \Pi \Delta a_{red,k} \right\} = \beta_{red,k} \quad \text{with} \quad \beta_{red,k} = \left\{ \beta_1, \beta_2, ..., \beta_k, \underbrace{0, 0, ..., 0}_{n_p - k} \right\}^T \tag{285}$$

finally, yielding the physical parameter change $\left\{ \Delta a_{red,k} \right\}$ after backward substitution, where the remaining $n_p - k$ parameter changes that do not contribute to the solution of the problem are zero.

The remaining dimensionless error e_k for truncation after k follows from:

$$e_k^2 = 1 - \sum_{i=1}^{k} \overline{\beta}_i^2 = \frac{\left\| r - SII\left\{ \Delta a_{red,k} \right\} \right\|^2}{\left\| r \right\|^2}$$ [286]

The idea of the error reduction ratio was presented in [FRI 98, JEN 99], but it is given here in the context of the truncated QR-decomposition. A detailed flow chart of the QR algorithm described is here given in [FRI 98].

2.6.3. Other solution methods

Other methods for solving the inverse problem are based on truncated Singular Value Decomposition (SVD) or iterative methods such as the Method of Conjugate Gradients [BAU 87, LOU 89]. The important aspect with the iteration-based methods is that the iteration is stopped at the right point. The general behavior is that the solution rapidly improves at the beginning but then becomes worse again. However, the truncated SVD and the iterative solution method always work with the full solution vectors and a proper subset selection is not possible in this way.

A combined two-step parameter reduction strategy with parameter pre-selection is described in [FRI 99a, FRI 99b, FRI 99c]. As a result, only those parameters are considered that yield a significant contribution to reducing the error in equation [267].

Bohle [BOH 05] has investigated the effects of outliers and concluded that Robust Regression is superior to the Least Squares Solution in that case.

The Minimum Rank Perturbation Theory (MRPT) presented by Zimmerman *et al.* in various publications [ZIM 94, ZIM 95, ZIM 99] follows the same aim as the parameter reduction approach described above, because damage is usually concentrated at certain locations, rather than spreading throughout the whole structure. The change of the matrix is determined in such a way that it has minimum rank rather than minimum matrix norm.

Instead of carrying out stepwise linear regression, non-linear optimization methods (as described for model updating) can be used to solve the regression problem. Global optimization methods such as genetic algorithm or Simulated Annealing methods, which are based on random search, can also be used. The Simulated Annealing method can be used to find the global minimum of the problem by including not only steps towards the minimum but also those that lead to larger function values within a certain probability. Local minima can be finally combined into the global minimum. The Simulated Annealing method is, however, very time-consuming.

2.6.4. *Variances of the parameters*

From the weighted normal equations, the covariance of the parameters Σ_a can be calculated and this provides useful information about the spread of values that can be expected [NAT 92, LIN 99, FRI 97, FRIS 95]. Large variances of the parameter values are the consequence of low sensitivity and/or high measurement errors and mean that low accuracy can be expected from single experiments. The parameter covariance matrix for a reduced parameter set with k significant parameters:

$$\Sigma_{a,k} = (S_{red,k}{}^T W\ S_{red,k})^{-1}\ , \text{ with } W = W_\varepsilon = \Sigma^{-1} \qquad [287]$$

is only calculated by those k columns of S which are associated with the k parameters which are actually taken into account.

On the assumption that the measurement errors are uncorrelated random quantities having equal variances so that $\Sigma_\varepsilon = \sigma^2 I$, where I is the identity matrix and σ^2 is the variance of the measurement error:

$$\Sigma_{a,k} = \hat{\sigma}^2 (S_{red,k}{}^T S_{red,k})^{-1} \qquad [288]$$

$\hat{\sigma}^2$ can be estimated from:

$$\hat{\sigma}^2(k) = \frac{\varepsilon_k{}^T \varepsilon_k}{n-k} = \frac{\|\varepsilon_k\|^2}{n-k} \qquad [289]$$

where:

$$\varepsilon_k = r - S_{red,k} \Pi \Delta a_{red,k}$$

n and k are the number of measurement sets and the number of parameters actually taken into account, respectively. The condition number of $S^T S$ plays an important role. Large condition numbers lead to large co-variances. Usually the parameter subset selection improves the condition number. Also, large measurement errors increase σ^2 and hence the co-variances. The diagonal elements of Σ_a represent the parameter variances, which provide us with information about their statistical uncertainty.

For the regularized solution, the parameter covariance matrix is [WAL 97]:

$$\Sigma_{a,k} = (S^T W\ S + \gamma^2 I)^{-1} (S^T W\ S) (S^T W\ S + \gamma^2 I)^{-1} \qquad [290]$$

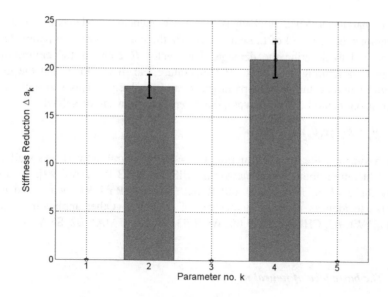

Figure 2.25. *Identification result with parameters and standard deviation for the five-dof oscillator with two damage sites*

Figure 2.25 shows the damage identification result for the five-dof oscillator shown in Figure 2.19 where two steel columns have been removed between storeys 1–2 and 3–4 in this example. Removal of one of the five steel columns means a 20% reduction in stiffness. The data used here were measured FRFs. Further details about this example can be found in [FRI 97, JEN 99]. The estimation result using QR-decomposition clearly identifies that only two parameters are here significant and these are the correct parameters 2 and 4, the other three parameters being set to zero. The reduction of 20% has been approximately reproduced by the parameter estimation. The estimated standard deviations, square roots of the variances, are shown as error bars.

2.7. Neural network approach to SHM

The forward problem of calculating the dynamic changes from the system parameters can be written as:

$$r = \Phi(\theta_d, \theta_e) \tag{291}$$

where the vector θ_d contains information about the damage parameters (i.e. a crack length or a stiffness reduction) and the vector r represents the changes of the dynamic parameters such as changes of eigenfrequencies, mode shapes or spectral distributions, etc. Instead of the parameters θ_d, the matrix correction parameters Δa can be used to describe the damage. The vector θ_e contains the environmental parameters describing the effects of temperature, etc. It is shown that the inverse problem of finding the damage parameters can be solved by classical optimization or regression methods. Alternatively, the inverse problem can be solved:

$$\theta_d = \Phi^{-1}(r,\theta_e) = \Psi(r,\theta_e) \qquad\qquad [292]$$

by approximating the inverse function $\Phi^{-1} = \Psi$ by neural network. As well as the general literature about neural networks [BIS 95, NAB 02, HAY 99], we follow closely the work of Kirkegaard and Rytter [KIR 94, RYT 97], who used neural networks to identify damage in mechanical structures. Other applications can be found in [WU 92, CHU 96, LOP 00, WOR 93, WOR 04, MAN 02, STA 04a].

2.7.1. *The basic idea of neural networks*

Artificial neural networks are computational models inspired by the neuron architecture and operation of the human brain. The paper of McCulloch and Pitts [MCC 43] from 1943 is usually considered as the pioneering work in this field. An artificial neural network is an assembly (network) of a large number of highly connected processing units, the so-called neurons or nodes. The neurons are connected by unidirectional communication channels, the so-called connections; see Figure 2.26.

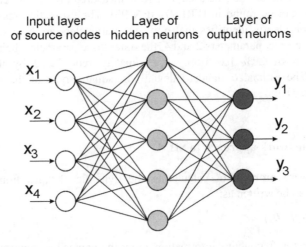

Figure 2.26. *A perceptron network with input, output and one hidden layer*

The strength of the connections between the neurons is represented by weights, which are numerical values; see Figure 2.27. Knowledge is stored in the form of the complete collection of weights. Each node has an activation value that is a function of the sum of inputs received from the preceding layer and the weights.

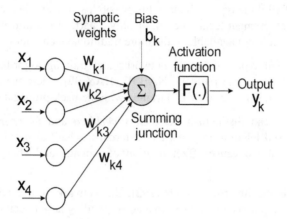

Figure 2.27. *Principle of the generation of a node output from multiple inputs via summation and processing by an activation function with two layers only*

Neural networks are capable of self-organization and learning. Most neural networks have some sort of training rules to adjust the weights by using some well defined patterns of input and output. One of the characteristics of neural networks is the capability of producing correct, or nearly correct, outputs when presented with partially incorrect or incomplete inputs. Furthermore, neural networks are capable of performing an amount of generalization from the patterns on which they are trained.

2.7.2. *Neural networks in damage detection, localization and quantification*

The basic strategy for developing a neural network-based approach to be used in connection with vibration-based monitoring of engineering structures is to train a neural network to recognize different damage scenarios from the measured response of the structure, strains, modal parameters, etc. For instance, a neural network might be trained with natural frequencies and mode shape components as input and the corresponding damage state as output.

Therefore, the training of a neural network with appropriate data containing the information about the cause and effect is a key requirement of a neural-network based damage assessment technique. This means that the first step is to establish the training sets which can be used to train the network in such a way that the network can recognize the behavior of the damaged as well as undamaged structure from measured quantities. Ideally, the training sets should contain data for the undamaged as well as for the damaged structure in various damage states. These data can be obtained by measurements, model tests, through numerical simulation or as a combination of all three types of data. This possibility of using in a neural network-based damage assessment technique all, or only a part, of the information obtained is a capability that is not available in the more traditional techniques.

Many different types of neural networks and training algorithms have been proposed (see for example [BIS 95, NAB 02, HAY 99]). Here we discuss the so-called multilayered perceptron (MLP) trained with the back-propagation algorithm.

Basically we can distinguish between *supervised* and *unsupervised* learning modes. Supervised learning requires an "external teacher" to provide the network information about the cause–effect relation of damage and change of dynamic behavior.

To establish the inverse relation [292], the network is trained by assigning labelled pairs of input vectors $x^{(s)}=r^{(s)}$ to a corresponding target vector $t^{(s)}$. The target vector contains the information about the damage parameters θ_d in a certain encoded and normalized way. For practical purposes it is necessary not only to provide one training set for one pair made up of an input and a target vector $r^{(s)} \rightarrow t^{(s)}$ but to generate different inputs with superimposed random noise: $(r^{(s)}+\Delta r^{(s)}) \rightarrow t^{(s)}$. This enables the network to recognize a certain damage pattern if the measured input is polluted by noise and does not look exactly like those presented during the learning phase. Basically, the training procedure is an optimization task bringing the network output or $y^{(s)}$ (which follows from the network input) as close as possible to the predefined target vectors $t^{(s)}$ by adjusting the free parameters of the neural network:

$$\min \sum_{s=1}^{s_{\max}} \left\| y^{(s)} - t^{(s)} \right\|^2 \qquad [293]$$

The free parameters are the synaptic weights and biases, which will be explained below. Once the network is trained, it can be presented with the currently measured data sets. The network then delivers a damage vector as output and this allows to judge the present state of damage. While the training phase can be very time-consuming, this calculation is very fast and allows an on-line evaluation. The pattern recognition character of this approach is obvious.

The training sets can be generated, for instance, through experimental tests or by a computational model, such as finite element analysis, where r is measured/calculated for a well-defined damage vector θ or Δa, respectively. If the training sets are generated by a computational model and if this model is inaccurate, it is obvious that the same kind of trouble whether we use optimization or a neural network for the solution of the damage identification problem will be encountered. Hence, the quality of the model is a key requirement when using model-based damage identification.

In contrast to *supervised learning*, it is not always possible to obtain such labeled training sets where a typical dynamic behavior is assigned to a defined damage scenario. In this case, the notion of *unsupervised learning* is referred. Very often, it is only possible to measure the behavior of a system in the undamaged state. These unlabeled training data are presented to the network as inputs. The network outputs are certain characteristic features, which the network extracts from the input data, such as the characteristics of a probability distribution. Depending on the type of network, a self-organization of the network is performed in one way or another. These extracted features are characterized by the fact that the dimension of the feature space is usually much smaller than that of the original data. For a more detailed discussion of unsupervised learning methods, see [BIS 95, NAB 02, HAY 99].

2.7.3. Multi-layer Perceptron (MLP)

One widely used type of network is the MLP. It belongs to the class of layered feed-forward nets with supervised learning [KIR 94, RYT 97, BIS 95, NAB 02, HAY 99, WOR 04, WOR 93, STA 04a]. An MLP network is made up of one or more hidden layers between the input and the output layer, as shown in Figure 2.26. Each layer consists of a number of nodes, which are connected to all the nodes in the previous and the subsequent layers. During the training phase, activation flows are only allowed in one direction, a feed-forward process, from the input layer to the output layer through the hidden layer [RYT 97]. The input vectors feed each of the nodes in the first layer, the output of this layer feed into each of the second layer nodes and so on:

$$y_k = F\left(\sum_{i=1}^{n_l} w_{ki} x_i + b_k\right)$$ [294]

y_k is the node output following from the inputs x_i of the previous layer multiplied by the weights w_{ki} which characterize the strength of the connection; see Figure 2.27. The bias or threshold b_k introduces a shift of the zero-point of the activation function input. The output of this node can then be the input for the next layer.

Several types of activation functions can be used [BIS 95, NAB 02, HAY 99]. The non-linear logistic sigmoid function:

$$F(\alpha) = \frac{1}{1 + \exp(-\alpha)}$$ [295]

generates values between zero and 1 with a smooth transition between these values. Linear or other types of non-linear transfer functions are also in use.

In the case of three layers (input, output, and one hidden layer), we pass the n_{in} inputs x_i to the hidden layer, which has n_h neurons. This produces outputs of the hidden layers $x_{k,h}$:

$$x_{k,h} = F\left(\sum_{i=1}^{n_{in}} w_{ki,1} x_i + b_{k,1} \right); \; k = 1,...,n_h \qquad\qquad [296]$$

which are again taken as input for output layer:

$$y_k = F\left(\sum_{i=1}^{n_h} w_{ki,2} x_{i,h} + b_{k,2} \right), \; k = 1,...,n_{out} \qquad\qquad [297]$$

This can be further expanded to more hidden layers. The weights of the connections between the layers are initialized by random values at the start of the training process. During this process, representative examples of input–output patterns are presented to the network. Each presentation is followed by small adjustments of the weights and the bias values if the computed output is not correct. A classical approach to the adjustment is the so-called back propagation algorithm of Rumelhard and McClelland [RUM 86], which is a gradient method to minimize the error between the output and the target vectors. If there is any systematic relationship between input and output, and the training examples are representative of this, and if the network topology is properly chosen, then the trained network will often be able to generalize beyond learned examples. For the practical implementation of MLP, the NETLAB toolbox is used. This toolbox is written in MATLAB and described in [NAB 02].

In [RYT 97], a Radial Basis Network (RBF) is used; see also [BIS 95, NAB 02, HAY 99] for detailed explanation and programming of this network type. In their study, Rytter and Kirkegaard [RYT 97] observe that this type of network failed completely.

2.8. A simulation example

2.8.1. Description of the structure

The example given in Figure 2.28 is a simulation study to demonstrate the behavior of different methods under various conditions such as incomplete measurement data and noise. The system under consideration is a simply supported beam, which is divided into 20 elements. The total length of the beam is 0.4 m with a beam cross section of $20mm \times 20mm$. The material is steel with density of 7850 kg/m^3 and

Young's modulus of $E = 210,000$ MPa. The damage is simulated by a loss of bending stiffness in elements 8 and 12. The corresponding stiffness reductions are $\Delta EI/EI = 0.4$ (for beam element no. 8) and 0.3 (for beam element no. 12).

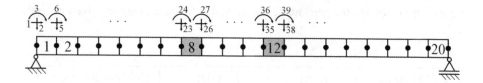

Figure 2.28. *Beam with 20 finite elements and damage at two elements (no. 8 and no. 12)*

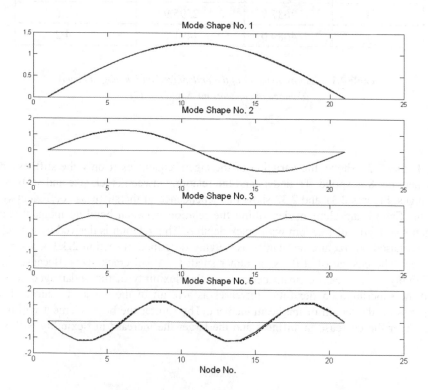

Figure 2.29. *Bending modes 1 to 4 of the beam with and without damage (Mode no. 5 is the fourth bending mode). It can be seen that changes of the modes are relatively small*

The mode shapes with and without the damage are shown in Figure 2.29. Especially for mode 3, there is almost no deviation. As can be seen, the changes are rather small. The case of two damage locations with almost symmetric distribution is a more complicated case compared to one-sided damage, because the one-sided damage causes a stronger asymmetry after the damage. The corresponding eigenfrequencies and the percentage deviation due to the damage are given in Table 2.1.

Mode no. i	$f_{0,i}$ [Hz]	$f_{d,i}$ [Hz]	Relative difference [%]
1	293.2	279.9	−4.5
2	1172.5	1148.6	−2.0
3	2637.9	2595.9	−1.6
4	4689.0	4491.2	−4.2

Table 2.1. *Eigenfrequencies of the undamaged and damaged beam (damage at elements no. 8 and no. 12)*

Figure 2.28 shows the variation of the eigenfrequencies if only the stiffness at element no. 8 is varied. To understand the relation between crack size and stiffness changes, Figures 2.31 and 2.32 show the dependence of the stiffness coefficient k_{33} of the finite beam element describing the relation between bending moment and rotational degree of freedom with crack depth a. This relation is derived by a crack model based on fracture mechanics as roughly described in section 2.3.3.4. As can be seen, the change of stiffness is rather small for small crack sizes. Because the flexibility of the beam element depends on the flexibility of the undamaged beam and the superimposed flexibility increases as a result of the crack, the curve also depends on the length of the beam element. The shorter the beam element will be, the larger the decrease in stiffness and the larger the increase in flexibility will be notices.

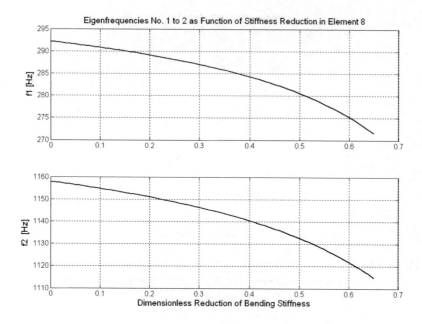

Figure 2.28. *The first two eigenfrequencies as function of bending stiffness reduction in element no. 8*

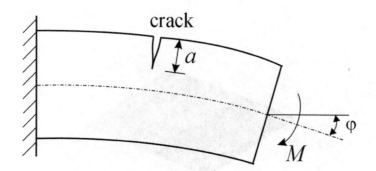

Figure 2.29. *Beam element with a crack, showing the relationship between the bending moment and the rotation*

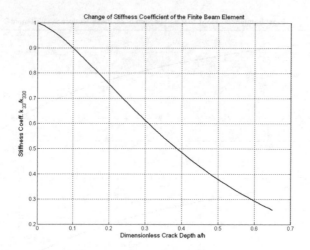

Figure 2.30. *Variation of the stiffness coefficient k_{33} describing the relationship between bending moment and rotational degree of freedom with crack depth a related to beam height h*

The 20 elements result in a model with 21 nodes (three dofs per node) with stiffness and mass matrices K and M of dimension 63×63. The change of stiffness has a very local character with respect to changes in the stiffness matrix. The change of the stiffness at the two beam elements no. 8 and no. 12 yields the following changes in the stiffness matrix $\Delta K = K_d - K_0$, as displayed graphically in Figure 2.31.

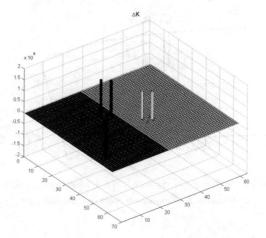

Figure 2.31. *Change of the 63×63 stiffness matrix ΔK due to changes in beam elements 8 and 12*

2.8.2. *Application of damage indicator methods*

Starting with the modal force residuals (see section 2.5.7):

$$\varepsilon_i = \Delta f_i = \left(K - \lambda_{i,d} M \right) \varphi_{i,d} = -\Delta K \; \varphi_{i,d}$$

we can localize the errors in ΔK visually from their plot; see Figure 2.34. Three modes are investigated. The black bars are the force residuals resulting from the complete displacement vector including all horizontal and vertical displacements and the rotations. Especially in the middle and lower subplots, it can be seen that the large values are connected with the changes of elements 8 and 12. They correspond to the residual force components 23, 26 (belonging to element 8) and 35, 38 (belonging to element 12), respectively.

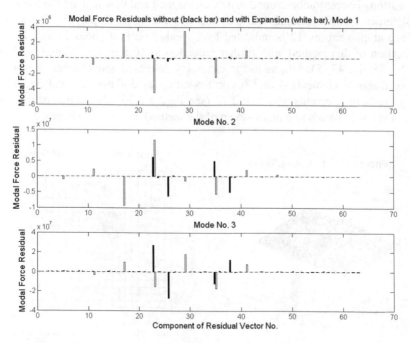

Figure 2.32. *Influence of small expansion errors on the modal force residuals*

Now, the transverse beam displacements of each second node starting with dof no. 5 as measurement information are retained. This means that 10 transverse displacements are still used from the total number of 63 dofs: 5, 11, 17, 23, 29, 35, 41, 47, 53 and 59. The missing information has to be reconstructed using the expansion techniques described above. The problem here is that the expansion of the modes of the damaged system can only be performed using the modes from the undamaged system, which introduces errors in the expansion process. Figure 2.32 also demonstrates how strong the influence is on the relatively small expansion errors on the modal force residuals. The white bar is the result of expanding the incomplete measurements and feeding them into the modal force residual calculation. As can be seen, the two vectors have almost nothing in common. The result is more or less the same if dynamic or SEREP expansion is applied. This negative effect of errors is actually increased if noise is added. The input residual formulations are commonly known to be very sensitive with respect to disturbances in the input data [FRI 86].

Figure 2.33 shows the result of the Error Matrix Method as presented by Ewins and Sidhu [EWI 00] (see section 2.5.2). Here, the first four bending modes are used as well, but all 21 transverse displacements are used in the first step. It can be seen that a strong concentration around matrix elements 8 and 9, which are the transverse displacements corresponding to beam element no. 8, is displayed. The smaller damage at element no. 12 is indicated by a weaker peak at nodes 12 and 13. The application of this method with further reduction to 10 transverse dofs (5, 11, 17, 23, 29, 35, 41, 47, 53, 59), as in the examples discussed above, leads to peaks at matrix diagonal elements 4 and 6 corresponding to dof nos. 23 and 35, which belong to beam elements nos. 8 and 12 (see Figure 2.36). The superposition of a noise level of 1% leads to a break-down of the method in this case (Figure 2.37).

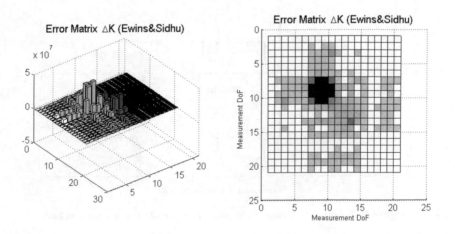

Figure 2.33. *21 × 21 matrix **ΔK** after Ewins and Sidhu [EWI 00] derived from measurement of all 21 transverse displacements and four bending modes*

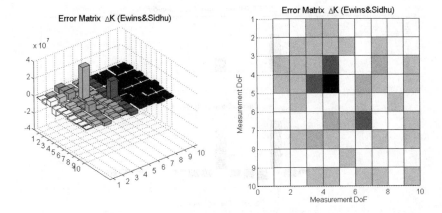

Figure 2.34. *10 × 10 matrix **ΔK** after Ewins and Sidhu [EWI 00] derived from measurement of 10 transverse displacements*

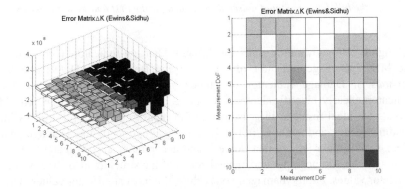

Figure 2.35. *10 × 10 matrix **ΔK** after Ewins and Sidhu [EWI 00] derived from measurement of 10 transverse displacements, four bending modes and 1% noise*

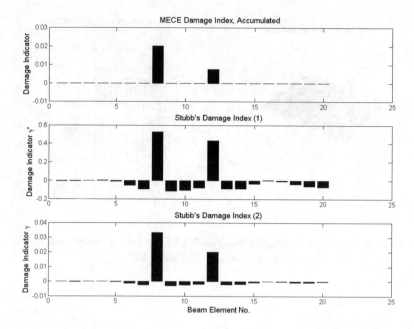

Figure 2.36. *MECE Damage Indicator and Damage Indicator according to Stubbs for the full dataset without noise, accumulated for the first four bending modes*

Figure 2.36 shows the results for different Damage Indicators (see section 2.5): the MECE Damage Indicator of Ladeveze and Reynier [LAD 89] and the two versions of Damage Indicator of Stubbs *et al.* [STU 95]. All three Damage Indicators clearly identify elements 8 and 12 as the damage location. The MECE reveals the ideal result in the case of the full measurement set (all dofs) with four bending modes and without noise. Due to assumption of the Damage Indicator Method of Stubbs *et al.* that some quantities remain approximately unchanged (although they do not), the undamaged zones of the structure also show some small indicator values, but the damaged zones show the most significant values. The more interesting cases of incomplete measurement with and without noise are closer to practical application and will be examined in the following section. If the expanded measurement sets have to be used instead of the full error-free set, the expansion errors lead to some spread of the Damage Indicators to the neighboring elements. These results are shown in Figure 2.37. Again the MECE indicator yields the best results, keeping the indicator values small where the structure is not damaged. Finally, adding normally distributed random noise to displacement with a standard deviation of 1% of the maximum displacement of the maximum modal displacement yields the situation depicted in Figure 2.38. Also in the case of noisy data, the MECE method brings out the clearest results.

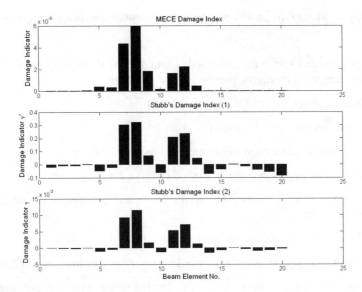

Figure 2.37. *MECE Damage Indicator and Damage Indicator according to Stubbs for the expanded dataset without noise, accumulated for the first four bending modes*

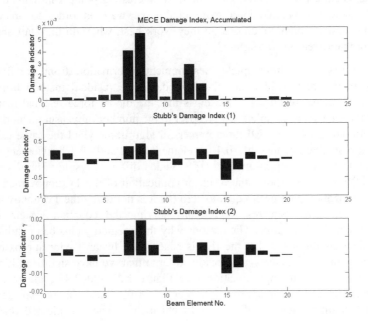

Figure 2.38. *MECE Damage Indicator and Damage Indicator according to Stubbs for the expanded dataset with 1% noise, accumulated for the first four bending modes*

2.8.3. *Application of the modal force residual method and inverse eigensensitivity method*

This section deals with those methods localizing the damage by determining the stiffness correction parameters from an inverse problem:

– modal force residual method (section 2.5.7),

– inverse eigensensitivity method (section 2.5.6),

– modal kinetic energy sensitivity method (section 2.5.8.1),

– modal strain energy sensitivity method (section 2.5.8.2).

The solution for the parameter vectors is found by application of different numerical procedures, such as:

– Tykhonov regularization method (section 2.6.1),

– parameter subset selection by forward/backward regression (section 2.6.2),

– parameter subset selection by QR-decomposition and forward regression (section 2.6.3).

In the author's opinion, it does not make sense to compare numerical result within an accuracy of several digits. The most interesting point is how the different methods generally behave in a more qualitative way, in order to answer the questions: under which conditions do they work best, when do they fail and what are the probable reasons for failure?

In the first case of complete measurement information from the first four bending modes and without noise, the modal force residual method brings out perfect and correct results. The reason is that the modal force residual method is exact if the data set is complete with respect to the number of measurement dofs and if the data are noise-free. All three numerical algorithms yield the same excellent result for a stiffness change of –0.4 for element no. 8 and –0.3 for element no. 12. The inverse eigensensitivity method (IES) yields different results. The upper right picture of Figure 2.39 shows the result of application of the Tykhonov method but with regularization parameter equal to zero (so it is not really the Tykhonov method but a simple Least Squares Regression) solving the equation system for all parameters at the same time. The reasons why this method yields bad results under perfect data conditions is that the IES is based on a linear Taylor approximation. Since the parameter changes are already so large that we have significantly moved away from the linear approximation (see Figure 2.28 and 2.42), the results are strongly influenced by the approximation errors of the linear Taylor expansion and the large condition number of the coefficient matrix. The two methods based on parameter selection find the correct elements but they overestimate their values, and this is also due to the linearization error. The size of the overestimation can be easily understood when looking at Figure 2.40. The effect of the linearization errors can be easily compensated by improving the two localized parameters 8 and 12 iteratively. This will lead to the correct parameter values. The threshold for the QR-decomposition is set to 5% throughout all calculations. This means that an additional parameter is only selected if this parameter can improve the error by more than 5% of the overall error. Figure 2.41 shows the correlation measure κ^2 of the different columns of the sensitivity matrix and the residual vector. Columns/elements 8 and 12 show highest correlation with the residuals vector. This can be used as a preselection step in more complex situations in order to get a reduction of parameters before entering the phase of repeated solution of equation systems during the forward/backward selection procedure. A threshold value can be set and parameters j for which $\kappa_j^2 \leq \kappa_{thres}^2$ are deleted as potential candidates for describing the damage.

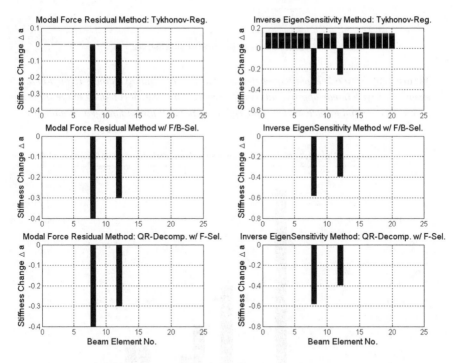

Figure 2.39. *Comparison of modal force residual method and inverse eigensensitivity method using different solution techniques, full data set, no noise and regularization parameter zero*

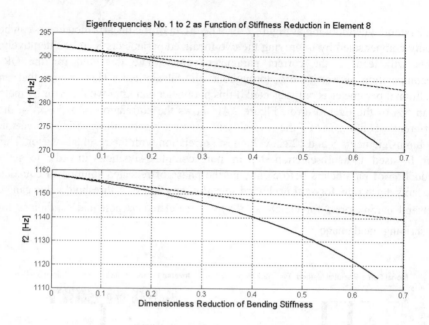

Figure 2.40. *First two eigenfrequencies as a function of bending stiffness reduction in element no. 8 with linear approximation of the curve. The tangent is placed at the stiffness reduction equal to zero, which is the reference point of the undamaged structure*

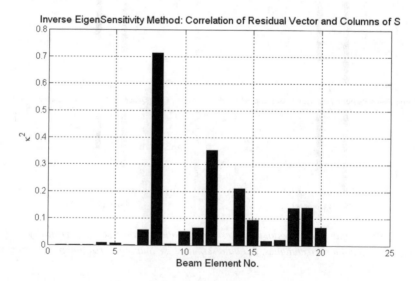

Figure 2.41. *Correlation measure of the different columns of the sensitivity matrix and the residual vector. Elements 8 and 12 show highest correlation with the residuals*

Figure 2.42 repeats the test run, shown in Figure 2.40, but with a regularization parameter which is non-zero. The QR-decomposition is not affected by this, however, and the effect on the solution of the other methods is as follows.

The solution using the IES method with the full set of 21 parameters is improved. The parameters belonging to the undamaged elements (which should be zero) are now smaller than before. Also, the values belonging to the damaged elements are smaller now. This is the effect of the regularization term trying to minimize the changes.

The MFRM solution combined with the Tykhonov method becomes worse than before because the minimization of the changes leads to a reduction of the two damage parameters 8 and 12 and also an increase of the non-zero parameters corresponding to the undamaged elements. The reason is that the attempt to reduce the overall norm of the solution vector primarily leads to a reduction of the large solution components (due to the quadratic dependence in the Euclidian norm). The small increase of the other parameters leads to much smaller values in the Tykhonov penalty term.

The Modal Force Residual Method combined with forward/backward selection suppresses the "unimportant" parameters but the two parameters 8 and 12 are also slightly reduced due to the Tykhonov regularization term. The same can be observed for the IES with forward/backward selection and Tykhonov regularization.

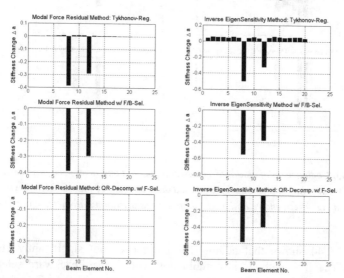

Figure 2.42. *Comparison of modal force residual method and inverse eigensensitivity method using different solution techniques, full data set and no noise, regularization parameter non-zero*

The case shown in Figure 2.43 is the same as before, but, instead of the full data set of 63 vector components, only the 10 transverse displacements are used.

The MFRM is known to be very sensitive to errors in the input data due to correlated residuals. This correlation can yield string bias on the estimates. Here, the data errors are coming from the expansion process. As can be seen, a major outcome is that the results of the MFRM worsen significantly. The MFRM in connection with QR-decomposition (lower left chart) suppresses the undesired components, but it does not hit the right elements (7 and 11 instead of 8 and 12). The outcome of the IES method is much better here, especially for the QR-decomposition and the forward/backward selection, although less information is used.

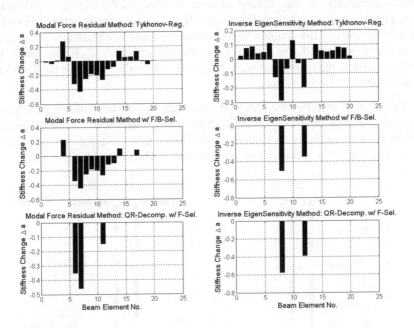

Figure 2.43. *Comparison of modal force residual method and inverse eigensensitivity method using different solution techniques, expanded data set from 10 measured dofs without noise on eigenfrequencies and mode shapes, regularization parameter non-zero*

Figure 2.44 shows that adding 1% noise does not significantly change the situation. The IES methods in combination with forward/backward selection and regularization, or with the QR-decomposition, turn out to be the most robust. Figure 2.45 shows the case with expansion and 1% noise but now *without regularization* (regularization parameter is equal to zero). As can be seen, the consequence is that the IES solution with the full parameter set (upper right chart) becomes completely useless, because the amplification of the noise is so dominant. This shows the importance of the regularization under noise conditions. Also, the IES method with forward/backward selection suffers from the lack of regularization; compare Figure 2.44 and 2.47.

Figure 2.46 displays the result of the weighted modal force residual method, as described in section 2.5.7.3, instead of the normal MFRM. As can be seen, this improves results for this method significantly, see left column of Figure 2.46 in comparison to the left column of Figure 2.44. The reason is that the premultiplication with the flexibility matrix can partially decouple the errors in the residual vector although the same expanded measurement sets are used, which results in an improved parameter estimates. Finally, Figure 2.47 shows the results with added noise (1% standard deviation related to the largest eigenvector component). The left column is again important and it can be seen that the noise worsens the results, although the solution is still better compared to the non-weighted case.

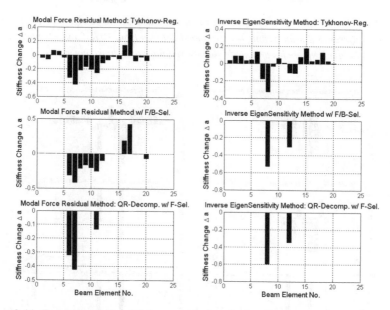

Figure 2.44. *Comparison of modal force residual method and inverse eigensensitivity method using different solution techniques, expanded data set from 10 measured dofs and 1% noise on eigenfrequencies and mode shapes, regularization parameter non-zero*

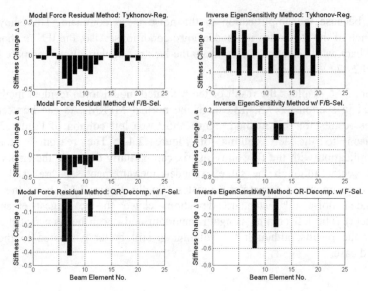

Figure 2.45. *Comparison of modal force residual method and inverse eigensensitivity method using different solution techniques, expanded data set from 10 dof and 1% noise on eigenfrequencies and mode shapes, regularization parameter zero*

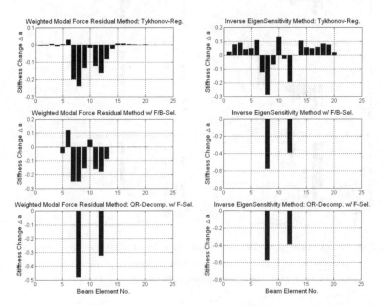

Figure 2.46. *Comparison of weighted modal force residual method and inverse eigensensitivity method using different solution techniques, reduced and expanded data set (10 dof) and without noise, regularization parameter non-zero*

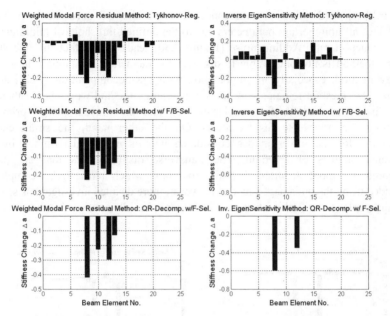

Figure 2.47. *Comparison of weighted modal force residual method and inverse eigensensitivity method using different solution techniques, expanded data set from 10 dof and 1% noise on eigenfrequencies and mode shapes, regularization parameter non-zero*

2.8.4. *Application of the kinetic and modal strain energy methods*

This section is devoted to the results obtained from the modal kinetic energy method and the modal strain energy method. In contrast to the energy indicator methods described in section 2.5.8, which also use the strain energy to indicate element changes, an equation system based on energy sensitivities is built up and solved to get the parameter values of the predicted changes. The results are shown in Figure 2.48 for the ideal case under full availability of all vector components and no noise (and with a non-zero regularization parameter, which is only relevant for the upper and middle rows of the figure). The correct damage locations can be found with slightly different parameter values. We have to keep in mind that everything here is based on a linear Taylor approximation and hence in a first step attention must be paid on possible linearization errors, which could be eliminated through further iterations. Figure 2.49 considers the reduction of available measurement information to 10 transverse displacements and reconstruction of the missing dofs by expansion. The expanded full vector is required in order to calculate the element kinetic and strain energies. As an important outcome, it can be seen that the expansion process has a strong influence on the quality of the localization result. The best solution is provided by the modal kinetic energy methods with the forward/backward selection and especially with the QR-decomposition which eliminates all unimportant parameters due to the defined threshold of minimum 5% error reduction rate. Adding noise (Figure 2.50) leads to the fact that the strain energy method is no longer able to discover the element 12, although the kinetic energy method, especially in combination with QR-decomposition, yields stable results. The question remains why the kinetic energy method works better than the strain based method. A first explanation is that the kinetic energy primarily addresses the transverse displacements. On the other hand, the strain energy expression for the lower modes considered here is mainly determined by the rotational degrees of freedom. The transverse displacements that are partly measured are more accurate than the rotations, which have to be completely expanded using the undamaged model. However, more detailed investigations seem to be worthwhile to clarify this situation, which cannot be fully explained at the moment.

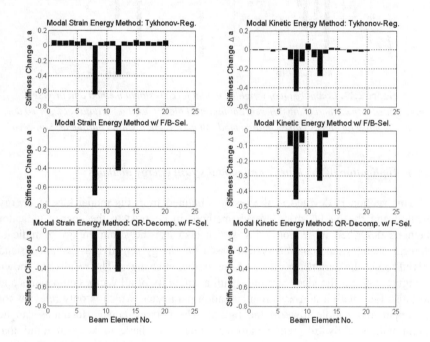

Figure 2.48. *Comparison of modal strain energy method and modal kinetic energy method for different solution techniques, full data set, no noise, regularization parameter non-zero*

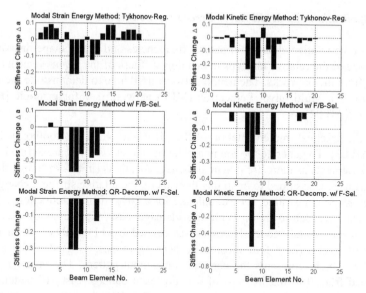

Figure 2.49. *comparison of modal strain energy method and modal kinetic energy method for different solution techniques, expanded data set from measurement of 10 dofs, no noise, regularization parameter non-zero*

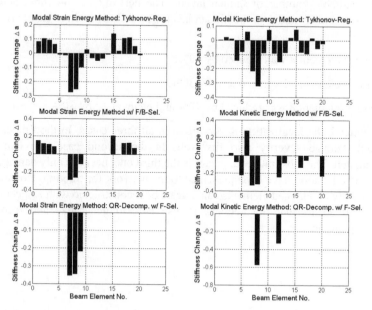

Figure 2.50. *comparison of modal strain energy method and modal kinetic energy method for different solution techniques, expanded data set from measurement of 10 dofs, 1% noise on eigenfrequencies and mode shapes, regularization parameter non-zero*

2.8.5. *Application of a Multi-Layer Perceptron neural network*

Again the relative changes in natural frequencies and mode shape data are used as inputs, and the relative changes in bending stiffness of the beam elements are the outputs from the networks. The neural network used here to demonstrate the general procedure is a Multi-Layer Perceptron (MLP) network; see for example [BIS 95, HAY 99, NAB 02].

For the practical implementation of the MLP, the NETLAB toolbox was used; see [NAB 02]. To generate the training patterns, target vectors *t* are defined to have as many elements as there are potential damage locations (in this example, there are 20). If damage is assumed to be in the *j*th element, the relative stiffness change is set as the corresponding value, for example 0.2 for a 20% stiffness reduction. All other components are set to zero. With this damage case the (relative) changes of eigenfrequencies and mode shapes are calculated by FE analysis for a defined number of modes (here four incomplete bending modes, as used in the previous examples) which represents the corresponding input pattern. The damage severity has varied from 0 to 0.5 in steps of 0.1. For each damage case, 10 patterns were generated with superimposed random noise (1% related to the maximum mode shape value). This is recommended so that the network sees a number of random variations of the same pattern. Altogether, there are 1200 training sets. One has to try out a suitable number of hidden layers. The network has 44 input nodes, 20 output nodes and 75 nodes in the hidden layer. The following three cases are shown here:

– case 1: element 8 was reduced by 40%,

– case 2: element 12 was reduced by 30%,

– case 3: element 8 was reduced by 40% and element 12 by 30%.

As noticed, the localization is very good in all three cases. In the first case, the estimated damage severity also fits well. In the second case, the damage is overestimated. The most interesting case is the third one, because only damage cases with a single location have been presented to the network for training. However, the new case of two damaged elements can at least be reconstructed with respect to their locations. It seems that the network is also able to generalize and reconstruct untrained patterns.

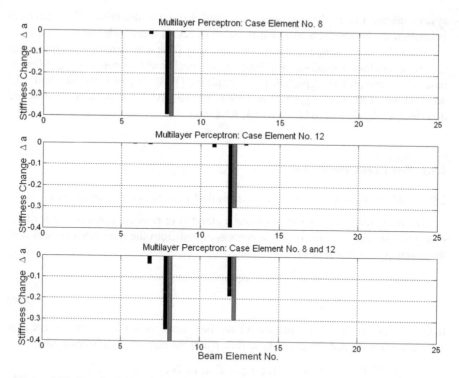

Figure 2.51. *Result of damage identification with the MLP neural network. The black bars are the outcome of the network, while the grey bars represent the true damage cases*

2.9. Time-domain damage detection methods for linear systems

In the control and automation literature, many interesting approaches to monitoring general dynamic systems which can be adapted for the purpose of SHM, are found in [GER 98, GUS 00, BAS 88, BAS 93, WIL 76, PAT 89, PAT 91, FRA 94, LJU 99]. Some of these algorithms rely on an analytical redundancy concept based on quantitative mathematical models, which represent the undamaged system. Other overviews can be found in [NAT 97, DOE 96, SOH 04, WAL 90].

As discussed above, in any model-based approach, residuals have to be defined that quantify the deviation of the actual system from its nominal, undamaged state. The model of the undamaged structure is then continuously tested against the measured data of the real structure and, in the case of damage, some change in the statistics of the residuals indicating abnormal behavior is expected to be observed. In order to produce such residuals, a Kalman Filter (KF) is used [FRI 03a]. The KF was originally developed as an algorithm for the optimal estimation of unknown system states. Later, the KF was also used for fault detection. The statistical evaluation of the residuals makes it possible to set up thresholds, which enables to say if an observed change is significant, resulting from a change of the structure or simply a random variation. Applications to mechanical systems using filter or observer techniques are reported, for example by [WAL 90, SOE 91, SEI 95b, SEI 96, FRI 90, FRI 95]. Approaches based on AR, ARX or ARMA models are described in [LJU 99, SOH 01, PIO 93, AND 97].

2.9.1. *Parity equation method*

2.9.1.1. *Monitoring of deterministic systems with measured input and output*

The basic idea of the parity relations method is to provide a proper check of the parity (consistency) of the measurements acquired from the monitored system with the nominal (reference) model.

Let us start with the measurement equation at time instant k:

$$y_k = C\,z_k + D\,u_k \tag{298}$$

then write down the same equation for the next time instant $k + 1$, where the new state z_{k+1} is substituted for z_k and u_k so that, using the state-space system model:

$$\begin{aligned} y_{k+1} &= C\,z_{k+1} + D\,u_{k+1} = C\,(A\,z_k + B\,u_k) + D\,u_{k+1} \\ &= CA\,z_k + CB\,u_k + D\,u_{k+1} \end{aligned} \tag{299}$$

We can proceed with $k + 2$ and substitute the new state again for the previous ones:

$$\begin{aligned} y_{k+2} &= C\,z_{k+2} + D\,u_{k+2} = C\,(A\,z_{k+1} + B\,u_{k+1}) + D\,u_{k+2} \\ &= CA^2\,z_k + CAB\,u_k + CB\,u_{k+1} + D\,u_{k+2} \end{aligned} \tag{300}$$

and so on. For a general future time instant $k + i$ it is found that:

$$\begin{aligned} y_{k+i} &= C\,z_{k+i} + D\,u_{k+i} \\ &= CA^i\,z_k + CA^{i-1}B\,u_k + CA^{i-2}B\,u_{k+1} + \cdots + CB\,u_{k+i-1} + D\,u_{k+i} \end{aligned} \tag{301}$$

Now all measurement and input vectors for all time instants k to $k + i$ are put together in one equation system:

$$
\begin{bmatrix} y_k \\ y_{k+1} \\ y_{k+2} \\ \vdots \\ \vdots \\ y_{k+i} \end{bmatrix} = \begin{bmatrix} C \\ CA \\ CA^2 \\ \vdots \\ \vdots \\ CA^i \end{bmatrix} z_k + \begin{bmatrix} D & 0 & 0 & \cdots & 0 & 0 \\ CB & D & 0 & \cdots & 0 & 0 \\ CAB & CB & D & \cdots & 0 & 0 \\ \vdots & \vdots & \vdots & \cdots & \vdots & 0 \\ \vdots & \vdots & \vdots & \cdots & D & 0 \\ CA^{i-1}B & CA^{i-2}B & CA^{i-3}B & \cdots & CB & D \end{bmatrix} \begin{bmatrix} u_k \\ u_{k+1} \\ u_{k+2} \\ \vdots \\ \vdots \\ u_{k+i} \end{bmatrix} \quad [302]
$$

or in short:

$$
Y_{k,i} = J_i\, z_k + K_i\, U_{k,i} \qquad [303]
$$

The matrix J_i is the observability matrix:

$$
J_i = \begin{bmatrix} C \\ CA \\ CA^2 \\ \vdots \\ CA^i \end{bmatrix} \qquad [304]
$$

Now, we define:

$$
R_{k,i}^{*} = Y_{k,i} - K_i U_{k,i} = J_i z_k \qquad [305]
$$

The states are non-measurable and therefore they have to be eliminated. This goal can be achieved by constructing a matrix S in such a way so that:

$$
S^T J_i = 0 \qquad [306]
$$

This means that the columns of S are orthogonal to the columns of J. Then, the residual vector R is defined according to:

$$
R_{k,i} = S^T R_{k,i}^{*} = S^T (Y_{k,i} - K_i U_{k,i}) = S^T J_i z_k = 0 \qquad [307]
$$

so that the residual $R_{k,i}$ is zero, independently of the unknown state z_k.

In order to find such a matrix S the Singular Value Decomposition (SVD) [GOL 96] of the matrix J is calculated:

$$
J = U\, \Sigma\, V^T = \sum_{j=1}^{n} \sigma_j\, u_j v_j^T \qquad [308]
$$

where $\sigma_1 \geq \sigma_2 \geq \sigma_3 \geq \cdots \geq \sigma_v > 0 = \sigma_{v+1} = \cdots = \sigma_n$ are the singular values, which are ordered by their magnitudes. If the matrix has no full rank, the singular values can also be zero. The vectors u and v are left and right singular vectors, which are normalized to 1 and are orthogonal:

$$u_i^T u_j = 0 \text{ and}$$

$$v_i^T v_j = 0 \text{ for } i \neq j$$

Let us assume that J has rank v and we premultiply J by the transposed right singular vector u_{v+l}, $l = 1, 2,..., n - v$. Due to the orthogonality and the fact that $\sigma_{v+l} = 0$, we get:

$$u_{v+l}^T J = \sum_{j=1}^{n} \sigma_j u_{v+l}^T u_j v_j^T = \sigma_{v+l} u_{v+l}^T u_{v+l} v_{v+l}^T = 0 \qquad [309]$$

This means that the vectors u_{v+l}, $l = 1, 2,..., n - v$ span the left kernel space of the matrix J. Putting all the vectors u_{v+l}, $l = 1, 2,..., n - v$ into the matrix U_0:

$$U_0 = \begin{bmatrix} u_{v+1}, & u_{v+2}, & \cdots, & u_n \end{bmatrix} \qquad [310]$$

It is observed that $U_0^T J = 0$, therefore the desired matrix S can be chosen as the matrix U_0 or a subspace, spanned only by some u-vectors:

$$S = U_0 = \begin{bmatrix} u_{v+1}, & u_{v+2}, & \cdots, & u_n \end{bmatrix} \qquad [311]$$

Once a proper matrix S is found, the monitored system can be checked by measuring sequences of outputs and inputs. If a system change reflected by a change of the system matrix ($A \rightarrow \tilde{A} = A + \Delta A$) occurs from equation [304], it follows that:

$$\tilde{J}_i = \begin{bmatrix} C \\ C(A+\Delta A) \\ C(A+\Delta A)^2 \\ \vdots \\ C(A+\Delta A)^i \end{bmatrix} = J_i + \Delta J_i \qquad [312]$$

so that:

$$S^T \tilde{J}_i = S^T (J_i + \Delta J_i) = S^T \Delta J_i \neq 0 \qquad [313]$$

and hence the residual $R_{k,i}$ is no longer zero, indicating a change of the system:

$$R_{k,i} = S^T (Y_{k,i} - K_i U_{k,i}) \neq 0 \qquad [314]$$

Note that also the matrix C can change: $\tilde{C} = C + \Delta C$. In practice, this criterion has proven to be very sensitive with respect to system changes so that very small changes can be detected while significant changes of the eigenfrequencies cannot be observed at this early stage of damage. When dealing with real data from measurement, the residual is usually not exactly zero but small. If the matrix set (A, B, C, D) is not known, it has to be identified from data y and u stemming from the original, undamaged system. Methods for this identification step can be found in [LJU 99, JUA 85, JUA 94].

2.9.1.2. Monitoring of stochastic output-only systems

Going back to equation [302] and adapting it to the new situation of a stochastic input w:

$$
\begin{bmatrix} y_k \\ y_{k+1} \\ y_{k+2} \\ \vdots \\ \vdots \\ y_{k+i} \end{bmatrix} = \begin{bmatrix} C \\ CA \\ CA^2 \\ \vdots \\ \vdots \\ CA^i \end{bmatrix} z_k + \begin{bmatrix} 0 & 0 & 0 & \cdots & \cdots & 0 \\ C & 0 & 0 & \cdots & & 0 \\ CA & C & 0 & \cdots & & 0 \\ \vdots & & & & & 0 \\ \vdots & & & & & \\ CA^{i-1} & CA^{i-2} & \cdots & \cdots & C & 0 \end{bmatrix} \begin{bmatrix} w_k \\ w_{k+1} \\ w_{k+2} \\ \vdots \\ \vdots \\ w_{k+i} \end{bmatrix} + \begin{bmatrix} v_k \\ v_{k+1} \\ v_{k+2} \\ \vdots \\ \vdots \\ v_{k+i} \end{bmatrix} \qquad [315]
$$

Now the vector y_{k-1}^T from the right is multiplied by:

$$
\begin{bmatrix} y_k \\ y_{k+1} \\ y_{k+2} \\ \vdots \\ \vdots \\ y_{k+i} \end{bmatrix} y_{k-1}^T = \begin{bmatrix} C \\ CA \\ CA^2 \\ \vdots \\ \vdots \\ CA^i \end{bmatrix} z_k y_{k-1}^T \cdots
$$

$$
+ \begin{bmatrix} 0 & 0 & 0 & \cdots & \cdots & 0 \\ C & 0 & 0 & \cdots & & 0 \\ CA & C & 0 & \cdots & & 0 \\ \vdots & & & & & 0 \\ \vdots & & & & & \\ CA^{i-1} & CA^{i-2} & \cdots & \cdots & C & 0 \end{bmatrix} \begin{bmatrix} w_k \\ w_{k+1} \\ w_{k+2} \\ \vdots \\ \vdots \\ w_{k+i} \end{bmatrix} y_{k-1}^T + \begin{bmatrix} v_k \\ v_{k+1} \\ v_{k+2} \\ \vdots \\ \vdots \\ v_{k+i} \end{bmatrix} y_{k-1}^T \qquad [316]
$$

and the expectation operator is applied:

$$
\begin{bmatrix}
E\{ y_k \, y_{k-1}^T \} \\
E\{ y_{k+1} \, y_{k-1}^T \} \\
E\{ y_{k+2} \, y_{k-1}^T \} \\
\vdots \\
\vdots \\
E\{ y_{k+i} \, y_{k-1}^T \}
\end{bmatrix}
=
\begin{bmatrix}
C \\
CA \\
CA^2 \\
\vdots \\
\vdots \\
CA^i
\end{bmatrix}
E\{ z_k \, y_{k-1}^T \} \ldots
$$

$$
+
\begin{bmatrix}
0 & 0 & 0 & \cdots & \cdots & 0 \\
C & 0 & 0 & \cdots & & 0 \\
CA & C & 0 & \cdots & & 0 \\
\vdots & & & & & 0 \\
\vdots & & & & & \\
CA^{i-1} & CA^{i-2} & \cdots & \cdots & C & 0
\end{bmatrix}
\begin{bmatrix}
E\{ w_k \, y_{k-1}^T \} \\
E\{ w_{k+1} \, y_{k-1}^T \} \\
E\{ w_{k+2} \, y_{k-1}^T \} \\
\vdots \\
\vdots \\
E\{ w_{k+i} \, y_{k-1}^T \}
\end{bmatrix}
+
\begin{bmatrix}
E\{ v_k \, y_{k-1}^T \} \\
E\{ v_{k+1} \, y_{k-1}^T \} \\
E\{ v_{k+2} \, y_{k-1}^T \} \\
\vdots \\
\vdots \\
E\{ v_{k+i} \, y_{k-1}^T \}
\end{bmatrix}
$$

[317]

The correlation expressions are correlation matrices of the output vectors:

$$
E\{ y_{k+v} \, y_{k-1}^T \} = R_{yy}(v+1) \tag*{[318]}
$$

with time delay $(v + 1)$, the correlation of the state and the output yields:

$$
E\{ z_k \, y_{k-1}^T \} = R_{zy}(1) \tag*{[319]}
$$

and the correlation of the stochastic disturbance w and the output y:

$$
E\{ w_k \, y_{k-1}^T \} = 0 \tag*{[320]}
$$

$$
E\{ v_k \, y_{k-1}^T \} = 0 \tag*{[321]}
$$

These expressions are both equal to zero because the output y_{k-1} does not depend on the future disturbance w_k. The correlation matrices provide long-time averaging of the data series gained from the structure to be monitored.

With the correlation matrices, we get:

$$
\begin{bmatrix}
R_{yy}(1) \\
R_{yy}(2) \\
R_{yy}(3) \\
\vdots \\
\vdots \\
R_{yy}(i+1)
\end{bmatrix}
=
\begin{bmatrix}
C \\
CA \\
CA^2 \\
\vdots \\
\vdots \\
CA^i
\end{bmatrix}
R_{zy}(1) \tag*{[322]}
$$

If the vector y_{k-1}^T from the right and a larger vector, containing a larger number of time delays $(1,2,\ldots,\ j)$: $\begin{bmatrix} y_{k-1}^T & y_{k-2}^T & \cdots & y_{k-j}^T \end{bmatrix}$, are multiplied , a straightforward calculation as before yields the relation:

$$
\begin{bmatrix}
R_{yy}(1) & R_{yy}(2) & R_{yy}(3) & \cdots & R_{yy}(j) \\
R_{yy}(2) & R_{yy}(3) & R_{yy}(4) & \vdots & R_{yy}(j+1) \\
R_{yy}(3) & R_{yy}(4) & R_{yy}(5) & & R_{yy}(j+2) \\
\vdots & \vdots & \vdots & & \vdots \\
\vdots & \vdots & \vdots & \vdots & \vdots \\
R_{yy}(i+1) & R_{yy}(i+2) & R_{yy}(i+3) & \cdots & R_{yy}(i+j)
\end{bmatrix}
\qquad [323]
$$

$$
= \begin{bmatrix} C \\ CA \\ CA^2 \\ \vdots \\ \vdots \\ CA^i \end{bmatrix} \begin{bmatrix} R_{zy}(1) & R_{zy}(2) & R_{zy}(3) & \cdots & R_{zy}(j) \end{bmatrix}
$$

The left-hand side of the last equation is called Block–Hankel matrix:

$$
H_{i,j} = \begin{bmatrix}
R_{yy}(1) & R_{yy}(2) & R_{yy}(3) & \cdots & R_{yy}(j) \\
R_{yy}(2) & R_{yy}(3) & R_{yy}(4) & \vdots & R_{yy}(j+1) \\
R_{yy}(3) & R_{yy}(4) & R_{yy}(5) & & R_{yy}(j+2) \\
\vdots & \vdots & \vdots & & \vdots \\
\vdots & \vdots & \vdots & \vdots & \vdots \\
R_{yy}(i+1) & R_{yy}(i+2) & R_{yy}(i+3) & \cdots & R_{yy}(i+j)
\end{bmatrix}
\qquad [324]
$$

The matrix on the right-hand side can be written as follows. Defining:

$$
G = R_{zy}(1) = E\left\{ z_k\ y_{k-1}^T \right\}
\qquad [325]
$$

the next time delay can be expressed as:

$$
R_{zy}(2) = E\left\{ z_k\ y_{k-2}^T \right\} = E\left\{ z_{k+1}\ y_{k-1}^T \right\} = E\left\{ (Az_k + w_k)\ y_{k-1}^T \right\} = E\left\{ Az_k\ y_{k-1}^T \right\} = AG
$$

so that:

$$
\begin{bmatrix} R_{zy}(1) & R_{zy}(2) & R_{zy}(3) & \cdots & R_{zy}(j) \end{bmatrix} = \begin{bmatrix} G & AG & A^2G & \cdots & A^{j-1}G \end{bmatrix}
\qquad [326]
$$

The Block–Hankel matrix can now be expressed by means of the matrices (A,C) and the state-output correlation:

$$H_{i,j} = \begin{bmatrix} C \\ CA \\ CA^2 \\ \vdots \\ \vdots \\ CA^i \end{bmatrix} \begin{bmatrix} G & AG & A^2G & \cdots & A^{j-1}G \end{bmatrix} = J_i \, Q_j \qquad [327]$$

which is finally:

$$H_{i,j} = \begin{bmatrix} CG & CAG & CA^2G & \cdots & CA^{j-1}G \\ CAG & CA^2G & & & \vdots \\ CA^2G & CA^3G & & & \\ \vdots & & & & \vdots \\ CA^iG & CA^{i+1}G & CA^{i+2}G & \cdots & CA^{(i+j-1)}G \end{bmatrix} \qquad [328]$$

On the other hand, the correlation matrices $R_{yy}(\nu)$ appearing in the Block–Hankel matrix can be calculated practically from the channel-wise output correlation using the empirical correlations to replace the expectation by a finite summation:

$$\hat{R}_{yy}(\nu) = \frac{1}{n_t - \nu - 1} \sum_{\mu=1}^{n_t - \nu} y_{\mu+\nu} \, y_{\mu}^T \qquad [329]$$

where n_t is the length of the time series of the output data. Equations [327] and [328] can also be used to identify the set of matrices (A, C, G). This is shown in the following section.

Applying the SVD to the Hankel matrix:

$$H_{i,j} = U_H \, \Sigma_H \, V_H^T = (U_H \, \Sigma_H^{1/2})(\Sigma_H^{1/2} \, V_H^T) = J_i \, Q_j \qquad [330]$$

the SVD can be used for a decomposition of H (which is available from measurement data) into the observability matrix J and the matrix Q. This also enables C and G to be extracted from the first blocks of J and Q.

Now, in order to define a residual for the detection of system changes/damage, let us write:

$$R_{i,j} = S^T H_{i,j} \qquad [331]$$

which is zero in the nominal undamaged case. This can be achieved by using the SVD setting:

$$S = U_{H0} \qquad [332]$$

where U_{H0} contains only the left-hand singular vectors which are associated with the singular values equal to zero (as shown above in the case of deterministic inputs).

$$R_{i,j} = S^T H_{i,j} = S^T J_i Q_j = 0 \qquad [333]$$

As before, in the case of deterministic inputs u, the residual is zero in the nominal case if:

$$S^T J_i = 0 \qquad [334]$$

The idea here is to collect the output data from the structure to be monitored, calculate the Hankel matrix based on the empirical correlation functions, equation [329], and use the S-matrix resulting from data of the nominal undamaged structure. As soon as there is any damage, the system changes: $A \rightarrow A + \Delta A$, $C \rightarrow C + \Delta C$, $G \rightarrow G + \Delta G$ and consequently the Hankel matrix changes $H_{i,j} \rightarrow H_{i,j} + \Delta H_{i,j}$ as can be seen from equation [333] so that the new residual will indicate that a system change has occurred if:

$$R_{i,j} = S^T (H_{i,j} + \Delta H_{i,j}) = S^T \Delta H_{i,j} \neq 0 \qquad [335]$$

2.9.1.3. The damage detection method

To derive a damage detection method from the measured Hankel matrix, the ideas of Basseville is introduced; see [BAS 88, BAS 93, BAS 00, FRI 04b]. The damage indicator can be defined as:

$$D_n^2 = \zeta_n^T \hat{\Sigma}^{-1} \zeta_n \qquad [336]$$

with:

$$\zeta_n = vec\left(S^T H_{\alpha,\beta,n}\right) \qquad [337]$$

where ζ_n is the residual and $\hat{\Sigma}$ is an estimate of the residual covariance matrix. Note that the upper limits for the blocks of the Hankel matrix are α and β instead of i and j. $vec(...)$ is the stack operator rearranging the columns of a $m \times n$ matrix into one vector of length $m \cdot n \times 1$. To determine the residual ζ_n and the residual covariance matrix $\hat{\Sigma}$, the following steps are necessary [FRI 04b]; see also Figure 2.52:

I. Learning Phase (undamaged structure)

1. Determine the Hankel matrix $H_{\alpha,\beta,0}$ of the undamaged structure. The index 0 indicates the undamaged structure.

2. Determine the left kernel space S^T of the Hankel matrix $H_{\alpha,\beta,0}$, using singular value decomposition (SVD).

3. The product $vec\left(S^T H_{\alpha,\beta,n}\right) = \zeta_n$ defines the residual ζ_n. If the Hankel matrix is composed of the data for the undamaged structure, the residual ζ_n should be close to zero. If damage occurs, ζ_n should be significantly different from zero.

4. Determine an estimate of the residual covariance matrix $\hat{\Sigma}$.

$$\hat{\Sigma} = \frac{1}{n-1}\sum_n \zeta_n \zeta_n^T \qquad\qquad [338]$$

II. Detection Phase (undamaged or damaged structure)

1. Recall the covariance matrix $\hat{\Sigma}$ and the matrix S of the Learning Phase.

2. Compute the Damage Indicator values D_n using the actual measurement data.

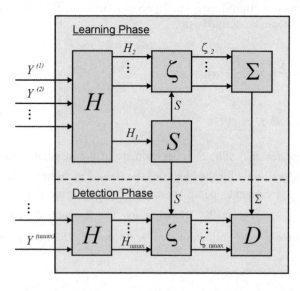

Figure 2.52. *Damage detection method using subsequent measurement sets* $Y^{(1)}, Y^{(2)}, ...Y^{(n)}, ... Y^{(nmax)}$

2.9.2. *Kalman filters*

Given the discrete-time state-space description with sampling interval Δt ($t_k = k\Delta t$, $k = 0,1,2,\ldots$):

$$z_{k+1} = A\,z_k + B\,u_k + w_k \qquad\qquad [339]$$

where A is the $2m \times 2m$ system matrix, B is the input matrix and u is the n_u-dimensional input vector. The stochastic disturbance is described by the additional random input w.

The second equation of the state-space description is the measurement equation:

$$y_k = C\,z_k + v_k \qquad\qquad [340]$$

describing the relation between the state variables x and the n_m-dimensional vector of outputs y by the $n_m \times 2n$ measurement matrix C. v is an additional immeasurable random disturbance called measurement noise. The processes in the sequence, process noise w_k and measurement noise v, are assumed to be Gaussian white noise random processes, uncorrelated to each other and with the initial state x_0. Both have zero mean and covariance matrices Q and R, respectively (see also section 2.3.2.4):

$$E\{v_k\} = 0 \quad ; \quad E\{v_k v_l^T\} = R_k\,\delta_{kl} \qquad\qquad [341]$$

$$E\{w_k\} = 0 \quad ; \quad E\{w_k w_l^T\} = Q_k\,\delta_{kl} \qquad\qquad [342]$$

(which then makes it possible to calculate the steady-state Kalman gain $K_{g,\infty}$).

The matrices A, B and C can be determined by physical modeling or identified by means of the ERA method, developed by Juang and Pappa [JUA 85] (see also Juang [JUA 94]). In the stochastic case of unknown inputs, the stochastic subspace identification method [PEE 00a, VAN 93] can be used.

The Kalman filter [JAZ 70, LEW 86, GEL 74] determines estimates of the unknown states z of the system with minimum variance, given the state space model (A,B,C) and the noise covariance matrices R and Q. Starting with the initial conditions z_0, P_0 for the initial state and covariance, respectively:

$$\hat{z}_0 = z_0 \quad ; \quad P_0 = P_{z_0} \qquad\qquad [343]$$

the following steps have to be carried out, see for instance [LEW 86]:

– **Prediction**: time update (effect of system dynamics)

State estimate and error covariance:

$$\hat{z}_{k+1/k} = A\,\hat{z}_k + B\,u_k \qquad\qquad [344]$$

$$P_{k+1/k} = A P_k A^T + Q_k \qquad [345]$$

– **Correction**: measurement update (effect of measurement y_k)

Corrected state estimate and output estimate:

$$\hat{z}_{k+1} = \hat{z}_{k+1/k} + K_{g,k+1}\left(y_{k+1} - C\,\hat{z}_{k+1/k}\right) \qquad [346]$$

$$\hat{y}_{k+1} = C\,\hat{x}_{k+1} \qquad [347]$$

Kalman gain matrix:

$$K_{g,k+1} = P_{k+1/k} C^T \left(C\,P_{k+1/k}\,C^T + R_{k+1}\right)^{-1} \qquad [348]$$

Error covariance:

$$P_{k+1} = \left(I - K_{g,k+1}\,C\right) P_{k+1/k} \qquad [349]$$

The time update makes a prediction for the state vector for the next time step on the basis of the model only. The uncertainty of this prediction is described by the covariance matrix $P_{k+1/k}$. Q_k determines the part of the uncertainty coming from the unknown disturbances. The correction part takes the latest measurement data y_{k+1} into account. The predicted state $\hat{z}_{k+1/k}$ is corrected by the difference $y_{k+1} - C\,\hat{z}_{k+1/k}$, also called innovations, weighted by the Kalman gain matrix. The property of optimality is important for damage detection. If the system changes, for example, from A to $A + \Delta A$, but the Kalman filter is still designed for the old nominal system A and now the measurement data are coming from the changed system $A + \Delta A$, this mistuning will lead to a state estimation that is no longer optimal. The residuals will show a correlation and also lead to an increase of the variances that can be used as indication for the system change.

2.9.2.1. *Steady state Kalman filter*

For a large number of time steps k $(k \to \infty)$ and constant noise covariances for system disturbances $Q_k = Q$ and measurement noise $R_k = R$, the state covariance matrix converges to a steady state:

$$\lim_{k \to \infty} P_{k+1/k} = P_\infty \qquad [350]$$

so that $P_{k+1/k} = P_{k/k-1} = P_\infty$. Putting this into equation [345], an algebraic Riccati equation is obtained for the steady-state covariance matrix P_∞ :

$$P_\infty = A\left[P_\infty - P_\infty C^T (C\,P_\infty C^T + R)^{-1}\,C\,P_\infty\right]A^T + Q \qquad [351]$$

which then allows to calculate the steady-state Kalman gain $\boldsymbol{K}_{g,\infty}$:

$$\boldsymbol{K}_{g,\infty} = \boldsymbol{P}_\infty \boldsymbol{C}^T \left(\boldsymbol{C} \, \boldsymbol{P}_\infty \, \boldsymbol{C}^T + \boldsymbol{R} \right)^{-1} \qquad [352]$$

The state estimation using a constant gain matrix is then:

$$\hat{\boldsymbol{z}}_{k+1} = \hat{\boldsymbol{z}}_{k+1/k} + \boldsymbol{K}_{g,\infty} \left(\boldsymbol{y}_{k+1} - \boldsymbol{C} \, \hat{\boldsymbol{z}}_{k+1/k} \right) \qquad [353]$$

2.9.2.2. *Statistical evaluation of the residuals and hypothesis testing*

The general scheme of information flow is shown in Figure 2.53. The difference between the vectors of measurement and filter output estimate is called the residual vector (of dimension n_m); see also Figure 2.54:

$$\boldsymbol{\varepsilon}_k = \boldsymbol{y}_k - \hat{\boldsymbol{y}}_k \, ; \; k = 1, 2, \dots, n_t \qquad [354]$$

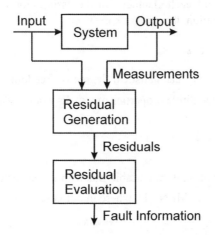

Figure 2.53. *Basic structure of residual generation and evaluation to get information about system faults (damage)*

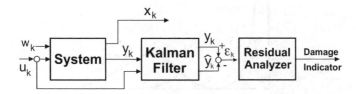

Figure 2.54. *Scheme of damage detection using a Kalman filter*

The statistics of the residuals can be used to classify the "distance" between the filter output based on the model representing the undamaged system and the measurements y_k from the structure. One assumes that the measurement noise v_k and the process noise w_k are Gaussian white noise with zero mean, thus the residuals ε_k also have a Gaussian distribution with zero mean.

The covariance matrix of the residuals is $C_{\varepsilon\varepsilon} = E\{\varepsilon\,\varepsilon^T\}$ and can be estimated by:

$$\hat{C}_{\varepsilon\varepsilon} = \frac{1}{n_t - 1}\sum_{k=1}^{n_t}\varepsilon_k\,\varepsilon_k^T \qquad [355]$$

The *normalized* residuals:

$$\bar{\varepsilon}_k = \hat{C}_{\varepsilon\varepsilon}^{-1/2}\varepsilon_k \qquad [356]$$

have a *unit variance* for each channel. On the assumption that the residual vector has a Gaussian distribution, the scalar expression:

$$V_k = \varepsilon_k^T\hat{C}_{\varepsilon\varepsilon}^{-1}\,\varepsilon_k = \bar{\varepsilon}_k^{-T}\bar{\varepsilon}_k \qquad [357]$$

has a χ^2 distribution with the statistical degree of freedom equal to the number of outputs n_m. Due to the additive properties of χ^2 distributed variables, the sum:

$$V = \sum_{k=1}^{n_t}V_k = \sum_{k=1}^{n_t}\bar{\varepsilon}_k^{-T}\bar{\varepsilon}_k \qquad [358]$$

is also a χ^2-distributed random variable but with the statistical degree of freedom $n_V = (n_t - 1)\cdot n_m$ [WIL 76, MEN 91], where in our case $n_t \gg n_m$.

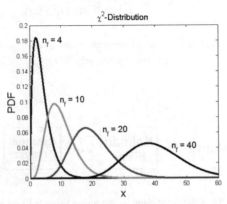

Figure 2.1. χ^2 *distribution for different degrees of freedom n_f*

Subsequently, data sets of time length n_t can be measured and tested against critical threshold values. These critical values can be calculated or taken from tables as the upper and lower tail values for the two-sided χ^2 test: $\chi^2_{\alpha/2}$ and $\chi^2_{1-\alpha/2}$ [MEN 91]. The critical values have to be based on n_V degrees of freedom.

For $V = \chi^2$ one wish to test the:

null-hypothesis $H_0 : V = V_0 = n_V$ (structure is undamaged)   [359]

against the:

alternative hypothesis $H_a : V \neq V_0 = n_V$ (structure is damaged)   [360]

If the test statistic $\chi^2 = V$ falls into the interval $\chi^2_{1-\alpha/2} < \chi^2 < \chi^2_{\alpha/2}$, the measured data do not provide sufficient evidence to reject the null-hypothesis and the structure is assumed to be undamaged. If the test statistic V is outside this region (rejection region: $\chi^2 < \chi^2_{1-\alpha/2}$ or $\chi^2 > \chi^2_{\alpha/2}$), the null hypothesis has to be rejected and the structure is assumed to be damaged. If $\alpha = 1\%$ is chosen, the error probability of V lying outside the above range because the null-hypothesis $V = \chi^2 = V_0$ is rejected, while the system is undamaged, is 1%, given the fact that it also the probability of false alarm (Type I error). Alternatively, other α values can be chosen.

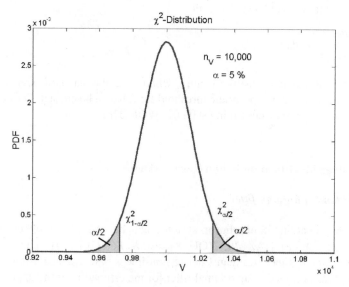

Figure 2.56. χ^2 *distribution for residual test variable V with n_v*
degrees of freedom and $\alpha = 5\%$

2.9.3. *AR and ARX models*

Sohn *et al.* [SOH 01] have demonstrated the use of the Auto-Regressive (AR) and AR with Exogenous input (ARX) models. The basic idea is to identify a single-input single-output ARX reference model from data sets of discrete time series y_0 representing the undamaged structure in a first phase (the k indicates the discrete time instant t_k and $(k-j)$ refers to t_{k-j} etc.):

$$y_0(k) = \sum_i \alpha_i y_0(k-i) + \sum_j \beta_j e_{y0}(k-j) + \varepsilon_{y0}(k). \qquad [361]$$

The model is defined by the coefficients α_i and β_j representing the dynamic behavior of the undamaged system (reference state). The time series $e_{y0}(k)$ is mainly caused by unknown external inputs and $\varepsilon_{y0}(k)$ is the residual error after fitting the ARX model. In the second stage this model is tested against new data sets y:

$$\varepsilon_y(k) = y(k) - \sum_i \alpha_i y(k-i) - \sum_j \beta_j e_y(k-j) \qquad [362]$$

and the resulting residual error $\varepsilon_y(k)$ is statistically evaluated. As long as the feature extracted from the residuals lies within a defined range of the statistical variation, there is no evidence that the structure has changed its physical properties. This method provides a level I test. The ratio:

$$\frac{\sigma(\varepsilon_y)}{\sigma(\varepsilon_{y0})} > h, h > 1,$$

is defined as a damage sensitive feature, where σ is the standard deviation of the residual time series. An appropriate threshold for h has to be chosen. Further aspects of outlier analysis are discussed in [SOH 01, WOR 02b].

2.10. Damage identification in non-linear systems

2.10.1. *Extended Kalman filter*

The identification of non-linear systems can be carried out by the extended Kalman filter (EKF) [JAZ 70, YOU 84, LEW 86, GEL 74]. The EKF is an extension of the ordinary Kalman filter to non-linear systems. The original Kalman filter (KF) was designed as an optimal filter for the estimation of the unknown states of a linear system under stochastic disturbances. Furthermore, the EKF can be used for the estimation of unknown parameters. The state vector to be reconstructed z is augmented by the unknown parameters p so that the new state-space vector is now defined as follows:

$$z = \begin{bmatrix} z_x \\ z_v \\ z_p \end{bmatrix} = \begin{bmatrix} x \\ \dot{x} \\ p \end{bmatrix} \qquad [363]$$

Also, in the case where the mechanical structure is linear, the estimate algorithm becomes non-linear as a result of the multiplicative coupling of parameters and state variables.

In the case where the system is time invariant during the data acquisition phase, the parameters are constant so that $\dot{p} = 0$.

Now, the state-space formulation of the augmented state space equation takes the form:

$$\dot{z} = \begin{bmatrix} \dot{z}_x \\ \dot{z}_v \\ \dot{z}_p \end{bmatrix} = \begin{bmatrix} z_v \\ M^{-1}(z_x, z_p, t) \left[g(z_x, z_v, z_p, u, t) - f(t) \right] \\ 0 \end{bmatrix} \qquad [364]$$

The augmented state space vector z has dimension $2n_f + n_p$. The linear system can be treated in the same way. The EKF yields estimates of the first two statistical moments of z: the mean $\hat{z}(t) = E\{z(t)\}$ and the covariance matrix $P = E\{(z - \hat{z})(z - \hat{z})^T\}$.

The derivation of the filter equations can be found in [JAZ 70, LEW 86, GEL 74] and will not be studied here. The EKF consists of a predictor and a corrector part. The predictor makes use of the model and forecasts the state of the system for the next time step by numerical integration of the equation of motion. The corrector takes the measurement y_{k+1} into account and compares it with the prediction, calculating the residual $y_{k+1} - C_m \hat{z}_{k+1/k}$ and weighting this residual by the Kalman gain matrix K_g:

Prediction:

$$\hat{z}_{k+1/k} = \varphi(\hat{z}_k, u_k, t_{k+1}, t_k) \qquad [365]$$

$$P_{k+1/k} = A_k^* P_k A_k^T + Q_k \qquad [366]$$

Correction:

$$\hat{z}_{k+1} = \hat{z}_{k+1/k} + K_{g_{k+1}} \left(y_{k+1} - C \, \hat{z}_{k+1/k} \right) \qquad [367]$$

$$P_{k+1} = \left(I - K_{g_{k+1}} C \right) P_{k+1/k} \qquad\qquad [368]$$

Kalman:

$$K_{g_{k+1}} = P_{k+1/k} C^T \left(C \, P_{k+1/k} C^T + R \right)^{-1} \qquad\qquad [369]$$

Initial conditions: $\hat{z}_0$, P_0 normally distributed.

The EKF is given in a time-discrete formulation. Starting from a state z_k and a corresponding covariance matrix P_k, the predictor yields a forecast for the next time step at time t_{k+1} solely on the basis of the model. The mapping φ, which is in general a non-linear function, follows from the integration of the continuous state space equation f with z_k as initial condition:

$$\hat{z}_{k+1/k} = \hat{z}_k + \int_{t_k}^{t_{k+1}} f\left(\hat{z}, u, \tau\right) d\tau = \varphi\left(\hat{z}_k, u_k, t_k, t_{k+1}\right) \qquad\qquad [370]$$

The prediction of the covariance matrix $P_{k+1/k}$ can be derived by a local linearization for each individual time step t_k about the nominal trajectory z_k [4]. The linearization yields:

$$A_k = \left. \frac{\partial f\left(z, u, t\right)}{\partial z} \right|_{z=\hat{z}_k} \qquad\qquad [371]$$

and the transition from continuous to discrete time yields:

$$A_k^* = e^{A_k \Delta t} = I + A_k \Delta t + A_k^2 \frac{\Delta t^2}{2!} + \dots = \left. \frac{\partial \varphi}{\partial z} \right|_{z=\hat{z}_k} \qquad\qquad [372]$$

Matrix A_k is the Jacobian of the vector $f\left(z, u, t\right)$ and it corresponds to the linear system matrix. A_k^* is the analogous expression for the discrete time.

As mentioned earlier, the covariance matrices R and Q describe the statistical properties of the measurement noise v_k and the system noise w_k, where v_k and w_k are both zero-mean uncorrelated Gaussian white noise processes.

---

[4] Note that this is not a global linearization about a fixed reference point. In fact the linearization is carried out for each individual time step about the actual system state.

The relation between the measurement variables y_k and the state variables z_k has been assumed to be linear and time-invariant.

The recursion is started with initial values $\hat{z}_0, P_0$. If the initial states $\hat{z}_0$ are very uncertain, we have to start with large covariances P_0. Usually, P_0 is chosen to be a diagonal matrix. For more details see [JAZ 70, YOU 84, GEL 74, LEW 86].

The result is the state of the system including the estimated parameters together with the covariance matrix, where the variances of the states and parameters are obtained as the diagonal elements of the covariance matrix.

The innovation sequence can be calculated from:

$$\gamma_k = y_k - C\,\hat{z}_{k/k-1} \tag{373}$$

2.10.2. Localization of damage using filter banks

A filter bank is an arrangement of parallel filters, each one with a different model that is driven by the same input $u(t)$; see Figure 2.57. Each of the filters represents a model with a special type of damage or fault. The 0-model represents the undamaged system; see for instance [KIN 89]. In the case of a structure with cracks, one filter consists of a model with one special location of the crack. The depth of the crack is a free parameter that also has to be identified. More general types of defects can be considered, for example unbalance due the loss of mass, bearing defects, etc. The filter-bank scheme is well suited to parallel processing. The parallel models and filters with their respective different damage types and locations produce residuals γ_k, $k = 0,1,...,n_t$ which can be examined by statistical methods. It should be expected that the model producing the smallest residuals fits best to the real measurement:

$$J = \sum_{k=1}^{n_t} \gamma_k^T\,\Sigma_\gamma^{-1}\,\gamma_k \tag{374}$$

where J is a weighted least squares time domain sum. The model which produces the smallest value of J is adopted.

In other words, one is searching for the model with the most likely type of damage characterized by maximum *a posteriori* probability. The logarithmic likelihood function [WAL 97] under the assumption of Gaussian noise is:

$$\ln L = -\frac{n_m\,n_t}{2}\ln 2\pi - \frac{n_t}{2}\ln \det \Sigma_\gamma - \frac{1}{2}\sum_{k=1}^{n_t} \gamma_k^T\,\Sigma_\gamma^{-1}\,\gamma_k \tag{375}$$

where n_m is the number of vector components of γ and n_t the number of time steps.

The covariance matrix Σ_γ can be estimated by [WAL 97]:

$$\hat{\Sigma}_\gamma = \frac{1}{n_t} \sum_{k=1}^{n_t} \gamma_k \gamma_k^T \qquad\qquad [376]$$

According to the Bayes rule and normalization, the likelihood function can be used to evaluate the *a posteriori* probabilities of an assumed damage type or location.

The EKF for parallel state and parameter estimation is applied where the unknown parameter is a characteristic quantity describing the extent of damage (for example, the reduction of stiffness $\Delta EI / EI_0$ or the crack depth).

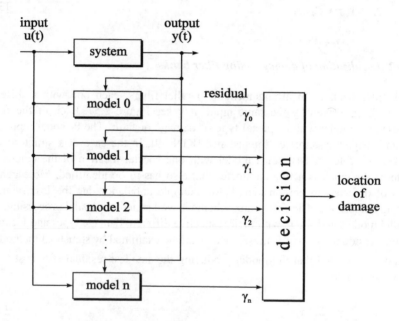

Figure 2.57. *Scheme of a filter bank*

2.10.3. *A simulation study on a beam with opening and closing crack*

The system under consideration (Figure 2.58) is a simply supported beam with a closing and opening crack (in element no. 3) and therefore non-linear [FRI 95]. The beam changes locally its stiffness depending on the current state of deformation:

$$M\ddot{x} + C\dot{x} + K(x)x = f(t)$$

with:

$$K(x) = K_0 + \Delta K(x)$$

where the change matrix $\Delta K(x)$ is not constant but depends on the state (especially the displacements) of the system.

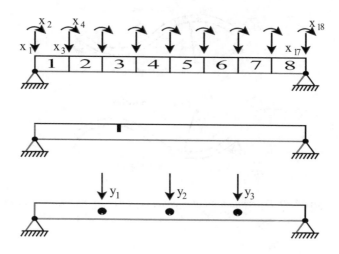

Figure 2.58. *Beam with a transverse crack and three measurement positions*

If the beam element with the crack has a positive curvature, the crack is open and the stiffness is reduced; otherwise the crack is closed and the stiffness is assumed to be the same as in the undamaged case. The displacements and rotations of the beam sections are described by the Hermitian polynomials as shape functions. The curvature of the beam can be calculated with the second derivatives of the displacements (Figure 2.59).

The dynamic behavior can be described by a state (x) dependent element stiffness matrix corresponding to element 3, making the global stiffness matrix K also state dependent.

The beam structure is divided into eight elements and excited by a random force at dof no. 7; three outputs are measured (displacements at dof nos. 5, 9 and 13). The transverse crack is located in the middle of element no. 3. The position of the three sensors shows that the crack behavior cannot be observed directly by the measurement alone. Fortunately, however, it is not possible to observe the crack indirectly by the implemented Kalman filter reconstructing the unmeasured states of the system. Thus we can determine whether the crack is open or closed for any time step. Figure 2.60 shows the random input, the three responses and the opening and closing of the crack indicated by the sign of the curvature in element no.3. The full state estimated by the EKF makes it possible to reconstruct the switching of the crack opening and closing, respectively.

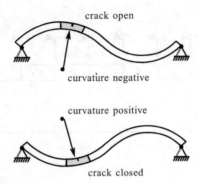

Figure 2.59. *Determination of opening or closing of the crack by means of local beam curvature*

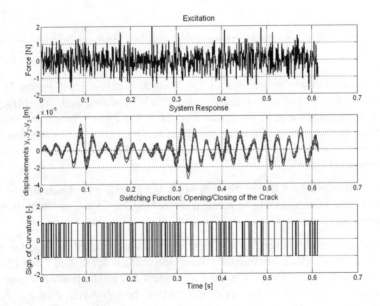

Figure 2.60. *Random excitation, responses and sign of curvature to describe the opening and closing of the crack (+ 1: crack closed, − 1: crack open)*

The localization of the crack is performed by nine parallel EKFs, where each filter contains a model with damage at a different location (Figure 2.57). The 0-model represents the undamaged system. The filters estimate the system states as well as the losses of stiffness in a combined state and parameter estimation. The application of the filter bank together with the EKF yields the time series of the residuals for all possible crack locations that are evaluated by equations [374] and [375]. This yields a distribution of likelihood values, which clearly indicates that beam element no. 3 contains the crack (Figure 2.61), and this is the true location.

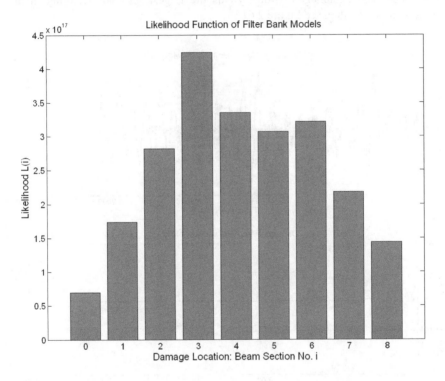

Figure 2.61. *Values for the likelihood function evaluated for different potential crack locations with maximum for model 3*

Figure 2.62 is the result for the EKF with the crack in element no. 3. It shows the comparison of the "measured" noisy displacement y_1 and the EKF estimation $\hat{y}_1$ of this displacement, which are almost identical. The third curve is the difference between the two (the residual), which is displayed as a random signal close to the zero-line in the upper subplot. The lower subplot of Figure 2.62 demonstrates the convergence of the EKF for crack position no. 3 starting with the initial value of r_0

= 0.001 (no crack). After about 250 steps, the parameter has converged to the true value. Finally, the loss of stiffness $\hat{r} = (\Delta EI / EI)_3 = 0.202$ (true value $r = 0.200$) was estimated on the basis of a time series of 2048 samples with a satisfying accuracy of 1%. During the periods when the crack is closed the sensitivity of the displacements according to the crack depth is zero and the EKF does not change the value of r.

Filter 3 with the crack at the correct position yields residuals with the smallest covariances. The undamaged 0-filter yields the largest residuals resulting in the lowest likelihood value; see Figure 2.61.

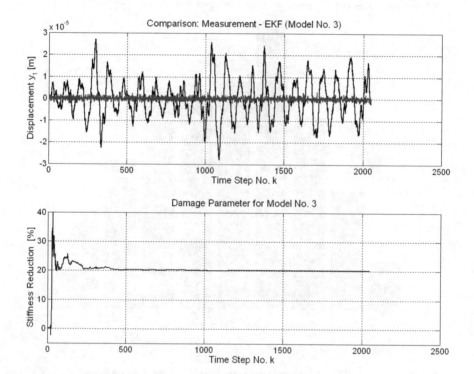

Figure 2.62. *Upper: "measured" and estimated output, residuals for crack location at element no. 3, Lower: the fast convergence of the estimated damage parameter to the true value of 0.2*

For lower levels of noise, it is shown in Figure 2.63 that the decision among the different models is much more obvious.

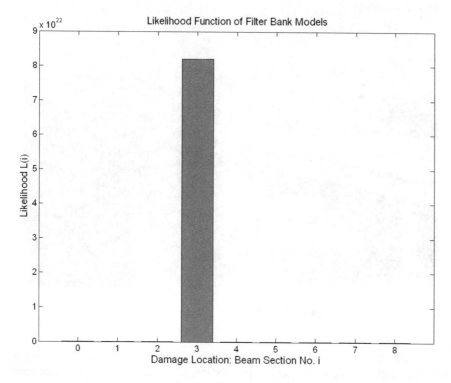

Figure 2.63. *Values for the likelihood function evaluated for different potential crack locations for lower measurement noise levels*

2.11. Applications

2.11.1. *I-40 bridge*

2.11.1.1. *Description of the I-40 bridge testing*

The bridge on Interstate 40 ("I-40 bridge") over the Rio Grande in New Mexico, USA, consisted of two separate spans for each traffic direction divided into three identical, almost structurally independent sections. Each section was made up of three spans (Figure 2.64). Los Alamos National Laboratories (LANL) performed several tests on the eastern section of the bridge carrying eastbound traffic. Because the bridge was to be razed, several states of damage could be introduced, simulating regularly observed fatigue cracking.

Figure 2.64. *I-40 highway bridge [FAR 96]*

The geometry of the tested bridge section is shown in Figure 2.65. The bridge was made up of a concrete deck supported by two plate girders and three steel stringers. The loads from the stringers were transmitted to the plate girders by floor beams. Details are described by Farrar *et al.* [FAR 94, FAR 96].

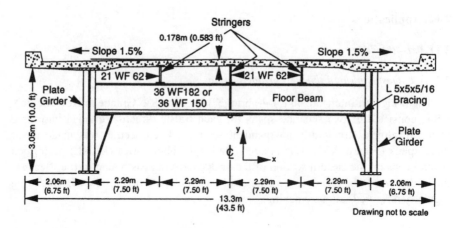

Figure 2.65. *Typical cross section of the tested bridge [FAR 96]*

The bridge section was supported by three pairs of concrete piers, where the eastern piers restricted all the dofs, while the other piers restricted all but the longitudinal DOF, in order to enable thermal expansion. LANL measured the structural response at 26 accelerometer locations (Figure 2.67). The bridge was excited by a shaker system in the range from 2 to 12 Hz with maximum force amplitude of 8900 N.

After the measurements at the undamaged bridge section, four states of damage were introduced into the plate girder near point N7. First, a 0.6 m long and 1 cm wide cut was inserted at the mid-height of the girder plate (Figure 2.66). In the second state this cut was expanded to the bottom of the plate. In the third state, half of each side of the bottom girder flange was cut. In the final state of damage the flange was cut through completely.

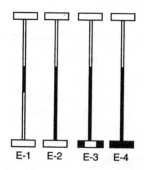

Figure 2.66. *Damage scenarios*

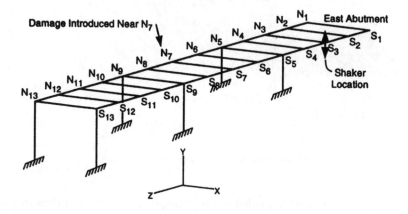

Figure 2.67. *Mesh grid along the bridge section [FAR 96]*

2.11.1.2. *Computational model*

A 3586-dof finite element (FE) model of the tested bridge section, made up of 480 shell and 600 beam elements, was developed. For modeling, the non-commercial Matlab-Toolbox "Matfem" was used [LIN 97a].

In the first stage of modeling, the actual geometric dimensions and material properties of the main structural components were used. Details of the construction such as bracing and cross-bracing had been neglected to reduce the model's number of dofs. The concrete slab and the girder plates were modeled with four-node shell elements and the other structural units shown in Figure 2.68 with two-node beam elements.

As boundary conditions, all translational and rotational dofs at the bottom of the piers were restricted. The abutment was simulated by restricting all dofs except the rotational DOF about the *x*-axis at the eastern lower girder flanges.

A model update was performed minimizing the objective function. The parameters given for optimization were the Young's moduli and the shear moduli of both steel and concrete.

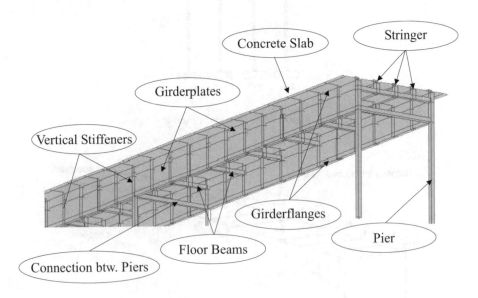

Figure 2.68. *Structural units of the FE model*

The updating algorithm yielded the eigenfrequency and MAC values in Tables 2.2 and 2.3, which show a remarkably good coincidence between the model and the real bridge. All MAC values of the first three bending and torsional modes are greater than 0.98, where a MAC of more than 0.9 means a significant correlation between the compared mode shapes. The adjacent modes 4 and 5 of the measurement correspond to the modes 8 and 7 of the model, but this was not considered to be critical because of the high MAC values obtained.

Mode	f meas. [Hz]	f model [Hz]	Diff. [%]
1	2.479	2.459	−0.81%
2	2.949	3.029	2.70%
3	3.494	3.532	1.08%
4	4.077	4.137	1.45%
5	4.167	4.075	−2.21%
6	4.636	4.552	−1.82%

Table 2.2. *Comparison of eigenfrequencies*

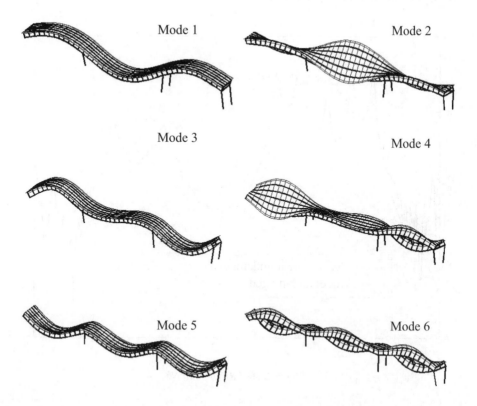

Figure 2.69. *Calculated mode shapes of the bridge*

	Mod 1	Mod 2	Mod 3	Mod 4	Mod 5	Mod 6	Mod 7	Mod 8	Mod 9	Mod 10
Meas 1	0.866	0.000	**0.997**	0.001	0.001	0.000	0.000	0.006	0.000	0.000
Meas 2	0.002	0.466	0.002	0.561	**0.995**	0.000	0.002	0.000	0.009	0.339
Meas 3	0.102	0.000	0.000	0.000	0.000	**0.993**	0.006	0.000	0.000	0.000
Meas 4	0.032	0.002	0.012	0.002	0.000	0.000	0.000	**0.988**	0.010	0.007
Meas 5	0.003	0.012	0.000	0.132	0.000	0.012	**0.983**	0.001	0.001	0.029
Meas 6	0.002	0.502	0.001	0.122	0.004	0.000	0.000	0.016	**0.981**	0.652

Table 2.3. *MAC values of measurement- and model mode shapes*

Furthermore, the frequency response functions of measurement and model (Figure 2.72) show good agreement, so the assumption of a realistic reflection of the real properties by the model's behavior can be maintained. Note that the peaks in higher frequency regions adequately match the real peaks, although they were not considered in the model updating.

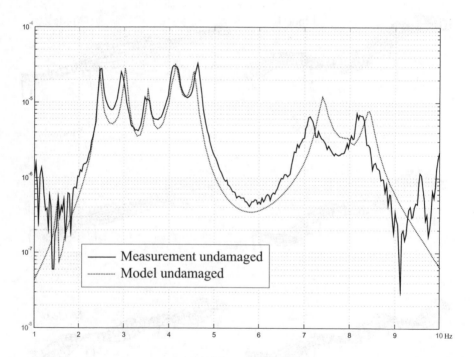

Figure 2.70. *FRF of point N7*

2.11.1.3. *Damage identification*

The first sequence in the damage identification algorithm is the calculation of the residual vector r. In this example, r has 58 elements resulting from the residuals of the six eigenvalues and the residuals of the first bending and torsional mode with 26 elements. The second step is the determination of the finite element sensitivities, where the changes in eigenvalues and eigenvectors are caused by changes in the stiffness of the finite element. The sensitivity matrix S is of size 58×1080 and the resulting under-determined equation system leads to the ill-posed inverse problem with a non-unique solution that has been discussed above.

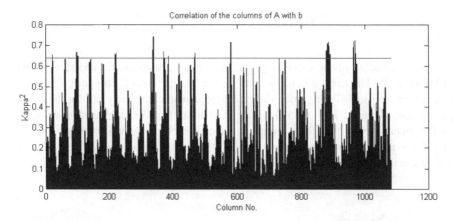

Figure 2.71. *Correlation of S and r*

To bypass that problem, ridge-regression and a parameter selection by a correlation threshold $\kappa_{th}^2 = 0.64$ are performed (Figure 2.71). The 1080 parameters are reduced to 34, which are the only ones considered in further calculations [FRI 99b]. After this first selection, the second one follows, where the error reduction ratio of each parameter with respect to the residual vector r is defined as the initial criterion. This selection leads to a subset of ten parameters.

The results of the parameter selection and the solution for their changes are shown in Figure 2.72, where the damaged shell elements around 340 and the damaged beam elements around 580 are extracted. A small error contribution is detected at 223, which is an undamaged element. To avoid this, the boundaries for error reduction should have been chosen more restrictively. However, the low parameter value at 223 indicates that it can be assumed that there is no serious damage here.

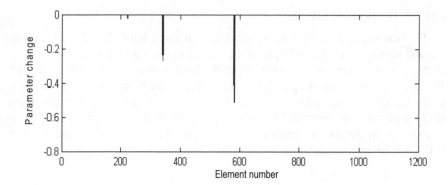

Figure 2.72. *Extracted parameters*

The association of the received parameter changes with the FE model is shown in Figure 2.73. The darkest color is used for the beam elements 580 and 581, which simulate the region of the girder flange where it was completely cut through. The girder plate elements 340 and 341 beside the cut are found to be the most damaged shell elements. Note that these locations exactly encircle the real damage. A final model update led to a reduction in stiffness of nearly 100% for beam elements 580 and 581 and of about 30% for shell elements 340 and 341.

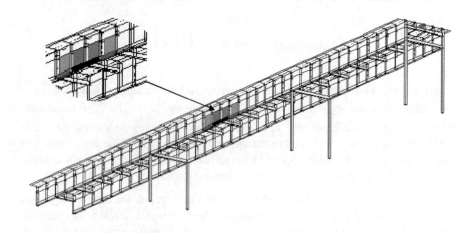

Figure 2.73. *Detected damage adapted to the FE model*

Damage scenario 4 constituted the main center of interested in this section, but adequate results were also obtained from scenarios 2 and 3. In case of scenario 3, the actual location matched exactly as in scenario 4, while in scenario 2, damage was detected in the two elements alongside the true location. For scenario 1, damage localization was not unique, because the changes in dynamic behavior due to the damage could not be distinguished from the measurement uncertainties.

2.11.2. Steelquake structure

2.11.2.1. Description of the structure

The structure corresponds to a two-storey frame as shown in Figure 2.74. It was investigated during the European COST F3 action as a benchmark example for comparing different algorithms. The main dimensions are 8 m × 3 m × 9 m. The storeys are made up of corrugated sheets supporting a concrete slab, forming a composite with orthotropic elastic characteristics, and are connected by welded vertical and horizontal steel girders. Details are shown in [PAS 98].

Figure 2.74. *"Steelquake" test structure at ELSA-JRC, Ispra, Italy [PAS 98]*

The vertical steel girders were stiffened by cross bracings in the plane parallel to the wall (see Figure 2.74). In the background can be seen the reaction wall that supports the four pistons (not visible in the photo) which will deform the structure (on each frame and on each storey). There are several standard profiles of steel girders in use (Figure 2.77). The columns consist of HE300B, the storeys of IPE400 on the long side and IPE300 on the short side. Bracings are made of L60 × 30 × 5 profiles. A detailed description can be found in [PAS 98]. The produced crack is shown in Figure 2.78.

Figure 2.75. *Details of the lower storey*

Figure 2.76. *Crack after the "steelquake"*

2.11.2.2. *Description of the computational model*

A 1476-dof finite element model of the tested structure, consisting of 104 four-node shell (white) and 172 two-node beam (grey) elements, was developed; see Figure 2.79. For modeling, the non-commercial Matlab Toolbox "Matfem" was used [LIN 99].

In the first stage of modeling the actual geometric dimensions and material properties of the steel girders were assumed. The concrete slab/corrugated sheet combinations were modeled by orthotropic shell elements. The column beam elements were fixed to the floor by grounded springs. Details of the construction such as bracings were modeled as beam elements.

A model update was performed minimizing the objective function:

$$f(\boldsymbol{p}) = \sum_i w_{\lambda_i} \left(\frac{\lambda_i^{meas} - \lambda_i^{mod}(\boldsymbol{p})}{\lambda_i^{mod}(\boldsymbol{p})} \right)^2 + \sum_i w_{MAC_i} \left(1 - MAC_{ii}(\boldsymbol{p})\right)^2 \qquad [377]$$

considering the MAC values and the eigenfrequencies of the first ten modes ($\lambda_i = \omega_i^2$). w_λ and w_{MAC} are individual weighting coefficients. The parameters $\boldsymbol{p}$ for the optimization run were the Young's moduli and the shear moduli of both steel and concrete, as well as the moment of inertia of the bracing and stiffnesses of the grounded springs.

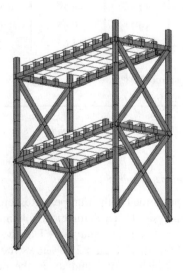

Figure 2.77. *FE model*

The updating algorithm yielded the eigenfrequencies and MAC-values in Table 2.4, which show a remarkably good coincidence between the model and the real structure. All MAC values of the first ten modes are greater than 0.98. The adjacent modes 4 and 5, as well as modes 9 and 10, are interchanged, but this is not considered to be critical because of the high MAC values obtained. The eigenfrequency deviations are less than 4% in the mean, although there are large deviations at the 4th and the 10th modes.

Comparison of Eigenfrequencies:
measurement vs model, undamaged case

Mode	$f_{meas,0}$ [Hz]	$f_{mod,0}$ [Hz]	Diff. [%]	MAC
1	3.13	3.19	1.92	1.00
2	3.93	3.87	−1.53	0.99
3	6.13	6.21	1.31	1.00
4	9.69	10.77	11.15	0.99
5	10.82	10.56	−2.40	0.99
6	12.27	12.12	−1.22	1.00
7	13.05	13.17	0.92	1.00
8	17.70	17.56	−0.79	0.99
9	19.03	19.21	0.95	0.99
10	21.41	18.06	−15.65	1.00

Mean frequency error: 3.98%

Mean MAC value: 0.99

Table 2.4. *Updating results*

2.11.2.3. *Description of the measurements*

The measurements were performed at the European Research Center ELSA-IRC, Ispra, Italy. For the modal analysis, 15 accelerometers, a hammer, a DAT recorder, an alimentation box for the accelerometers and an FFT analyzer were used. Details can be found in [PAS 98, WOR 02a]. Figure 2.78 shows the sensor locations and the measurement directions, which were found to be the optimal set for modal analysis of the first ten modes.

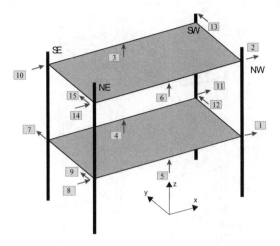

Figure 2.78. *Measurement sensor locations*

2.11.2.4. *Results of the damage identification procedure*

The damage identification algorithms were applied to the steelquake structure with different amounts of information. Before choosing the appropriate information in terms of mode shapes and eigenfrequencies, a detectability study was performed. The detectability study points out which damage locations can be found when using a certain selection of the overall available information. Modes 1, 5, 6 and 7 had the largest shifts in eigenfrequencies and MACs after damage.

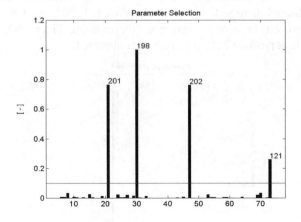

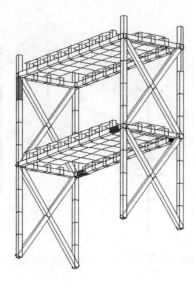

Figure 2.79. *Results of the damage localization with inverse eigensensitivity method, using modes 1, 5, 6 and 7*

The damage localization results from the inverse eigensensitivity method using modes 1, 5, 6 and 7 as illustrated in Figure 2.79. In the graph, one can see the selected parameters which most significantly contributed to reduce the error of the remaining equation system. The height of the bars shows the normalized error reduction; see [FRI 99c]. The four finally selected parameters were extracted out of a parameter set containing about 70 parameters, which all previously satisfied the preselection criterion consisting of a combination of correlation and sensitivity [FRI 99a]. The indicated damage locations at elements 121, 202 and 198 correspond to locations where cracks actually occurred (for example Figure 2.76) in the real structure, while at position 201 no damage was observed.

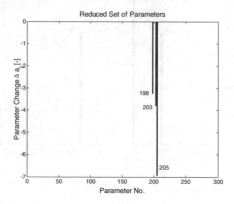

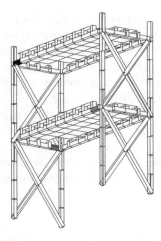

Figure 2.80. *Damage localization results with the MFR method, using modes 1 and 5*

Similar results to those of the eigensensitivity method were obtained by the modal force residual (MFR) method. The damage localization results shown in Figure 2.80 were obtained by only using modes 1 and 5. This method is very sensitive to the amount of information provided and to model inaccuracies and noise in the measurement data, which could be observed in simulation runs.

Figure 2.81 shows the results for the modified MFR method.

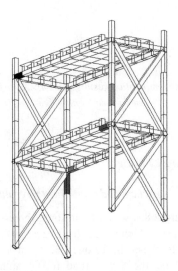

Figure 2.81. *Damage localization results with the modified MFR method*

The only direct method presented here is the MECE error localization technique. The results for damage localization using this method are shown in Figure 2.82. The overall error calculated for mode 1 and mode 5 in the undamaged comparison between the measurement and the model data was very small comparing to the overall error calculated for the measurement data of the damaged state and the undamaged model data. This indicates that the actual damage had the most severe effects on those two modes. The distribution of the overall error on the structural members identifies the damaged locations of the structure, although the degree of accuracy is not on the same level as in the other methods investigated, which use pre-selection. Final updating led to the following result: the bending stiffness of elements 121, 198, and 202 was reduced from 100% to about 5%.

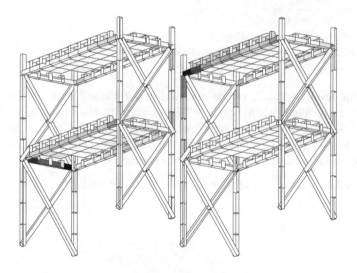

Figure 2.82. *Damage localization results with MECE, using mode 1 only (left) and mode 5 only (right)*

2.11.3. *Application to the Z24 bridge*

2.11.3.1. *Description of the structure and experimental results of the SIMCES-project*

The Z24 bridge (Figure 2.85) was a 60 m long, three-span pre-stressed concrete girder bridge with two lanes, which connected Utzenstorf and Koppigen by crossing the Swiss highway A1. Because the bridge had to make way for a huge railway project, it was provided by the Swiss road traffic authorities for the European SIMCES (System Identification to Monitor Civil Engineering Structures) project

[MAE 03a, MAE 03b, PEE 01b] for extensive investigations about condition monitoring. Toward the end of the one-year project, some artificial damage was introduced, where the damage scenarios were chosen to coincide with the most frequently observed real structural damage.

Figure 2.83. *The Z24 bridge (KU Leuven)*

Figure 2.84. *Implementation of hydraulic jacks to simulate the settlement of the pier*

Here, a damage scenario was investigated in which the Koppigen pier was lowered ("settled") by means of hydraulic jacks stepwise up to 95 mm (see Figure 2.86). During the measurements, the stiffness of the pier was raised to the original value by adding steel. This scenario simulated the undercutting of this pier, which is a dangerous, hardly detectable kind of damage that frequently occurs in reality.

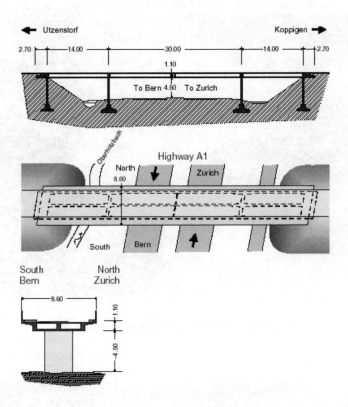

Figure 2.85. *Z24-bridge: elevation view, top view and cross section (KU Leuven)*

2.11.3.2. *Model updating*

For the model updating process, modal data identified with the Matlab Toolbox "Macec" (KU Leuven) from "Output-Only" measurement data, consisting of five eigenfrequencies and mode shapes, was used. These data were collected during the SIMCES project.

The initial model consisted of 330 four-node shell elements and approximately 2000 dofs. The model was based on technical drawings of the bridge and, to circumvent unnecessary sources of error, all measurement locations coincided with

FE model nodes. The unknown or uncertain material properties and the boundary conditions were roughly estimated; they are the outcome of the updating process. In addition, the stiffness of the connection between the bridge and the piers deflected by about 12° was uncertain. All parameters were physically meaningful, so the model could be used to predict the changes in the dynamic behavior within the framework of damage identification. The validation of the model was performed by using the normalized eigenfrequency deviation as well as the Modal Assurance Criterion (MAC) between the model and the reference measurement:

$$\varepsilon_{freq} = \sum_i \left(\frac{\omega_{i,model}^2 - \omega_{i,meas}^2}{\omega_{i,meas}^2} \right)^2 \qquad\qquad [378]$$

$$MAC_{ij} = \frac{\left(\varphi_{i,model}^T \varphi_{j,meas} \right)^2}{\varphi_{i,model}^T \varphi_{i,model} \cdot \varphi_{j,meas}^T \varphi_{j,meas}} \cdot 100\% \quad 0\% \leq MAC_{ij} \leq 100\% \quad [379]$$

The low MAC values (Tables 2.5 and 2.6) of the initial model show that some essential properties of the bridge – in spite of the detailed reproduction of the geometry shown in the technical drawings – were not reflected by the model. MAC values higher than 90% can be considered as satisfactory and only mode 1 actually reached this goal. However, the eigenfrequencies seem to roughly correspond to the measured ones, insofar as a model update using this initial model was started. The update parameters were the stiffnesses of the springs which had been used to model the boundary conditions (bridge bearings and connection of the piers to the ground) as well as the Young's modulus of the bridge material and the pier material. In all, 17 (partially very uncertain) parameters were obtained for updating and, in order to find suitable initial values, a Genetic Algorithm was applied [GOL 89]. The optimization was interrupted after only a few generations and an SQP-algorithm [MAT 02] was then used to find the optimum more rapidly. In addition, a mesh refinement was advantageous.

	Mode 1	Mode 2	Mode 3	Mode 4	Mode 5
Meas. [Hz]	3.9	5.0	9.8	10.3	12.7
Model [Hz]	3.5	6.0	9.2	10.0	12.3

Table 2.5. *Eigenfrequencies of initial model and reference measurement data*

MAC values [%]		Model							
		Mode 1	Mode2	Mode 3	Mode 4	Mode 5	Mode 6	Mode 7	Mode 8
Meas.	Mode 1	**99.5**	0.0	0.0	0.1	0.0	0.2	0.1	0.0
	Mode 2	0.3	**53.0**	4.8	52.3	7.0	1.2	0.0	0.6
	Mode 3	0.7	26.2	**48.9**	1.1	0.1	6.0	1.4	3.4
	Mode 4	0.0	14.5	**46.9**	23.0	0.4	2.2	1.4	2.9
	Mode 5	0.1	0.0	0.2	1.0	0.5	**70.4**	12.4	0.4

Table 2.6. *MAC values of initial model and reference measurement data (expanded)*

This approach led to a greatly increased model quality, but still some significant deviations in modes 3 and 4 remained. To improve the correlation of these modes, a remodeling of the connections between the bridge and the piers, which are very stiff in the real structure, was performed. A final updating considering these stiffness parameters resulted in very satisfactory MAC values with a minimum at 91.4% as well as eigenfrequency deviations with a maximum of 5.5% (Figure 2.88, Tables 2.7 and 2.8). These deviations are within the range of the different reference measurement sets identified from the output-only data. The updated model consisted of 650 shell elements and 3800 dofs.

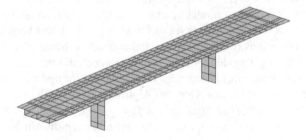

Figure 2.86. *FE-Model of the Z24-bridge after updating*

	Mode 1	Mode 2	Mode 3	Mode 4	Mode 5
Meas. [Hz]	3.9	5.0	9.8	10.3	12.7
Model [Hz]	3.9	5.3	9.8	10.4	12.0

Table 2.7. *Eigenfrequencies of the updated model and reference measurement data*

MAC values [%]	Model							
	Mode 1	Mode2	Mode 3	Mode 4	Mode 5	Mode 6	Mode 7	Mode 8
Meas. Mode 1	**99.6**	0.0	0.0	0.0	0.2	0.0	0.0	0.0
Mode 2	0.1	**97.0**	1.6	1.0	0.0	0.0	0.6	0.2
Mode 3	0.2	3.0	**96.2**	2.9	0.2	0.3	0.5	0.0
Mode 4	0.0	2.3	4.9	**94.8**	0.2	1.1	0.7	0.4
Mode 5	0.0	0.2	0.1	0.3	**91.4**	0.0	0.6	0.2

Table 2.8. *MAC values of the updated model and reference measurement data (expanded)*

2.11.3.3. Damage identification

Damage identification procedures were applied to the damage scenario described above, in which the Koppigen pier was "settled" by up to 95 mm. The experimental modal data, after introduction of the damage, were also extracted from output-only measurement data, based on the results provided by Professor de Roeck's group at the KU Leuven. Brief overviews of the effects due to the damage are shown in Tables 2.9 and 2.10.

	Mode 1	Mode 2	Mode 3	Mode 4	Mode 5
Undamaged structure [Hz]	3.9	5.0	9.8	10.3	12.7
Damaged structure [Hz]	3.7	4.9	9.2	9.7	12.0

Table 2.9. *Comparison of measured eigenfrequencies before and after damage*

MAC values [%]	Damaged structure				
	Mode 1	Mode2	Mode 3	Mode 4	Mode 5
Undamaged Mode 1	**99.8**	0.3	0.4	0.0	0.7
structure Mode 2	0.0	**98.8**	0.0	0.2	0.0
Mode 3	0.9	0.3	**87.0**	0.1	0.1
Mode 4	0.0	0.2	23.7	**84.2**	1.9
Mode 5	0.1	0.1	3.0	8.5	**89.7**

Table 2.10. *MAC values of measured eigenvectors before and after damage*

The simulated undercutting of the pier produced cracks on the lower side of the bridge near the pier, leading to a reduced stiffness in this bridge section [MAE 03b, MAE 01]. The stiffness reduction can be successfully identified with both damage identification methods, although the results are qualitatively slightly different. The grey elements in Figures 2.89 and 2.90 illustrate the identified damage: the darker the color, the greater the indicated damage.

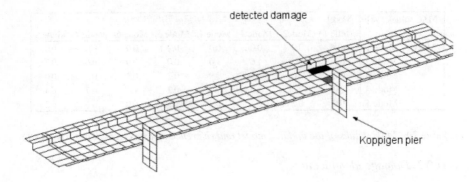

Figure 2.87. *Identification result using Inverse Eigensensitivity Method*

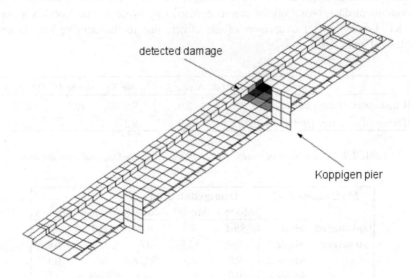

Figure 2.88. *Identification result using modal kinetic energy method*

2.11.4. *Detection of delaminations in a CFRP plate with stiffeners*

2.11.4.1 *Description of the problem*

For advanced composite materials like Carbon Fiber Reinforced Plastics (CFRP), impact by foreign objects is a major problem. Impacts can result in internal damage which is often difficult to detect but can drastically reduce the strength and stability of the structure. The damage detection methods used in this chapter are embedded in a framework of an intelligent structure. The structure is able to excite itself by a piezo-ceramic actuator (muscles) and the sensors (nerves) measure the dynamic response with respect to this excitation. The computer (brain) is able to decide upon the health state of the structure: "damaged" or "not damaged".

Figure 2.89 shows the plate, which is equipped with five surface-mounted piezo-ceramic (PZT) elements, one used as actuator (in the middle) and four as sensors. The overall size of the plate with stringers is approximately (490 mm × 490 mm × 1.9 mm) and its weight is 1.125 kg. The stringers are 36 mm high and 2.5 mm thick. The properties of the UD material are V_{Fiber} = 60%, $E_{longitudinal}$ = 142.6 GPa, $E_{transversal}$ = 9.65 GPa, v_{12} = 0.334, v_{13} = 0.328, v_{23} = 0.54 and G_{12} = G_{13} = 6.0 GPa. Plate and stringers consist of nine plies each.

Figure 2.89. *Composite plate with stringers and piezoelectric elements*

After the plate had been measured in the undamaged state, a delamination was introduced by an impactor mass with a kinetic energy of about 5 J, causing a small amount of damage (Figures 2.92 and 2.93), which was subsequently increased by a second 5 J impact. To obtain the size of the delaminations the CFRP plate was inspected by ultrasonic scan. After the first impact, a delamination of about 1 cm² in size was measured and, after the second impact a delamination of about 9 cm². Figure 2.91 shows the experimental setup for producing the delaminations. A steel bar (diameter 3.5 cm, with the end rounded (radius 1.75 cm), 20.25 cm long, weight 1.5 kg) was used. To introduce controlled delaminations the steel bar is guided by a tube when falling.

The energy of the impact is given by the selected height. The CFRP plate always impacts on the flat side. In this example, the CFRP plate was supported by small foam cubes at three points.

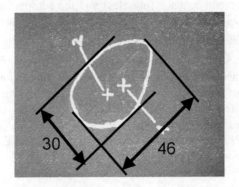

Figure 2.90. *Delamination area after two impacts*

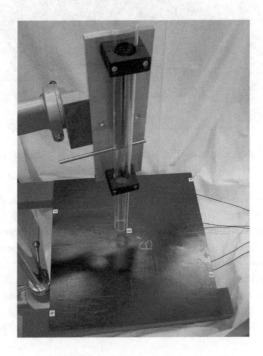

Figure 2.91. *Device to introduce impact damage to the CFRP plate with a tube to guide the impactor (steel bar, inside the tube)*

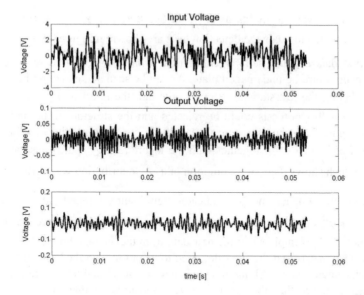

Figure 2.92. *Typical white noise excitation (input voltage) and responses from two sensors (output voltages)*

The plate was excited by broad band Gaussian noise with constant spectral density within the frequency range of interest using the 'built-in' actuator in the middle of the plate (Figure 2.94). The standard deviation of the input voltage was about 0.4 V. The voltage response of the four sensors was measured. The experimental set-up is described in [FRI 01c, FRI 02a]. The plate was supported by small foam cubes at three points. The sampling frequency was 2400 Hz and the measurement time was 120 s, which corresponds to 288,000 data samples.

2.11.4.2. *Application of Kalman filter for damage detection*

As a first step, a "learning test run", the matrices of the state-space model were identified from a data set of the undamaged system in order to provide a reference basis. With these, a stationary Kalman filter was set up and the residuals according to equation [354] were calculated. After letting transients decay, the number of time samples finally used for the residuals was n_t =273,000. From this set of residual data, the covariance matrix for the residuals was calculated from the dataset, equation [355], which served for all calculations of the normalized residuals, equation [356]. The result for $V = V_0$, equation [358], of this calibration run was 1,091,992, which is exactly equal to $n_V = (n_t - 1) \cdot n_m$ (exactly with respect to the number of digits displayed here), where the number of sensors (outputs) was $n_m = 4$. On the basis of n_V and a selected probability of false alarm $\alpha = 0.01$, we calculated the upper and lower thresholds as $\chi^2_{\alpha/2} = 1,095,806$ and $\chi^2_{1-\alpha/2} = 1,088,193$. The results of the following damage detection tests are shown in Figure 2.93.

The first dataset was used for calibration, while the next five sets were also measurements from the undamaged plate. As can be seen and as would be expected from the theory, the test statistic $\chi^2 = V$ falls into the interval $\chi^2_{1-\alpha/2} < \chi^2 < \chi^2_{\alpha/2}$ and thus the null-hypothesis could be rejected and the structure to be undamaged, assumed. Figure 2.94 shows a typical plot of the histogram of the V_k values (with degree of freedom $n_m = 4$), together with the theoretical χ^2-pdf. These are nearly identical, which shows that our assumption of a χ^2 distribution was correct.

After the first impact, the χ^2 indicator significantly jumped out of the region $\chi^2_{1-\alpha/2} < \chi^2 < \chi^2_{\alpha/2}$ into the rejection region and we assumed that there was damage to the structure. This implies that the new data from the damaged plate did not fit the old covariance matrix and the filter designed to be optimal on the basis of the model of the undamaged system. Although the changes of the modal data were very small, it is possible to see that the method is very sensitive so that even the smallest amount of damage could be observed. After the second impact, another jump of the χ^2 indicator could be observed.

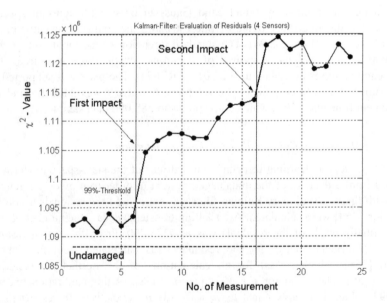

Figure 2.93. *Result of the damage detection for the CFRP plate before and after impact damage*

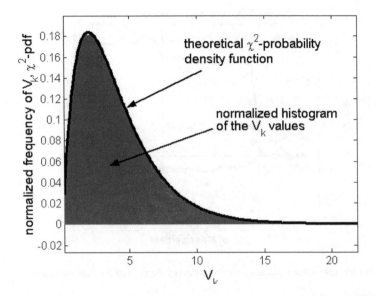

Figure 2.94. *Typical result for V_k for the undamaged structure. The measured normalized histogram of the residuals matches perfectly the theoretical χ^2 distribution*

When the number of data samples was reduced by one half (a measurement time of 60 s instead of 120 s), a jump into the rejection region after the impact could still be observed so that a statistical decision could be made, even though the jump was not as large as before. So, the number of data samples is an important parameter for the quality of the decision.

Using the information in the residuals for each output channel separately, it was possible to see how much each sensor "can see" of the change due to damage. The same formulas as before could be used, but V had to be calculated from the time series $\bar{\varepsilon}_{k,i}$ for each output channel i separately and the degrees of freedom had to be changed to $n_V = (n_t - 1)$. Figure 2.95 shows that the residuals of the four different output channels displayed quite different behavior. The indicators of all channels jumped into the rejection region. Channel 1 provided the most significant change. Further conclusions from this result have yet to be explored.

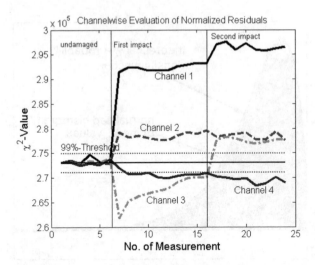

Figure 2.95. *Damage indicator: separate evaluation of the four channels*

Because of the large amount of needed data for a precise statistical evaluation, the structure is also able to identify a very small delamination, even though the eigenfrequencies, damping and mode shapes do not show any significant change.

Further investigations have to be carried out according to the size of the dataset, compensation of temperature effects, optimal placement of the sensors and actuators, etc. Future goals in connection with this method are also the localization and quantification of the damage, as well as application to more complex structures.

2.11.4.3. *Application of the stochastic subspace damage detection method*

As a first step, a calibration test run, the matrix S representing the left kernel space was obtained by orthogonalization with the Hankel matrix using the SVD as described above. The Hankel matrix was calculated with measurement data from the four response channels from the undamaged system. The size of the Hankel matrix is determined by α and β and the number of output channels (see equation [324] with α and β instead of i and j). It turned out that $\alpha = \beta > 100$ was a suitable choice in order to include the dynamic behavior of the structure. For use in subsequent runs, the matrix S was stored and kept constant while the Hankel matrix was recalculated each time using the new sensor output data. As can be seen in Figure 2.96, the first six measurements correspond to the undamaged plate. The damage indicator values are very small. Only the measurement noise causes a slight deviation from perfect orthogonality (which corresponds to zero). After these first measurements, a significant jump could be seen, indicating that the new data did not perfectly correspond with the matrix S. This was the point where the first impact on the plate took place. Although the changes of the modal data were very small, it can be seen that the method is very sensitive so that even the smallest amount of damage could be observed. The jump in the damage indicator from undamaged to 5 J impact was about 400, while the second jump after the second impact with the same energy but located close beside the first impact was about 650.

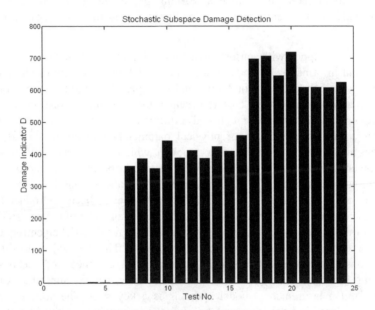

Figure 2.96. *Result of damage detection using the stochastic subspace damage detection method: the first six measurements sets taken in the undamaged state, measurement set 7 to 16 after the first 5 J impact and 17 to 24 after the second 5 J impact. Jumps of the damage indicator can be clearly identified*

Investigations using the stochastic subspace damage detection method showed that the estimated eigenfrequencies and damping coefficients for measurements nos. 1 to 26 did not change significantly.

2.12. Conclusion

In this chapter, SHM based on changes of the global vibration behavior has been discussed. The classical modal-based approach makes use of the low-frequency modes. These are global in nature, which corresponds, however, to a reduced sensitivity. The usefulness of these methods is based on the facts that the whole structure can be monitored with only a few sensors and the damage location does not need to be known in advance. It is not necessary for a sensor to be close to the damage location. The use of ambient excitation has the advantage that no additional actuation device is needed. On the other hand, ambient excitation often limits the frequency range. The change of the dynamic properties due to environmental changes has been recognized as a serious problem. Long-term monitoring of structures should therefore include a learning procedure as to how a structure behaves under certain environmental conditions so as to compensate for these influences.

Structures equipped with sensors, actuators and devices for the processing of the measurement information can be considered as "intelligent" structures that are able to perform a self-diagnosis to find structural damage. The "brain" is the computer, which can access a knowledge base represented by the computational model of the structure or any other relevant stored information. The model serves as an interpreter, translating changes of physical parameters into changes of dynamic properties and vice versa in the inverse problem, which allows level II/III problems to be solved. If appropriate damage models that describe the evolution of cracks, corrosion etc. are available, level IV predictions will also be possible. Three main elements of model-based damage assessment have been introduced: (1) measurement data, (2) reference model and (3) damage identification. Here, the changes of the dynamics have been expressed either as residuals of eigenfrequencies and mode shapes or as FRFs and time domain data. This has led us to different formulations of the inverse problem. To obtain a good reference model as basis for calculating important information, such as sensitivities, which cannot be extracted directly from measurements, model updating is a key step. The accuracy of the measurement data is very important because the uncertainty of the identification result is directly influenced by the measurement errors.

For complex structures such as bridges and building structures, there are usually a high number of unknown damage parameters. Thus, another key step in the procedure is to condense the large set of possible candidate damage parameters into a much smaller set by methods of subset selection. The strategy has been to find the smallest number of significant damage parameters that are needed to describe the changes of the dynamic properties with a given accuracy. This can improve the quality of the solution enormously compared to the solution with a full parameter set. Once those parameters have been localized, a non-linear optimization adjusting only this small subset of parameters makes it possible to quantify the damage, for example, in terms of stiffness changes.

Furthermore, the chapter has also dealt with model-free methods like some of the energy indicators and transmissibility ratios.

The time domain methods, namely the stochastic subspace fault detection and the Kalman filters, have been applied for damage detection (level I) of small delaminations in a stiffened composite plate. These methods showed an improved sensitivity to damage compared to a direct comparison of modal data. The reason for this may be that we are considering a broader frequency range. Usually, it is much more difficult to describe the higher frequency range by a computational model because of the more complex modes and modal overlap, especially if the structure is very complex (geometry, material, joints, etc.). Piezoelectric elements have been used as sensors and as actuators for higher frequency excitation.

Local methods based on ultrasonic wave propagation concentrate on a relatively small area of the structure. The wavelength is within the range of the damage size. Obviously, it is more difficult to find very small defects by global monitoring. Here, a sensitivity analysis helps judge the influence of the extent of the damage on the data used to define detection limits, e.g. by covariance analysis. Although desirable, it is not always a priority to detect damage at an incipient stage. In many applications, it is quite sufficient to be warned at an intermediate level or only to prevent the structure running into a critical situation.

For some applications, high sensitivity may be a critical requirement. The option here is to use local methods based on ultrasonic wave propagation with one or more dense local sensor networks. To increase the sensitivity of vibration-based methods, there is also a trend towards developing more damage-sensitive residuals and to expand the global methods into an intermediate frequency range. Modeling in the higher frequency range becomes more difficult as a result of strong modal overlap and the complexity of mode shapes, which also requires a finer sensor network. Some work should also be carried out on reducing this model dependence. Discussions about "global versus local methods" should be directed towards a combination of both worlds to take advantage of the complementary strengths of the two groups. Methods dealing with non-linear damage identification are still in the minority and are only being tested by simple examples. However, the use of global methods prevents the structure running into a critical condition. It needs to be investigated how combinations of global and local methods can yield improved solutions.

Finally, the success of SHM methods in practice will be determined by whether it is possible to develop robust sensor hardware that will survive the monitored object. Furthermore, the decisions made by the SHM system must have a sound statistical foundation. Too many false alarms, as well as missed indications of damage, will destroy the operator's confidence in such a system.

2.13. Acknowledgements

The author would like to thank his former and current PhD students, Dr D. Jennewein, K. Bohle and G. Mengelkamp for many discussions and their contributions, especially to the practical examples. Further thanks go to my skilled laboratory staff W. Richter and G. Dietrich. The measurement data of the I40 bridge were provided by Dr C.F. Farrar, LANL, NM. The author is very grateful for Dr Farrar's support. The investigation of a complex structure like this gave our whole method development a big push forward. The "Steelquake" structure was investigated during the EU COST action F3 chaired by Professor J.C. Golinval, University of Liege, Belgium. Many thanks to him and all colleagues involved in this project for the fruitful cooperation. The author would also like to thank Professor G. De Roeck, KU Leuven, and his co-workers, Dr B. Peeters and Dr J. Maeck, for their kind support and for the supply of the measurement data of the Z24 bridge. My enthusiastic students, P. Kraemer, R.T. Schulte and J. Zimmermann, have greatly contributed to bringing the manuscript into its final form by proofreading, word processing, graphics and discussion. Last but not least, I would like to thank my family for their endless patience and support during all the weekends needed to prepare the manuscript.

2.14. References

[AHM 97] AHMADIAN H., GLADWELL G.M.L., ISMAIL F., "Parameter Selection Strategies in Finite Element Model Updating", *Journal of Vibration and Acoustics*, Vol. 119(1), 1997, pp. 37-45.

[ALL 93] ALLEMANG R., BROWN D.L., "Experimental Modal Analysis", *Handbook of Experimental Mechanics*, Edition A Kobayashi, A., SEM, Bethel, CT, 1993, pp. 635-750.

[AND 97] ANDERSON P., "Identification of Civil Engineering Structures Using Vector ARMA Models", PhD Thesis, Dept. Building Technology and Structural Eng., Aalborg University, Denmark, 1997.

[AUW 03] VAN DER AUWERAER H., PEETERS B., "International Research Projects on Structural Health Monitoring: An Overview", *Structural Health Monitoring*, Vol. 2 (4), 2003, pp. 341-358.

[BAJ 93] BAJKOWSKI J., FRITZEN C.-P., SEIBOLD S., SÖFFKER D., "Verfahren zur Rißdetektion – Vergleich zweier modellgestützter Vorgehensweisen", in: Schwingungen in rotierenden Maschinen II, Irretier, Nordmann, Springer (Hrsg.), Vieweg Verlag, Braunschweig, Wiesbaden, 1993, pp. 55-63.

[BAL 02] BALAGEAS D.L. (Ed.), "Structural Health Monitoring 2002", Proc. 1st European Workshop on Structural Health Monitoring, ENS Cachan (Paris), France, July 10-12, 2002.

[BAR 78] BARUCH M., "Optimization Procedure to Correct Stiffness and Flexibility Matrices Using Vibration Data", *AIAA Journal*, 16(11), 1978, pp. 1208-1210.

[BAR 82] BARUCH M., "Correction of Stiffness Matrix Using Vibration Tests", *AIAA Journal*, 20(3), 1982, pp. 441-442.

[BAS 88] BASSEVILLE M., "Detecting Changes in Signals and Systems – A Survey", *Automatica*, Vol. 24(3), 1988, pp. 309-326.

[BAS 93] BASSEVILLE M., NIKIFOROV I.V., "Detection of Abrupt Changes: Theory and Applications", Prentice Hall, Englewood Cliffs, NJ [see also: http://www.irisa.fr/sigma2/kniga], 1993.

[BAS 00] BASSEVILLE M., ABDELGHANI M., BENVENISTE A., "Subspace-based fault detection algorithms for vibration monitoring", *Automatica*, Vol. 36, 2000, pp. 101-109.

[BAS 01] BASSEVILLE M., BENVENISTE A., GOURSAT M., *et al.*, "Output-Only Modal Subspace-Based Structural Identification: From Theory to Industrial Testing Practice", *J. of Dynamics, Measurement, and Control*, Transactions ASME, Vol. 123, Dec. 2001, pp. 668-676.

[BAU 87] BAUMEISTER J., "Stable Solutions of Inverse Problems", Friedr. Vieweg & Sohn Braunschweig/Wiesbaden, 1987.

[BEN 92] BEN HAIM Y., "Adaptive Diagnosis of Faults in Elastic Structures by Static Displacement Measurement: The Method of Selective Sensitivity", *Mechanical Systems and Signal Processing*, Vol. 6(1), 1992, pp. 85-96.

[BER 83] BERMAN A., NAGY E.J., "Improvement of a Large Analytical Model using Test Data", *AIAA Journal*, 21(8), 1983, pp. 1168-1173.

[BER 00] BERNAL D., "Damage Localization using Load Vectors", Europ. COST Conf. On System Identification & Structural Health Monitoring, Madrid, Edition A. Güemes, 2000, pp. 223-232.

[BER 02] BERNAL D., "Load Vectors for Damage Localization", *Journal of Engineering Mechanics*, ASCE, Vol. 128(1), 2002, pp. 7-14.

[BIS 95] BISHOP C.M., *Neural Networks for Pattern Recognition*, Oxford University Press, Oxford, 1995.

[BOH 03] BOHLE K., FRITZEN C.-P., "Schadenidentifikation und Model-Updating basierend auf Output-Only – Schwingungsmessdaten am Beispiel einer Autobahnbrücke", VDI-Schwingungstagung 2003, Magdeburg, VDI-Berichte 1788, 2003, pp. 105-118.

[BOH 05] BOHLE K., "Sensitivitätsbasierte Methoden zur modellgestützten Schadendiagnose mit Modaldaten", PhD Dissertation (in German), Dept. of Mechanical Engineering, University of Siegen, Germany, 2005.

[BOL 04] BOLLER C., STASZEWSKI W.J. (Eds.), "Structural Health Monitoring 2004", Proc. 2nd European Workshop on Structural Health Monitoring, Munich, Germany, July 7-9, 2004.

[BRI 00] BRINCKER R., ZHANG L., ANDERSON P., "Modal Identification of Output-Only Systems Using Frequency Domain Decomposition", Europ. COST Conf. On System Identification & Structural Health Monitoring, Madrid, Edition A. Güemes, 2000, pp. 273-282.

[CAO 98] CAO T., ZIMMERMAN D.C., JAMES G.H., "Identification of Ritz Vectors from Ambient Test Data", Proc. 16th Intl. Modal Analysis Conference (IMAC), Santa Barbara, CA, 1998, pp. 1609-1614.

[CAW 79] CAWLEY P., ADAMS R.D., "The location of defects in structures from measurements of natural frequencies", *J. Strain Analysis*, Vol. 14, 1979, pp. 49–57.

[CHA 97] CHANG F.-K. (Ed.), "Structural Health Monitoring – Current Status and Perspectives", Proc. Intl. Workshop on Structural Health Monitoring, Stanford, CA, Technomic Publ. Co. Inc., 1997.

[CHA 99] CHANG F.-K. (Ed.), "Structural Health Monitoring 2000", Proc. 2nd Intl. Workshop on Structural Health Monitoring, Stanford, CA, Technomic Publ. Co. Inc., 1999.

[CHA 01] CHANG F.-K. (Ed.), "Structural Health Monitoring – The Demands and Challenges", Proc. 3rd Intl. Workshop on Structural Health Monitoring, Stanford, CA, Technomic Publ. Co. Inc., 2001.

[CHA 03] CHANG F.-K. (Ed.), "Structural Health Monitoring 2003 – From Diagnostics and Prognostics to Structural Health Management", Proc. 4th Intl. Workshop on Structural Health Monitoring, Stanford University, Stanford, CA, Technomic Publ. Co. Inc., 2003.

[CHAU 94] CHAUDHRY Z., ROGERS C.A., "Smart Structures: On-line Health Monitoring Concepts and Challenges", Advanced Mater. and Process Technol. for Mechanical Failure Prevention, Proc. 48th Meeting of the Mechanical Failures Prevention Group, Wakefield, MA, USA, 1994, pp. 13–19.

[CHE 01] CHEDLIZE D., CUSUMANO J., CHATTERJEE A., "Failure Prognosis Using Non-linear Short-time Prediction and Multi-Time Scale Recursive Estimation", Proc. of DETC '01, ASME 2001 Design Engineering Techn. Conf., Pittsburg, Pennsylvania, USA, Paper DETC2001/VIB-21407, 10 pages (published on CD), 2001.

[CHE 03] CHEDLIZE D., "Dynamic System Approach to Material Damage Diagnosis", Proc. of DETC '03, ASME 2003 Design Engineering Techn. Conf., Chicago, Illinois, USA, Paper DETC2003/VIB-48452, 8 pages (published on CD), 2003.

[CHO 96] CHOI K., CHANG F.-K., "Identification of Impact Force and Location Using Distributed Sensors", *AIAA Journal*, 34, 1996, pp. 136-142.

[CHU 96] CHUKWUJEKWU OKAFOR A., CHANDRASHEKHARA K., JIANG Y.P., "Delamination Prediction in Composite Beams with Built-in Piezoelectric Devices Using Modal Analysis and Neural Network", Smart Mater. Struct. 5, 1996, pp. 338–347.

[COG 94] COGAN S., LALLEMENT G., BEN-HAIM Y., "Updating Linear Elastic Models with Modal-Based Selective Sensitivity", 12th IMAC, Honolulu, HI, USA, 1994, pp. 515-520.

[COR 99] CORNWELL P.J., DOEBLING S.W., FARRAR C.R., "Application of the Strain Energy Damage Detection Method to Plate-Like Strukturea", *Journal of Sound and Vibration*, Vol. 224, no. 2, 1999, pp. 359-374.

[DIR 88] DIRR B.O., HARTMANN D., SCHMALHORST B.K., "Cracked Cross Section Measurement in Rotating Machinery", in Structural Safety Evaluation Based on System Identification Approaches, Friedrich Vieweg & Sohn, Braunschweig/Wiesbaden, 1988, pp. 9-28.

[DOE 96] DOEBLING S.W., FARRAR C.R., PRIME M.B., SHEVITZ D.W., "Damage Identification and Health Monitoring of Structural Systems from Changes in Their Vibration Characteristics: A Literature Review", report no. LA-12767-MS, Los Alamos National Laboratory, 1996.

[DOY 97] DOYLE J.F., *Wave Propagation in Structures*, Springer, New York, NY, 1997.

[EIL 95] EILBRACHT G., LINK M., "Identification of Crack Parameters in Concrete Beams Using Modal Test Data", Proc. Int. Sympos. Non-Destructive Testing in Civil Engineering (NDT-CE), 1995, pp. 327-334.

[EYK 74] EYKHOFF P., "System Identification – Parameter and State Estimation" – John Wiley and Sons, Inc., New York, 1974.

[EWI 00] EWINS D.J., *Modal Testing: Theory, Practice and Applications*, Research Studies Press, Baldock, Hertfordshire, 2000.

[FAR 94] FARRAR C.R., BAKER W.E., BELL T.M., *et al.*, "Dynamic Characterization and Damage Detection in the I-40 Over the Rio Grande", Los Alamos National Laboratories, report no. LA-12767-MS, 1994.

[FAR 96] FARRAR C.R., DUFFEY T.A., GOLDMAN P.A., *et al.*, "Finite element analysis of the I-40 bridge over the Rio Grande", Los Alamos National Laboratories, Tech. report. no. LA-12979-MS, 1996.

[FAR 00] FARRAR C.R., CORNWELL P.J., DOEBLIN S.W., PRIME M.B., "Structural Health Monitoring Studies of the Alamosa Canyon and the I-40 Bridges", Report no. LA-13635-MS, Los Alamos National Laboratory, 2000.

[FAR 03] FARRAR C.R., SOHN H., HEMEZ F.M., *et al.*, "Damage Prognosis: Current Status and Future Needs", Los Alamos National Laboratory Report LA-14051-MS, 2003.

[FAR 04] FARRAR C.R., SOHN H., ROBERTSON A.N., "Application of Non-linear System Identification to Structural Health Monitoring", Proc. 2nd European Workshop "Structural Health Monitoring 2004", Munich, Germany, July, 7-9, 2004, pp. 59-67.

[FOX 68] FOX R.L., KAPOOR M.P., "Rates of Change of Eigenvalues and Eigenvectors", *AIAA Journal,* 6(12), 1968, pp. 2426-2429.

[FRA 94] FRANK P.M., "Fault Diagnosis in Dynamic Systems Using Analytical and Knowledge-based Redundancy – A Survey and some new Results", *Automatica,* Vol. 26(3), 1994, pp. 459-474.

[FRI 86] FRITZEN C.-P., "Identification of Mass, Damping and Stiffness Matrices of Mechanical Systems", *Journal of Vibration, Acoustics, Stress and Reliability in Design,* Transactions of the ASME, Vol. 108, Jan. 1986, pp. 9-16.

[FRI 90] FRITZEN C.-P., SEIBOLD S., "Identification of Mechanical Systems by Means of the Extended Kalman Filter", Proc. 3rd International IFToMM Conf. on Rotordynamics, Lyon, France, 1990, pp. 423-429.

[FRI 91] FRITZEN C.-P., ZHU S., "Updating of Finite Element Models by Means of Measured Information", *Computers & Structures,* Vol. 40, no. 2, 1991, pp. 475-486.

[FRI 92] FRITZEN C.-P., KIEFER T., "Localization and Correction of Errors in Analytical Models", Proc. 10th Int. Modal Analysis Conference (IMAC), San Diego, California, Feb. 1992, pp. 1063-1071.

[FRI 95] FRITZEN C.-P., SEIBOLD S., BUCHEN D., "Application of Filter Techniques for Damage Detection in Linear and Non-Linear Mechanical Structures", Proc. 13th Intl. Modal Analysis Conference (IMAC), Nashville, Tennessee, 1995, pp. 1874-1881.

[FRI 97] FRITZEN C.-P., JENNEWEIN D., KIEFER T., "Damage Detection based on Vibration Measurements and Inaccurate Models", paper no. DETC97/VIB-4157, Proc. DETC'97, ASME Design Eng. Techn. Conf., Sacramento, CA, 1997.

[FRI 98] FRITZEN C.-P., JENNEWEIN D., KIEFER T., "Damage Detection Based on Model Updating Methods", *Mechanical Systems & Signal Processing* 12 (1), 1998, pp. 163-186.

[FRI 99a] FRITZEN C.-P., BOHLE K., *Identification of Damage in Large Scale Structures by Means of Measured FRFs – Application to the I40-Bridge,* Damage Assessment of Structures, Gilchrist, M.D., Dulieu-Barton, J.M., Worden, K., Transtech Publ., 1999, pp. 310-319.

[FRI 99b] FRITZEN C.-P., BOHLE K., "Model-Based Health Monitoring of Structures – Application to the I40-Highway-Bridge", Proc. Identification in Engineering Systems IES99, Swansea, UK, 1999, pp. 492-505.

[FRI 99c] FRITZEN C.-P., BOHLE K., "Parameter selection strategies in model-based damage detection", Structural Health Monitoring 2000, Chang, F.-K., Technomic Publishing., 1999, pp. 901-911.

[FRI 00] FRITZEN C.-P., BOHLE K., STEPPING A., "Damage Detection in Structures with Multiple Cracks Using Computational Models", J.A. Güemes (Ed.), Proc. European

COST F3 Conf. on System Identification & Structural Health Monitoring, Madrid, Edition Güemes, 2000, pp. 191-200.

[FRI 01a] FRITZEN C.-P., BOHLE K., "Model-Based Damage Identification from Changes of Modal Data – A Comparison of Different Methods", Structural Health Monitoring 2001, Chang, F.-K. (Ed.), Proc. 2nd Intl. Workshop on Structural Health Monitoring, Stanford, CA, CRC-Press, 2001, pp. 849-859.

[FRI 01b] FRITZEN C.-P., BOHLE K., "Application of Model-Based Damage Identification to a Seismically Loaded Structure", *Smart Materials and Structures* (10), 2001, pp. 452-458.

[FRI 01c] FRITZEN C.-P., MENGELKAMP G., "Detection of Delaminations in Composite Materials Using a Smart Structures Concept", Proc. First European Workshop on Structural Health Monitoring, Paris, France, 2002, pp. 237-244.

[FRI 02a] FRITZEN C.-P., MENGELKAMP G., GÜEMES A., "A CFRP Plate with Piezoelectric Actuators and Sensors as Self-Diagnostic Intelligent Structure", Proc. ISMA 2002, Sept. 2002, Leuven, Belgium, pp. 185-191.

[FRI 03a] FRITZEN C.-P., MENGELKAMP G., "A Kalman Filter Approach to the Detection of Structural Damage", Proc. of the 4th Intl. Workshop on Structural Health Monitoring, Stanford University, Sept. 2003, pp. 1275-1284.

[FRI 03b] FRITZEN C.-P., MENGELKAMP G., GÜEMES A., "Detection of Small Delaminations in a CFRP Plate by Vibration-Based Structural Health Monitoring", Proc. of the 14th. Intl. Conference on Composite Materials, San Diego 14–18 July 2003, pp. 110-121.

[FRI 04a] FRITZEN C.-P., BOHLE K., "Damage Identification using a Modal Kinetic Energy Criterion and Output-Only Modal Data-Application to the Z-24 Bridge", in Structural Health Monitoring 2004, Proc. 2nd European Workshop Structural Health Monitoring, Munich, Germany, July 7-9, 2004, pp. 185-194.

[FRI 04b] FRITZEN C.-P., MENGELKAMP G., DIETRICH G., *et al.*, "Structural Health Monitoring of the ARTEMIS Satellite Antenna using a Smart Structure Concept", Proc. Second European Workshop on Structural Health Monitoring 2004, Munich, Germany, 2004, pp. 1075-1082.

[FRI 05] FRITZEN C.-P., MENGELKAMP G., "In-Situ Damage Detection and Localisation in Stiffened Structures", Proc. 23rd Intl. Modal Analysis Conference (IMAC), Orlando, FL, 2005, on CD-ROM.

[FRIS 95] FRISWELL M.I., MOTTERSHEAD J.E., *Finite Element Model Updating in Structural Dynamics*, Kluwer Academic Publishers, 1995.

[FRIS 97a] FRISWELL M.I., PENNY J.E.T., GARVEY S.D., "Parameter Subset Selection in Damage Location" in *Inverse Problems in Engineering*, Vol. 5, No. 3, 1997, pp. 189-215.

[FRIS 97b] FRISWELL M.I., PENNY J.E.T., "Is Damage Location Using Vibration Measurement Practical?", Proc. DAMAS '97, Structural Damage Assessment Using Advanced Signal Processing Procedures, Sheffield, UK, 1997, pp. 351-362.

[FRIS 02] FRISWELL M.I., PENNY J.E.T., "Crack Modelling for Structural Health Monitoring", *Structural Health Monitoring* 1(2), 2002, pp. 139-148.

[GEL 74] GELB A., *Applied Optimal Estimation*, The M.I.T. Press, Cambridge, MA, 1974.

[GER 91] GERTLER J.J., "Analytical Redundancy Methods in Fault Detection and Isolation", Proc. IFAC/IMACS Sympos. SAFEPROCESS '91, Baden-Baden, Sep. 1991, pp. 9-21.

[GER 98] GERTLER J.J., "Fault Detection and Diagnosis in Engineering Systems", Marcel Dekker, Inc., New York, Basel, Hong Kong, 1998.

[GIU 03] GIURGIUTIU V., HARRIS K., PETROU M., BOST J., QUATTLEBAUM J.B., "Disbond Detection with Piezoelectric Wafer Active Sensors in RC Structures Strengthened with FRP Composite Overlays", *Earthquake Eng. and Eng. Vibration*, Vol. 2(2), 2003, pp. 1-11.

[GLA 00] GLANTZ S.A., SLINKER B.K., *Primer of Applied Regression & Analysis of Variance*, McGraw Hill, New York, 2nd edition, 2000.

[GOL 89] GOLDBERG, D.E., "Genetic Algorithms in Search, Optimization and Machine Learning", Addison-Wesley Professional, 1989.

[GOL 96] GOLUB G.H., VAN LOAN C.F., *Matrix Computations*, Baltimore, Maryland, John Hopkins University Press, 1996.

[GOU 88] GOUNARIS G., DIMAROGONAS A.D., "A finite element of a cracked prismatic beam for structural analysis", *Computers & Structures* 28(3), 1988, pp. 309-313.

[GUD 82] GUDMUNDSON P. P., "Eigenfrequency Changes of Structures Due to Cracks, Notches or Other Geometrical Changes", *Journal of Mechanics and Physics of Solids* 30(5), 1982, pp. 339-353.

[GUD 83] GUDMUNDSON P. p., "The dynamic behavior of slender structures with cross-section cracks", *Journal of Mechanics and Physics of Solids* 31(4), 1983, pp. 329-345.

[GUS 00] GUSTAFSSON F., *Adaptive Filtering and Change Detection*, John Wiley & Sons Ltd, 2000.

[GUY 65] GUYAN J., "Reduction of Stiffness and Mass Matrices", *AIAA Journal*, Vol. 3, No. 2, 1965, pp. 380.

[GYS 90] GYSIN HP., "Comparison of Expansion Methods for FE Modelling Error Localization", Proc. 8th IMAC, Kissimmee, FL, 1990, pp.195-204.

[HAN 94] HANSEN pp.C., "Regularisation Tools: A Matlab Package for Analysis and Solution Of Discrete Ill-posed Problems", *Numerical Algorithms* 6, 1994, pp. 1-35.

[HAN 98] HANSEN pp.C., "Rank-Deficient and Discrete Ill-Posed Problems: Numerical Aspect of Linear Inversion", SIAM, Philadelphia, 1998.

[HAY 99] HAYKIN S., *Neural Networks – A Comprehensive Foundation*, 2nd Edition, Prentice Hall, Upper Saddle River, NJ, 1999.

[HE 94] HE J., LI Y.Q., "Relocation of Antiresonances of a Vibratory System by Local Structural Changes", *Modal Analysis*, Vol. 10, 1994, pp. 224-235.

[HEY 98] HEYLEN W., LAMMENS S., SAS P. P., "Modal Analysis Theory and Testing", Dept. Mech. Eng., Kath. Universiteit Leuven, Leuven, Belgium, 1998.

[HIM 72] HIMMELBLAU D.M., *Applied Non-linear Programming*, McGraw Hill, New York, NY, 1972.

[HU 01] HU N., WANG X., FUKUNAGA H., *et al.*, "Damage Assessment of Structures Using Modal Test Data", *Intl. J. of Solids and Structures* 38, 2001, pp. 3111-3126.

[HUT 02] HUTH O., "Ein adaptiertes Polyreferenz-Verfahren und seine Anwendung in der Systemidentifikation", PhD Dissertation, Bauhaus-Universität Weimar, 2002.

[IHM 03] IHM J.-B., "Built-in Diagnostics for Monitoring Fatigue Cracks in Aircraft Structures", PhD Dissertation, Stanford University, Stanford, CA, 2003.

[IHM 04] IHM J.-B., CHANG F.-K., HUANG J., DERRISO M., "Diagnostic Imaging Technique for Structural Health Monitoring", Proc. 2nd European Workshop "Structural Health Monitoring 2004", Munich, Germany, July, 7-9, 2004, pp. 109-116.

[ISE 84] ISERMANN R., "Process Fault Detection based on Modeling and Estimation Methods. A Survey", *Automatica*, Vol. 20, 1984, pp. 387-404.

[JAN 99] JANOCHA H. (Ed.), "Adaptronics and Smart Structures – Basics, Material, Design, and Applications", Springer-Verlag, Berlin, 1999.

[JAZ 70] JAZWINSKI A.H., "Stochastic Processes and Filtering Theory", *Mathematics in Science and Engineering*, Vol. 64, Academic Press, 1970.

[JEN 99] JENNEWEIN D., "Beitrag zur Verfahrensoptimierung bei der modellgestützten Schadensdiagnose an passiven und aktiven elastomechanischen Strukturen", Fortschritt-Berichte VDI, Reihe 11, Schwingungstechnik, Nr. 276, VDI Verlag-GmbH, Düsseldorf, 1999.

[JOH 02] JOHNSON T.J., ADAMS D.E., "Transmissibility as a Differential Indicator of Structural Damage", *J. Vibration and Acoustics*, Vol. 124, 2002, pp. 634-641.

[JOH 04] JOHNSON T.J., BROWN R.L., ADAMS D.E., SCHIEFER M., "Distributed Structural Health Monitoring With a Smart Sensor Array", *MSSP*, Vol. 18, 2004, pp. 555-572.

[JUA 85] JUANG J.N., PAPPA R., "An Eigensystem Realization Algorithm for Modal Parameter Identification and Model Reduction", *J. Guidance, Control and Dynamics* 8(5), 1985, pp.620-627.

[JUA 94] JUANG J.N., "Applied System Identification", Prentice Hall, Englewood Cliffs, 1994.

[KES 02] KESSLER S.S., "Piezoelectric Based In-Situ Damage Detection of Composite Materials for Structural Health Monitoring Systems", PhD Dissertation, MIT, Massachussetts, 2002.

[KIN 89] KING R., "Modellgestützte Überwachung kritischer Reaktionssysteme", Fortschritt-Berichte VDI, Reihe 8, Meß-, Steuerungs- und Regelungstechnik, Nr. 185, VDI Verlag-GmbH, Düsseldorf, 1989.

[KIR 94] KIRKEGAARD pp.H., RYTTER A., "Use of Neural Networks for Damage Assessment in a Steel Mast", Proc. 12th IMAC, Honolulu, Hawaii, USA, 1994, pp. 1128-1134.

[KRA 92a] KRAWCZUK M., "A Cracked Timoshenko Beam Finite Element", Proc.17th International Seminar on Modal Analysis, Vol. 1, Leuven, Belgium, 1992, pp. 67-81.

[KRA 92b] KRAWCZUK M; OSTACHOWICZ W.M., "Parametric Vibrations of Beam with Crack", *Archive of Applied Mechanics* 62, 1992, pp. 463-473.

[LAD 89] LADEVEZE pp., REYNIER M, "A Localization Method of Stiffness Errors for the Adjustment of FE Models", Proc. 12th ASME Conf. On Mechanical Vibration and Noise, Montreal, Canada, 1989, pp. 355-361.

[LAL 88] LALLEMENT G., "Localization Techniques", Proc. "Structural Safety Evaluation Based on System Identification Approaches", F. Vieweg & Sohn, Braunschweig, Wiesbaden, 1988, pp. 214-233.

[LAL 89] LALLEMENT G., PIRANDA J., FILLOD R., "Parametric Identification of Conservative Self Adjoint Structures", Proc. Int. Conf.: "Spacecraft Structures and Mechanical Testing", ESA SP-289, Noordwijk, The Netherlands, 1989, pp. 63-68.

[LAL 90] LALLEMENT G., PIRANDA J., "Localisation Methods for Parameter Updating of Finite Element Models in Elastodynamics", 8th International Modal Analysis Conference, Orlando, Florida, 1990, pp. 579-585.

[LAL 92] LALLEMENT G., COGAN S., "Reconciliation Between Measured and Calculated Dynamic Behaviors: Enlargement of the Knowledge Space", Proc. 10th IMAC, San Diego, CA, USA, 1992, pp. 487-493.

[LEM 01] LEMISTRE M., BALAGEAS D., "Structural Health Monitoring System based on Diffracted Lamb Wave Analysis by Multiresolution Processing", *Smart Materials and Structures*, Vol. 10, 2001, pp. 504-511.

[LEW 86] LEWIS F.L., "Optimal Estimation", John Wiley, New York, 1986.

[LIN 84] LINK M., "Finite Elemente in der Statik und Dynamik", B.G. Teubner, Stuttgart, 1984.

[LIN 91] LINK M., "Localization of Errors in Computational Models Using Dynamic Test Data", Structural Dynamics, Vol.1, Editions W.B. Krätzig *et al.*, Rotterdam, 1991, pp. 305-313.

[LIN 97a] LINK M., "Matfem User's Guide", University of Kassel, 1997.

[LIN 97b] LINK M., "Updating Analytical Models by Using Local and Global Parameters and Relaxed Optimisation Requirements", Mechanical Systems and Signal Processing 12(1), 1998, pp. 7-22.

[LIN 99] LINK M., "Updating of Analytical Models – Basic Procedures and Extensions", in J.M.M. Silva and N.M.M. Maia (eds.), Modal Analysis and Testing, 1999, pp. 281-304.

[LJU 99] LJUNG L., "System Identification: Theory for the User", 2nd edition, Prentice Hall, 1999.

[LOE 04] LÖHR M., DINKLER D., "Damage Detection in Structures using a Parameter Identification Method", Proceeding 2nd European Workshop in Structural Health Monitoring 2004, Edited by Boller C., Staszewski W.J., University of Sheffield, UK, July 7-9, 2004, pp. 237-242.

[LOU 89] LOUIS A. K., "Inverse und schlecht gestellte Probleme", B. G. Teubner, Stuttgart, 1989.

[LOP 00] LOPES V., PARK G., CUDNEY H.H., INMAN, D.J., "Impedance-Based Structural Health Monitoring with Artificial Neural Networks", *Journal of Intelligent Material Systems and Structures*, Vol. 11(3), 2000, pp. 206-214.

[LOP 01] LOPES V., MUELLER-SLANY H.H., BRUNZEL F., INMAN D.J., "A New Methodology of Damage Detection in Structures by Electrical Impedance and Optimization Technique", Proc. Diname 2001, IX Sympos. on Dynamic Problems in Mechanics, Florianopolis, SC, Brazil, 2001, pp. 311-316.

[LUO 00] LUO H., HANAGUD S., "Dynamics of Delaminated Beams", *Intl. Journal of Solids and Structures*, 37, 2000, pp.1501-1519.

[MAE 01] MAECK J., PEETERS B., DE ROECK G., "Damage identification on the Z24-bridge using vibration monitoring analysis", *Smart Materials and Structures*, 10(3), 2001, pp. 512-517.

[MAE 03a] MAECK J., DE ROECK G., "Description of Z24-benchmark", *MSSP*, 17(1), 2003, pp. 127-132.

[MAE 03b] MAECK J., DE ROECK G., "Damage assessment using vibration analysis on the Z24-bridge", *MSSP*, 17(1), 2003, pp. 133-142.

[MAE 03c] MAECK J., "Damage Assessment of Civil Engineering Structures by Vibration Monitoring", PhD Dissertation, KU Leuven, Leuven, Belgium, 2003.

[MAN 02] MANSON G., WORDEN K., ALLMAN D., "Experimental Validation of a Damage Severity Method", Proc. 1st Europ. Workshop on Structural Health Monitoring, Paris, France, DEStech Publ., Lancaster, PA, USA, 2002, pp. 845-852.

[MAT 02] MATLAB, Optimization Toolbox User's Guide, Version 2, The MathWorks Inc., 2002.

[MCC 43] McCULLOCH W.S., PITTS W., "A Logical Calculus of the Ideas Immanent in Nervous Activity", *Bull. Math. Biophys.*, Vol. 5, 1943, pp. 115-133.

[MEH 71] MEHRA R.K., PESCHON J., "An Innovations Approach to Fault Detection and Diagnosis in Dynamic Systems", *Automatica*, Vol. 7, 1971, pp. 637-640.

[MEN 91] MENDENHALL W., BEAVER R.J., "Introduction to Probability and Statistics", PWS-Kent Publishing Comp., Boston, 1991

[MIL 90] MILLER A.J., "Subset Selection in Regression", Chapman and Hall, London, New York, Tokyo, Melbourne, Madras, 1990.

[MOT 98] MOTTERSHEAD J.E., "On the Zeros of Structural Frequency Response Functions and Their Sensitivities", *MSSP*, Vol. 12(5), 1998, pp. 591-597.

[MUE 02] MÜLLER-SLANY H.H., BRUNZEL F., WEBER H.I., "Generation of Highly Accurate Low Order Updated FE Models For Damage Detection Using Optimization Procedures", Proc. Intl. Conf. On: Structural Dynamics Modelling, Madeira Island, Portugal, 2002, pp. 441-450.

[MUE 03] MÜLLER-SLANY H.H., "Möglichkeiten und Grenzen der Schwingungsdiagnose für mechanische Systeme auf der Basis numerischer Strukturmodelle", VDI-Schwingungstagung 2003, Magdeburg, VDI-Berichte 1788, 2003, pp. 71-104.

[NAB 02] NABNEY I.T., "NETLAB – Algorithms for Pattern Recognition", Springer, London, 2002.

[NAT 88] NATKE H.G., YAO J.T.P. (Eds.), "Structural Safety Evaluation Based on System Identification Approaches", Friedrich Vieweg & Sohn, Braunschweig/Wiesbaden, 1988.

[NAT 91] NATKE H.G., CEMPEL C., "Fault Detection and Localisation in Structures: A Discussion", *Mechanical Systems and Signal Processing*, Vol. 5(5), 1991, pp. 345-356.

[NAT 92] NATKE H.G., "Einführung in Theorie und Praxis der Zeitreihen- und Modalanalyse", Braunschweig, Vieweg, 3rd edition, 1992.

[NAT 93] NATKE H.G., TOMLINSON R., YAO J.T.P.(EDS), "Structural Safety Evaluation Based on System Identification Approaches", Fr. Vieweg, Braunschweig, Wiesbaden, 1993.

[NAT 97] NATKE H.G., CEMPEL C., "Model-Aided Diagnosis of Mechanical Systems", Springer Verlag, Berlin, 1997.

[NAU 01] NAUERZ A., FRITZEN C.-P., "Model-Based Damage Identification Using Output Spectral Densities", *Journal of Dynamic Systems, Measurement, and Control*, 123, 2001, pp. 691-698.

[NEL 76] NELSON R.B., "Simplified Calculation of Eigenvector Derivatives", *AIAA Journal* 14(9), 1976, pp.1201-1205.

[OCA 89] O'CALLAHAN J., AVITABILE PP., RIEMER R., "System Equivalent Reduction Expansion Process (SEREP)", Proc. 7th IMAC, Las Vegas, NV, USA, 1989, pp.29-37.

[OEL 98] OELJEKLAUS, M., "Ein Beitrag zur Systemidentifikation: Das Projektive Eingangsgrößenverfahren und das Regularisierte Ausgangsgrößenverfahren im Frequenzbereich für unvollständige Messungen", Habilitationsschrift, Curt-Risch-Institut, Universität Hannover, 1998.

[OEL 99] OELJEKLAUS M., "Projection Methods within Model Updating". In M. I. Friswell, J.E. Mottershead, A.W. Lees, editors, Identification in Engineering Systems, IES99, University of Wales, Swansea, 1999, pp. 325-335.

[OEL 00] OELJEKLAUS M., "Projection Methods within Model Updating", in *Inverse Problems in Engineering*, Vol. 8, 2000, pp. 119-141.

[OST 90] OSTACHOWICZ W.M., KRAWCZUK M., "Vibration analysis of a cracked beam", *Computers & Structures* 36(2), 1990, pp. 245-250.

[OST 01] OSTACHOWICZ W.M., KRAWCZUK M., "On Modelling of Structural Stiffness Loss Due to Damage", Proc. 4th Intl. Conf. On Damage Assessment of Structures (DAMAS 2001), Key Engineering Material, Vols. 204-205, Trans Tech Publications, Switzerland, 2001, pp.185-200.

[PAN 91] PANDEY A.K., BISWAS M., SAMMAN M., "Damage Detection from Changes in Curvature Mode Shapes", *Journal of Sound and Vibration* 145(2), 1991, pp. 321–332.

[PAN 94] PANDEY A.K., BISWAS M., "Damage Detection in Structures Using Changes in Flexibility", *Journal of Sound and Vibration* 169(1), 1994, pp. 3–17.

[PAR 00a] PARK G., CUDNEY H.H., INMAN D.J., "An Integrated Health Monitoring Technique Using Structural Impedance Sensors", *Journal of Intelligent Material Systems and Structures*, Vol 11(6), pp. 448-455, 2000.

[PAR 00b] PARK G., CUDNEY H.H., INMAN D.J., "Impedance-Based Health Monitoring of Civil Structural Components, *Journal of Infrastructure Systems*, Vol. 6(4), 2000, pp. 153-160.

[PAR 03] PARK G., SOHN H., FARRAR C.F., INMAN D.J., "Overview of Piezoelectric Impedance-Based Health Monitoring and Path Forward", *Shock and Vibration Digest*, Vol. 35(6), 2003, pp. 451-463.

[PAS 98] PASCUAL JIMINEZ R., MOLINA J., GOLINVAL J.C., "COST F3 Action Structural Dynamics Working Group 2", Short Term Scientific Mission Report, University of Liège, Belgium, July 1998.

[PAS 99] PASCUAL JIMINEZ R., "Model Based Structural Damage Assessment Using Vibration Measurements", PhD Thesis, University of Liège, Belgium, 1999.

[PAT 89] PATTON R., FRANK PP., CLARK R., "Fault Diagnosis in Dynamic Systems – Theory and Application", Prentice Hall, Hertfordshire, UK, 1989.

[PAT 91] PATTON R., "A Review of Parity-space Approach to Fault Diagnosis", Proc. IFAC/IMACS Sympos. SAFEPROCESS' 91, Baden-Baden, Sep. 1991, pp. 239-255.

[PEE 99] PEETERS B., DE ROECK G., "Reference-Based Stochastic Subspace Identification for Output-Only Modal Analysis", *Mechanical Systems and Signal Processing*, Vol. 13, no. 6, 1999, pp. 855-878.

[PEE 00a] PEETERS B., "System Identification and Damage Detection in Civil Engineering", PhD Thesis, Dept. Civil Eng., KU Leuven, Belgium, 2000.

[PEE 00b] PEETERS B., MAECK J., DE ROECK, G., "Dynamic Monitoring of the Z24-bridge – Separating Temperature Effects from Damage", Europ. COST Conf. On System Identification & Structural Health Monitoring, Madrid, Edition A. Güemes, 2000, pp. 377-396.

[PEE 01a] PEETERS B., DE ROECK, G., "Stochastic System Identification for Operational Modal Analysis: A Review", *Journal of Dynamic Systems, Measurement and Control* 123, 2001, pp. 659-667.

[PEE 01b] PEETERS B., MAECK J., AND DE ROECK G., "Vibration-Based Damage Detection in Civil Engineering: Excitation Sources and Temperature Effects", *Smart Materials and Structures*, 10(3), 2001, pp. 518-527.

[PEI 02] PEIL U., MEHDIANPOUR R., SCHARFF R., FRENZ M., "Life Time Assessment of Existing Bridges", Proc. First European Workshop on Structural Health Monitoring, Paris, France, 2002, pp. 999-1006.

[PIO 93] PIOMBO B., CIORCELLI E., GARIBALDI L., FASANA A., "Structures Identification Using ARMAV Models", Proc. 11th IMAC, Orlando, FL, 1993, pp. 588-592.

[POH 01] POHL J., HEROLD S., MOOK G., MICHEL F., "Damage Detection in Smart CFRP Composites Using Impedance Spectroscopy", *Smart Material and Structures*, Vol. 10, 2001, pp. 834-842.

[RIZ 90] RIZOS pp.F., ASPRAGATHOS N., DIMAROGONAS A.D., "Identification of crack location and magnitude in a cantilever beam from the vibration modes", *Journal of Sound and Vibration* 138(3), 1990, pp. 381-388.

[RUM 86] RUMELHARD D.E., McCLELLAND J.L., "Parallel Distributed Processing: Explorations in the Microstructure of Cognition", Vol.1, Foundations, Cambridge, MIT Press, 1986.

[RYT 93] RYTTER A., "Vibration Based Inspection of Civil Engineering Structures", PhD Thesis, Aalborg University, Denmark, 1993.

[RYT 97] RYTTER A., KIRKEGAARD pp.H., "Vibration Inspection Using Neural Networks", Proc. DAMAS '97, Structural Damage Assessment Using Advanced Signal Processing Procedures, Sheffield, UK, 1997, pp. 97-108.

[SAM 00] SAMPAIO R.P., MAIA N.M., SILVA J.M., RIBEIRO A.M., "On the Use of Transmissibility for Damage Detection and Localization", Europ. COST Conf. On System Identification & Structural Health Monitoring, Madrid, Edition A. Güemes, 2000, pp. 363-376.

[SCH 04] SCHMIDT H.-J., TELGKAMP J., SCHMIDT-BRANDECKER B., "Application of Structural Health Monitoring to Improve Efficiency of Aircraft Structure", Proc. 2nd European Workshop "Structural Health Monitoring 2004", Munich, Germany, July, 7-9, 2004, pp. 11-18.

[SEI 95a] SEIBOLD S., FRITZEN C.-P., "Identification Procedures as Tools for Fault Diagnosis of Rotating Machinery", *Int. Journal of Rotating Machinery*, Vol. 1(3-4), 1995, pp. 267-275.

[SEI 95b] SEIBOLD S., "Ein Beitrag zur modellgestützten Schadendiagnose bei rotierenden Maschinen", PhD Dissertation (in German), VDI Forschrittberichte, Reihe 11(219), VDI-Verlag, Düsseldorf, 1995.

[SEI 96] SEIBOLD S., FRITZEN C.-P., WAGNER D., "Employing Identification Procedures for the Detection of Cracks in Rotors", Modal Analysis, *The Intl. J. of Analytical and Experimental Modal Analysis*, Vol. 11, nos. 3 and 4, 1996, pp. 204-215.

[SOE 91] SOEFFKER D., BAJKOWSKI J., "Crack Detection of a Rotor by State Observers", Proc. 8th IFToMM World Congress on the Theory of Machines and Mechanisms, Prague, CZ, 1991, pp. 771-774.

[SOE 01] SOEFFKER D., "Monitoring and Control of Reliability Characteristics as a Base for Safe and Economical Operation of Technical Systems", Proc. 3rd Intl. Workshop on

Structural Health Monitoring, Stanford, CA, Technomic Publ. Co. Inc., 2001, pp. 784-793.

[SOH 99] SOHN H., LAW K.H., "Extraction of Ritz vectors from Vibration Test Data", Structural Health Monitoring 2000, Chang, F.-K., Technomic Publishing Co., 1999, pp. 840-850.

[SOH 01] SOHN H., FARRAR C., HUNTER N.F., WORDEN K., "Structural Health Monitoring Using Statistical Pattern Recognition Techniques", *Journal of Dynamic Systems, Measurement, and Control*, Transaction of the ASME, Vol. 123, Dec. 2001, pp.706-711.

[SOH 04] SOHN H., FARRAR C., HEMEZ F.M., SHUNK D.D., STINEMATES D.W., NADLER B.R., CZARNECKI J.J., "A Review of Structural Health Monitoring Literature: 1996-2001", Los Alamos National Laboratory Report, LA-13976-MS, Los Alamos, NM, 2004.

[SOH 04b] SOHN H., WAIT J.R., PARK G., FARRAR C.F., "Multiscale Structural Health Monitoring for Composite Structures", Proc. 2nd European Workshop "Structural Health Monitoring 2004", Munich, Germany, July, 7-9, 2004, pp. 721-729.

[STA 04a] STASZEWSKI W., WORDEN K., "Signal Processing for Damage Detection", in *Health Monitoring of Aerospace Structures – Smart Sensor Technologies and Signal Processing*, J. Wiley & Sons, Chichester, 2004, pp. 163-203.

[STA 04b] STASZEWSKI W., BOLLER C., TOMLINSON G. (EDS.), *Health Monitoring of Aerospace Structures – Smart Sensor Technologies and Signal Processing*, John Wiley & Sons, Chichester, 2004.

[STU 90] STUBBS N., OSGUEDA R., "Global Non-Destructive Damage Evaluation in Solids", Modal Analysis, *The Int. Journal of Analytical and Experimental Modal Analysis*, 5(2), 1990, pp. 81-97.

[STU 95] STUBBS N., KIM J.-T., FARRAR C.R., "Field Verification of a Nondestructive Damage Localization and Severity Estimation Algorithm", Proc. 13th Intl. Modal Analysis Conference, Nashville, TN, 1995, pp. 210-218.

[SEY 01a] SEYDEL R., CHANG F.-K., "Impact Identification of Stiffened Composite Panels: I. System Development", *Smart Materials and Structures* 10, 2001, pp. 354-369.

[SEY 01b] SEYDEL R., CHANG F.-K., "Impact Identification of Stiffened Composite Panels: II. Implementation Studies", *Smart Materials and Structures* 10, 2001, pp. 370-379.

[TAD 73] TADA H., PARIS P.-P.C., IRWIN, G.R., *The Stress Analysis of Cracks Handbook*, Hellertown, Pennsylvania: Del Research Corp., 1973.

[THE 90] THEIS W., *Längs- und Torsionsschwingungen bei quer angerissenen Rotoren*, VDI-Fortschrittsberichte, Reihe 11, Nr. 117, Düsseldorf, 1990.

[TEU 03] TEUGHELS A., "Inverse Modelling of Civil Engineering Structures Based on Operational Data", PhD Thesis, KU Leuven, Belgium, 2003.

[TRA 89] TRACY J., PARDOEN G., "Effect of Delamination on the Natural Frequencies of Composite Laminates", *Journal of Composite Materials* 23, 1989, pp. 1200-1215.

[TRU 97] TRUJILLO D.M., BUSBY R.B., *Practical Inverse Analysis in Engineering*, CRC Press, Boca Raton/New York, 1997.

[VAN 93] VAN OVERSCHEE pp., DE MOOR B., "Subspace Algorithm for the Stochastic Identification Problem", *Automatica* 29(3), 1993, pp. 649-660.

[WAL 90] WALLER H., SCHMIDT R., "The Application of State Observers in Structural Dynamics", *Mechanical Systems and Signal Processing*, 4(3), 1990, pp. 195-213.

[WAL 97] WALTER E., PRONZATO L., *Identification of Parametric Models*, Springer, Berlin, 1997.

[WAL 02] WALLASCHEK J., WEDMAN S., WICKORD W., "Lifetime Observer: An Application of Mechatronics in Vehicle Technology", *International Journal of Vehicle Design* 28 (2002), No. 1/2/3, 2002, pp. 121-130.

[WED 01] WEDMAN S., WALLASCHEK J., *The Application of a Lifetime Observer in Vehicle Technology*. Proc. 4th Intl. Conf. on Damage Assessment of Structures (DAMAS 2001), Key Engineering Materials, Vol. 204-205. Uetikon-Zürich: Trans Tech Publications, 2001, pp. 153-162.

[WET 86] WETHERILL B.G., *Regression Analysis with Applications,* Chapman and Hall Ltd, New York, 1986.

[WIL 76] WILLSKY A., "A Survey in Design Methods for Failure Detection in Dynamic Systems", *Automatica*, Vol. 12, 1976, pp. 601-611.

[WOL 03] WOLTERS K., SÖFFKER D., "Control of Damage Dependent Online Reliability Characteristics to Extend System Utilization", Proc. 4th Intl. Workshop on Structural Health Monitoring, Stanford, CA, DesTech Publ., 2003, pp. 796-804.

[WOR 93] WORDEN K., BALL A.D., TOMLINSON G.R., "Neural Networks for Fault Detection", Proc. 11th IMAC, Kissimmee, FL, USA, 1993, pp. 47-54.

[WOR 01] WORDEN K., MANSON G., ALLMAN D.J., "An Experimental Appraisal of the Strain Energy Damage Location Method", Proc. 4th Intl. Conference on Damage Assessment of Structures (DAMAS 2001), Key Engineering Material, vols. 204-205, 2001, pp. 35-46.

[WOR 02a] WORDEN K., MOLINA F.J., PASCUAL R., *et al.*, "COST Action F3 on Structural Dynamics: Benchmarks for Structural Health Monitoring", Proc. ISMA 2002, Vol.1, KU Leuven; Belgium, 2002, pp. 337-346.

[WOR 02b] WORDEN K., ALLEN D.W., SOHN H., *et al.*, "Extreme Value Statistics for Damage Detection in Mechanical Structures", Los Alamos National Laboratory Report LA-13903-MS, 2002.

[WOR 04] WORDEN K., DULIEU-BARTON J.M., "An Overview of Intelligent Fault Detection in Systems and Structures", *Structural Health Monitoring*, Vol. 3(1), 2004, pp. 85-98.

[WU 92] WU X., GHABOUSSI J., GARRETT J.H., "Use of Neural Networks in Detection of Structural Damage", *Computers and Structures*, Vol. 42(5), 1992, pp. 649-659.

[YOU 84] YOUNG pp., *Recursive Estimation and Time-Series Analysis*, Springer-Verlag, Berlin, 1984.

[ZAG 01] ZAGRAI A.N., GIURGIUTIU V., "Electro-Mechanical Impedance Method for Crack Detection in Thin Plates", *Journal of Intelligent Material Systems and Structures*, Vol. 12(10), 2001, pp. 709-718.

[ZIM 92] ZIMMERMANN D.C., SMITH S.W., Model Refinement and Damage Location For Intelligent Structures, in Tzou H.S. and Anderson G.L. (Eds.) "Intelligent Structural Systems", Kluwer Academic, Dordrecht, The Netherlands, 1992, pp. 403-452.

[ZIM 94] ZIMMERMANN D.C., KAOUK M., "Structural Damage Detection Using a Minimum Rank Update Theory", *ASME Journal of Vibration and Acoustics*, Vol. 116(2), 1994, pp. 222-231.

[ZIM 95] ZIMMERMANN D.C., SIMMERMACHER T., KAOUK M., "Structural Damage Detection using Frequency Response Functions", Proc. IMAC XIII, Nashville, TN, USA, 1995, pp. 179-184.

[ZIM 99] ZIMMERMANN D.C., "Looking into the Crystal Ball: The Continued Need for Multiple Viewpoints in Damage Detection, Damage Assessment of Structures", Gilchrist M.D., Dulieu-Barton J.M., Worden K., Trans Tech Publications, 1999, pp. 76-90.

[ZIM 01] ZIMMERMAN D.C., KIM H.M., BARTKOWICZ T.J., KAOUK M., "Damage Detection Using Expanded Dynamic Residuals", *Journal of Dynamic Systems*, Measurement, and Control, Vol. 123, Dec 2001, pp. 699-705.

[ZOU 96] ZOU Z., LINK M., "Identification of Delamination in Sandwich Plates Using Vibration Test Data", Proc. Identification in Engineering Systems, Swansea, UK, 1996, pp. 58-66.

Chapter 3

Fiber-Optic Sensors

3.1. Introduction

Fiber-optic technology started in the 1970s, for long-distance telecommunications, and it has experienced an exponential growth during the last three decades. Sensing applications are a small spin-off from this technology, taking advantage of developments in optoelectronic components and concepts.

In the design of fiber-optic sensors, many different techniques are employed, from very simple concepts, such as proximity sensors, to quite complex procedures, such as fiber-optic gyroscopes for inertial navigation systems. This chapter deals mainly with technologies related to strain and damage sensing, but the whole fiber-optic sensing field is covered in the books written by Udd [UDD 94], Culshaw & Dakin (1988–1997) and Lopez-Higuera [LOP 02]. The book by Measures [MEA 01] deals more specifically with fiber-optic strain monitoring, and it covers the theory and implementation of sensors in composite structures. Professor Measures is a pioneer in this field, as illustrated by an early article entitled 'Smart Structures with Nerves of Glass' [MEA 89]. Strong activity continues in Canada, mainly oriented to civil engineering applications, in the ISIS (Intelligent Systems for Innovative Structures) group (www.isiscanada.com), from whom a manual on the installation, use and repair of fiber-optic sensors can be obtained.

In addition, several review papers have appeared, with emphasis on different aspects of the technology and its applications. Highly recommended is the 60 page article by M. Todd [TOD 05], which gives the fundamentals of optical-based sensing for structural applications. Kersey [KER 97] offers a good discussion of the

Chapter written by Alfredo GÜEMES and Jose Manuel MENENDEZ.

most effective systems for static and dynamic strain sensing; while Grattan and Sun [GRA 00] have produced an excellent review including 167 references and covering the full spectrum of sensing techniques, including some that are not very common, such as LPGs, luminescent OFS and plastic fiber applications.

The understanding of some concepts from optics is required to work with fiber-optic sensors. Different models, with increasing levels of complexity, may be used to explain the different physical phenomena that can be observed featuring light and its interaction with the matter [SAL 91].

The easiest model is based on geometrical optics: light propagates like optical rays, and the selected path requires a minimum of time (the Fermat principle). Together with the definition of speed of light, and the concept of the refractive index, this theory explains the reflection and refraction of light.

The next level of complexity is the "scalar wave model": the propagation of light must satisfy Huygens equations. The reflection, refraction and diffraction of light are then explained.

Light can also be modeled as electromagnetic waves (**E** and **H** vectorial fields) satisfying the Maxwell equations. This model can explain the previous phenomena and other phenomena as well, such as the polarization of the light and the finite number of modes in waveguides.

The most complex model "Quantum Optics" reflects the wave–particle duality, and is required for study of the interaction with matter and the absorption/emission of light. The problem will be stated as a Schrödinger equation. Fortunately for our purposes, most of the phenomena we are going to deal with can be understood with the first two well known models.

Optical fibers (OF) are cylindrical dielectric waveguides for the propagation of light, made from high-purity, low-loss optical materials, usually silica (optical fibers made from plastic and other transparent materials are also commercially available). The refractive index of the core (about 1.46 for silica) is slightly higher than that of surrounding material or cladding, due to the presence of dopants. Internally traveling optical rays arriving at this interface with angles higher than the critical angle, as defined by Snell's law, will undergo a total reflection and continue to travel along the fiber, as paraxial rays, remaining locally confined in the core (see Figure 3.1). Only when the fiber is locally bent with a small radius can the light escape. The OF is externally protected from mechanical scratches by a plastic coating (the "buffer"), and frequently several optical fibers are bundled together and assembled with high strength mechanical fibers, such as Kevlar, to make a robust product known as optical cable, which can withstand rough industrial manipulation.

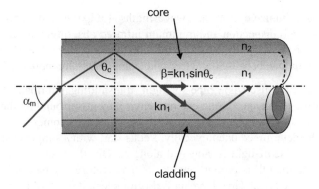

Figure 3.1. *Schematic representation of an optical fiber*

Optical fibers can be classified into two main groups:

– **Single mode OF**, of small core (about 10 µm, depending on the intended optical wavelength). The electromagnetic waves traveling in the core must satisfy the Maxwell equations; the cylindrical contour conditions only allow for a discrete number of solutions, *V*, dependent on the core diameter, and the wavelength:

$$V = \sqrt{2\pi^2 a^2 \frac{n_1^2 - n_2^2}{\lambda^2}}$$

[1]

When *V* is less than 2.4, only one mode (two orthogonal polarizations) propagates through the optical fiber, and this is called a single-mode OF.

Single-mode OF offers some advantages for optical communication, such as:

– optical attenuation is smaller for single-mode OF, due to the lower difference in refractive index between core and cladding, requiring a lower concentration of dopants.

– multimode OF will suffer group dispersion due to different modes traveling at different speeds.

– **Multimode OF** only has the advantage of a larger core (30 to 100 µm). The larger core makes simpler alignment with optical sources and connections. Multimode OF are preferred when used only as guides of light, as in many medical applications. For sensing purposes, they can only be used for intensity-based methods.

Optical power loss is extremely small with current OF, of the order of 0.03 dB per kilometer, meaning that over 100 km the optical power will only be reduced by a factor of two; this is an important requirement for telecommunications. Optical power loss is minimized near the wavelengths 1300 nm and 1550 nm, where minimum Rayleigh dispersion and minimum infrared absorption are found. From a sensing point of view, optical losses at the fiber are irrelevant, but 1550 nm is still the preferred window, because optoelectronic components are more easily available.

The optical fiber itself is a pipe for light, transmitting information, but it may also be sensitive to changes in the external environment surrounding the fiber, such as temperature, strain or chemical composition. Then, it becomes a sensor if these changes can be determined unequivocally. Let us start with a simple example. When a very narrow pulse of light is launched along an OF, the back-scattered radiation dispersed inside the OF is proportional to its temperature. By means of equipment called an OTDR (Optical Time Domain Reflectometer), it is possible to measure the light back-scattered along the fiber length, returning to the emitter after the time of flight; from the signal of intensity collected as a function of time, the temperature distribution can be derived. The fiber will then act simultaneously as a distributed sensor and as a cable for data transmission, all along its length. This seems to be an ideal approach, but accuracy and spatial resolutions are not good enough for many applications. An approach based on local sensors is more common, either by including in the optical path an external device (extrinsic sensor) or by modifying the fiber over a short length (about 1 cm) to make it sensitive to the strain or any other parameter. These are so-called intrinsic sensors, which are usually undetectable with the naked eye.

Common advantages of all kinds of optical fiber sensors arise from their small size and weight and their non-electrical nature, making them immune to electromagnetic interference and to electrical noise, also allowing them to work in explosive environments; they usually have a very high sensitivity and a wide operating temperature range. For smart structures, local intrinsic sensors are the most commonly used, because of their simplicity, the minimal perturbation of the host material and the accuracy of measurement. Multiplexing capability, or the possibility to "write" several sensors on the same optical fiber, is another highly desirable feature. A single optical fiber, 0.25 mm in diameter, can provide strain readings from 10 to 20 different points in the structure. When compared to the four leads required by each electrical strain gauge, a clear advantage can be seen.

In summary, fiber-optic sensors (FOS) have undergone rapid development during the last three decades, before attaining their current status. Criteria for assessing the potential suitability of any optical sensor will include the following:

– intrinsic in nature, for minimum perturbation and stability;

– localized, so it can operate remotely with insensitive leads;

– able to respond only to a strain field, discerning changes of direction;

– all-fiber for operational stability;

– single optical fiber, single ended for ease of installation and connection;

– sufficiently sensitive, with adequate measurement range;
– interrupt immune and capable of absolute measurements;
– amenable to multiplexing;
– easily manufactured, and adaptable to mass production.

3.2. Classification of fiber-optic sensors

A broad definition of a sensor refers to an artefact able to transform a certain physical or chemical parameter into readable information. In addition to the previous topological classification (local/distributed, intrinsic/extrinsic), a more basic classification may be made in terms of the optical parameters. The external action to be monitored must change some of the parameters which define the optical wave: intensity, phase, wavelength and polarization. Based on the type of change, different kinds of optical sensors have been developed.

3.2.1. *Intensity-based sensors*

These are the simplest FO devices, and consequently they were among the first tried; they are still in use as proximity sensors, for damage detection, cure monitoring and hydrogen detection. The components of the measurement system are a stable light source (white or monochromatic), an optical fiber (preferably multimode for higher power transmission) and a sensitive photodetector.

Commercially available proximity sensors are used for non-contact monitoring of the displacement of rotating shafts with an accuracy better than 3 µm. The distance between the cleaved fiber-optic end and the moving surface is related to the amount of light reflected by this surface, and captured again by the optical fiber (Figure 3.2).

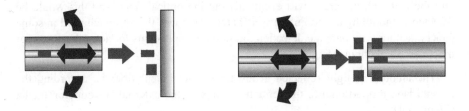

Figure 3.2. *Fiber-optic proximity sensors*

Microbending sensors were popular intrinsic sensors for pressure monitoring; they were based on intensity measurements (Figure 3.3). It is known that light is guided inside the optical fiber because its angle of incidence is higher than the critical reflection angle at the core–cladding interface, and this is always true except for very sharp bends. If the OF is placed between two rough surfaces, higher pressure promotes greater bending and, consequently, higher optical losses. If adequately calibrated, this would provide a measurement of the transverse load on the fiber, by using a local plate, or even the natural roughness of the composite fabrics. Early work was done using this concept, which has now been abandoned; fluctuations of optical power at the light source, connectors, temperature, etc., make the system rather unreliable. Nevertheless, microbending losses must be keep in mind when embedding optical fiber into composite fabrics, because they can reduce the optical signal.

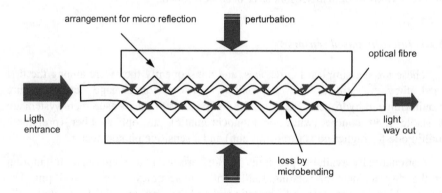

Figure 3.3. *Microbending sensor*

If a crack appears in a structure which has a continuous optical fiber embedded into the material, the crack will eventually cut the optical fiber, and this would be detected without any uncertainty by an OTDR. This was the approach used in some papers dealing with damage detection, but the concept seems to lack robustness for practical use in engineering.

The amount of light reflected at the tip of the optical fiber is, according the Fresnel law, proportional to the refractive index of the external surrounding media (Figure 3.4):

$$R = \frac{(n_{core} / n_{environment} - 1)^2}{(n_{core} / n_{environment} + 1)^2}$$

This phenomenon has been used for cure monitoring [ROD 99] (the refractive index of the resin is known to change with the extent of the cure, as shown at the bottom of Figure 3.4). However, the refractive index of the resin also varies with temperature, so it needs to be determined independently if useful results are to be obtained. In the figure, it is shown how this temperature measurement can be made very near to the fiber tip using an FBG (Fiber Bragg Grating).

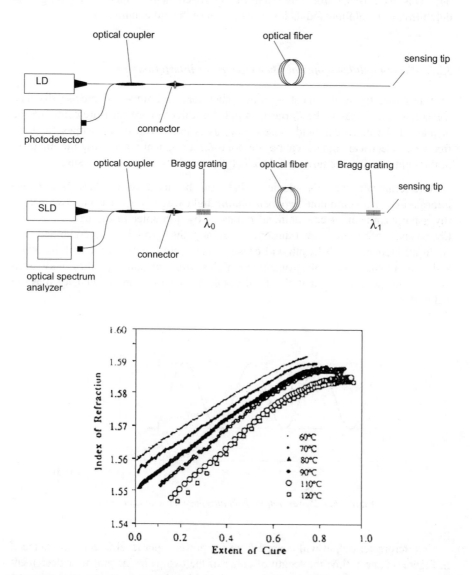

Figure 3.4. *Sketch of basic and improved cure monitoring systems using intensity based FOS*

A similar approach can be used for hydrogen leakage detection if the tip is coated with a hydrogen-sensitive material, such as palladium. With the future widespread usage of hydrogen as a green energy, this kind of low-cost sensor, with no sparking risk, has attracted great interest. A thin film of palladium is laid by vacuum deposition at the cleaved tip end of the fiber. Metallic palladium has a high refractive index, which decreases when it is converted to palladium hydride [GUE 04]. Unsolved issues for this kind of hydrogen sensor include its long-term durability, its resolution and its low time response in cold conditions.

3.2.2. Phase-modulated optical fiber sensors, or interferometers

Interferometry is the most accurate laboratory technique for precise distance measurements. It has to be borne in mind that direct phase measurements, or the display of the electrical field versus time, as is done with oscilloscopes for low-frequency electrical signals, cannot be done for an optical wave; only the intensity of the light, the average power of the electromagnetic field, can be measured.

Interferometers are the devices that can be used to produce this phase information. If a continuous *monochromatic* light wave is divided into two beams (by a partial mirror in conventional optics or by a coupler in the case of optical fibers) and these waves travel through different paths before being recombined, any slight difference in path-length will cause a delay between one wave and the other, and consequently their electromagnetic fields will not sum up; they may even oppose one another, so that the final result is that the output intensity may be reduced to zero.

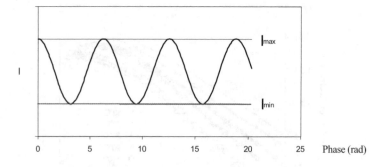

Figure 3.5. *Power output from interferometric measurements*

Interferometer output will change from the power input level to zero, as sketched in Figure 3.5, each time the length of either of the two paths increment or decrement by half a wavelength (380 nm, if the red light of a He–Ne laser is used). With the equipment correctly tuned, if the experiment is carefully carried out, changes in length of the order of 10 nm (the size of an oxygen molecule!) can be detected. This extreme sensitivity to any perturbation in the optical path of any of the two arms makes measurements difficult outside the laboratory environment, because a local temperature change alone may cause a drift of several maxima and minima. The sensor is the whole length of the optical path of one arm; it is an "averaging distributed sensor". The main drawback is that measurements are related to the path differences when the equipment is switched on and any interruption will mean that the reference is lost. In addition, there is no direct indication whether an increment or a decrement is occurring, and a quadrature technique is needed to identify this.

Figure 3.6 shows the most common optical fiber interferometric architectures, known as Mach–Zender, Michelson and Fabry–Perot, from the names of classical optical interferometry. In the Michelson and Fabry–Perot systems, the light is reflected at the end of the optical fibers, traveling back and recombining. The Fabry–Perot system is special in the sense that light travels most of the time through a single fiber and interference is caused between the waves reflected at a partial mirror and the waves that pass this mirror and are reflected back later. A detailed discussion on each system can be found in the bibliography [LOP 02].

For smart structures, a single optical fiber is preferred, because only one ingress point into the structure is required, as it is in the Fabry–Perot architecture. A lot of work has been done to develop a rugged sensor head. Soldering optical fibers after partial coating of the fiber ends was first used for intrinsic sensors. With the development of Bragg grating sensors, this line of work was almost completely abandoned. However, extrinsic Fabry–Perot heads, known as "microcavities", are still commercially available; a precision machined wafer is bonded to the end of the optical fiber, leaving a small air gap, creating miniature pressure and temperature sensors.

The Michelson arrangement is employed in "white light interferometry", a widely used technique for strain monitoring in civil engineering, which is commercially available from a Swiss company, under the name SOFO. The interference phenomena described earlier can only be seen when pure monochromatic light is used. The "spectral purity" of a source of light is related to its "coherence length", or maximum difference in path length between the two arms of the interferometer that still produces a visible interference. The coherence length is several meters for a laser source, but only about 50 μm for a typical LED (Light Emitting Diode) with a central wavelength of 1550 nm and a spectral width of 40 nm. For a Michelson interferometer illuminated with "white light", the interference fringes only appear when the two arms have the same length; for the commercial system, one of the arms is the "sensing arm" and the other is connected to some controlled stretching system. Changes in length are detected with an accuracy of 5 μm or better, giving the average strain. A drawback of this system is the lack of multiplexability.

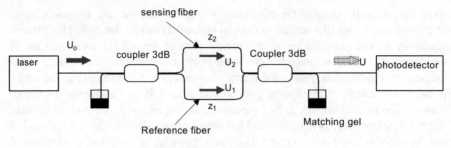

MACH-ZENDER INTERFEROMETER

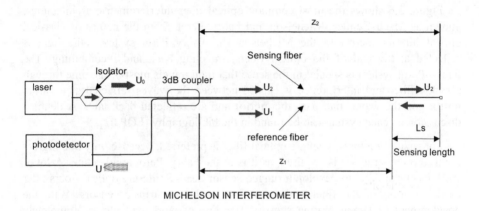

MICHELSON INTERFEROMETER

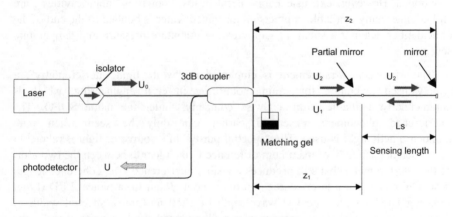

FABRY-PEROT INTERFEROMETER

Figure 3.6. *Typical interferometric arrangements*

3.2.3. *Wavelength based sensors, or Fiber Bragg Gratings (FBG)*

These are a kind of local, intrinsic, absolute, multiplexable, interruption immune fiber-optic strain sensors, which have been the main focus of attention since their discovery, at the beginning of the 1990s.

The basic idea is to engrave, at the core of the optical fiber and for a short length (about 1 cm), a periodic modulation of its refractive index. This will behave as a series of weak partially reflecting mirrors, which, by an accumulative phenomenon of repeated interferences called diffraction, will reflect back the optical wavelength that is exactly proportional to their spacing. The diffraction law, first established by Bragg and widely used for crystal structure analysis, simplifies under normal incidence to the simple equation:

$$\lambda_b = 2\bar{n}_e \Lambda_o \qquad\qquad [2]$$

where Λ_0 is the pitch, or period of the modulation, and $\bar{n}_e$ is the average refractive index. The Bragg grating behaves as a very narrow-bandwidth optical filter, as shown in Figure 3.7. When broadband light is traveling through the fiber, the grating ensures that a very narrow wavelength band is reflected back. If the grating is submitted to a uniform axial strain, or a uniform thermal increment is applied, the central wavelength of the spectrum reflected by the grating will shift as a result of changes in the pitch and the refractive index.

Any Optical Spectrum Analyzer (OSA) will be able to detect these changes, and transform it into readable information (Figure 3.8).

Furthermore, commercially available white light sources have a spectral width of around 40 to 60 nm. As the maximum deviation in wavelength is 10 nm for 10,000 microstrains, several Bragg gratings, centered at different wavelengths, can be written on the same optical fiber and interrogated at the same time. Multiplexing is easily implemented in the same optical fiber by wavelength multiplexing. This can be combined with multiplexing in the time domain, usually done by an optical switch which sequentially connects several optical fibers, as is also shown in Figure 3.8. The fact that the information is wavelength encoded makes the sensor very stable to aging, allowing absolute measurements of strain after long periods without recalibration, a common nightmare with standard strain gauges.

Optical spectrum analyzers are not suitable for every-day sensor systems because they are expensive and their wavelength scanning speed is too slow. Various techniques have been developed for the interrogators. Some are quite simple but are more limited in measurement resolution, dynamic range or multiplexing, and some are more complicated and provide better resolution but are more expensive or need stabilization. Not all of them are appropriate for commercial systems. A good discussion is found in [BYO 03].

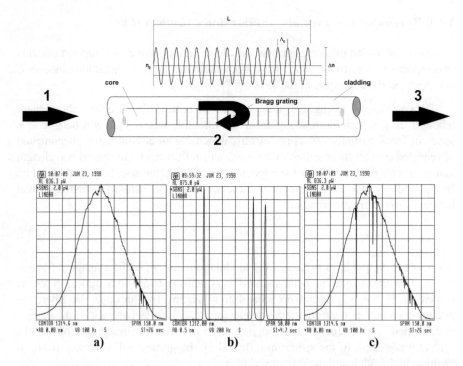

Figure 3.7. *Schematic of a fiber Bragg grating and operating principle. a) Intensity spectrum of a broadband source launched into the fiber. b) Spectra reflected back by three fiber Bragg gratings. c) Transmission spectrum after passing the three Bragg gratings*

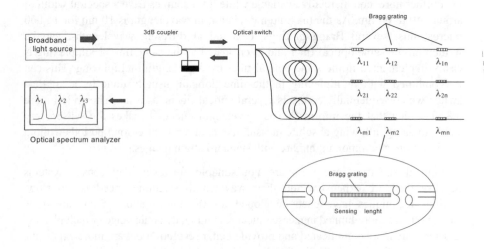

Figure 3.8. *A multiplexed reading system for Bragg gratings*

The simplest interrogators use linearly wavelength-dependent devices (couplers or filters). The Bragg wavelength shift is monitored by the detected power change. This structure is quite simple and has been commercialized, but it is not suitable for multiplexed sensors. In fact, this type of power detection method does not fully use one of the key advantages of FBG sensors – information is coded as wavelength change and the information is not affected by fluctuations in light power.

Most common interrogation systems are based on scanning the wavelength, by illuminating the FBGs with "white light" and collecting the reflecting signal through a tunable Fabry–Perot; other systems are based on illuminating the FBGs with a tunable laser. In both cases, resolution better than 1 microstrain is currently achieved, with a dynamic range up to 100 Hz, which is enough to monitor most of the static and vibrational problems in structures. For high-frequency phenomena, such as impacts and transient analysis, special devices working in the kHz range have been developed.

FBGs may be bought from several commercial sources, tuned to the selected wavelength. The only ability required on the part of the user is to make the optical connections, normally with fusion splicers, which require minor investment and training. This is the common approach for mechanical engineers, who are mainly interested in the use of this technology for strain measurement. Procedures for manufacturing FBGs require major investment and highly trained staff, and it is useful only for those interested in sensor development. Because of this, only a few comments have been made about it.

Dope optical fiber is photosensitive; its refractive index changes when it is exposed to intense UV light. This property is used to engrave the Bragg grating as a periodic pattern of refractive index. A sequence of very close light spots needs to be focused onto the core of the fiber, for a short time. The best way to produce this regular pattern is by diffraction of monochromatic light, either by holographic interference or by the phase mask procedure.

The holographic interference method has the advantage that the pitch of the Bragg grating can be easily changed by changing the angle of the two incident rays, but a highly coherent source is required. The phase mask procedure is easier to use, but different phase masks are required for different pitches.

3.3. The fiber Bragg grating as a strain and temperature sensor

3.3.1. *Response of the FBG to uniaxial uniform strain fields*

The equation that describes the drifting of the central wavelength of the reflectivity spectrum of a FBG loaded axially is:

$$\frac{\delta\lambda}{\lambda_o} = S_\varepsilon \Delta\varepsilon + S_T \Delta T \tag{3}$$

where S_ε and S_T are the coefficients for the fiber-optic strain and temperature sensitivity, respectively. The equation indicates that wavelength drift is essentially linear, with typical values for S_ε and S_T of $0.78 \times 10^{-6}\ \mu\varepsilon^{-1}$ and $6 \times 10^{-6}\ C^{-1}$ (the dependence on strain and temperature will be discussed later). Both coefficients reflect the change in the spacing and the average refractive index.

Concerning the strain dependence, the previous equation is only a simplification of the case $\varepsilon_1 = \varepsilon_1$, $\varepsilon_2 = \varepsilon_3 = -v\varepsilon_1$, others $= 0$ 5n the more general equation:

$$\frac{\delta\lambda}{\lambda_o} = \varepsilon_1 - \frac{\overline{n}_\varepsilon^2}{2} p_{ij} \varepsilon_{j1} \tag{3b}$$

where p_{ij} is a 6×6 matrix of photoelastic coefficients, which for homogeneous isotropic optical media such as silica is:

$$p_{ij} = \begin{bmatrix} p_{11} & p_{12} & p_{12} & 0 & 0 & 0 \\ p_{12} & p_{11} & p_{12} & 0 & 0 & 0 \\ p_{12} & p_{12} & p_{11} & 0 & 0 & 0 \\ 0 & 0 & 0 & \dfrac{p_{11}-p_{12}}{2} & 0 & 0 \\ 0 & 0 & 0 & 0 & \dfrac{p_{11}-p_{12}}{2} & 0 \\ 0 & 0 & 0 & 0 & 0 & \dfrac{p_{11}-p_{12}}{2} \end{bmatrix}$$

Typical values for the given coefficients are: $p_{11} = 0.113$ and $p_{12} = 0.252$, $n_e = 1.45$.

The validity of equation [3] can be experimentally verified in Figures 3.9 and 3.10. Possible deviations from linearity of the relationship strain/wavelength shift are typically less than the resolution of the optoelectronic demodulators. In the case of temperature, non-linearity is found for large temperature changes as a result of the dependence of the thermo-optical coefficient on temperature.

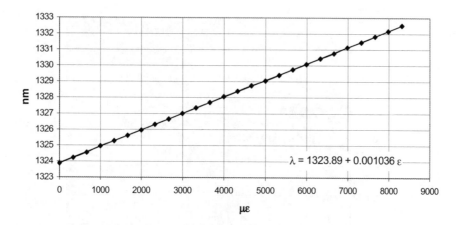

Figure 3.9. *Wavelength shift versus axial strain. The FBG was submitted to uniaxial strain*

3.3.2. Sensitivity of the FBG to temperature

Temperature influences the behavior of the FBG as a result of the linear thermal expansion of the grating, and also by changes in its refractive index. Both effects may be summarized by the following equation:

$$\frac{\delta \lambda_b}{\lambda_{b,o}} = (\alpha + \zeta)\delta T \tag{4}$$

Typical values for these coefficients are 5 to 5.5×10^{-7} [BUT 78, HOC 79] for the thermal expansion of silica, and one order of magnitude higher 8 to 10×10^{-6} for the thermo-optic coefficient [MEA 89, SAL 91]. Change in the refractive index is the main factor, and both coefficients are fairly constant over a wide temperature range.

These values suggest a drift of about 1 nm per 100°C for a grating with an unperturbed wavelength of 1300 nm. For accurate strain measurement, temperature effects need to be corrected, in the same way as they are for common strain gauges.

The first exposure of a grating to high temperatures (over 100°C) causes a partial loss of reflectivity of the grating, and also a permanent drifting in the central wavelength. To get a repeatable response at high temperatures, gratings need to be annealed. Figure 3.10 shows the response from room temperature to 700°C of gratings previously annealed at this maximum temperature. Behavior is almost linear, but for this very large temperature range a fourth-order fit was found to adjust better (coefficients are dependent on the wavelength used, around 1300 nm in this case).

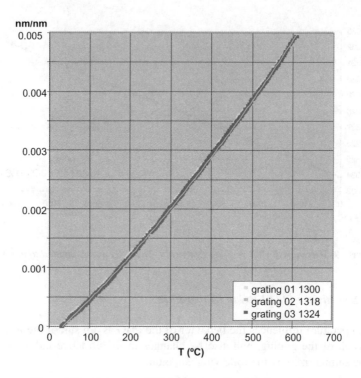

Figure 3.10. *Behavior of FBG under cooling-heating cycles*

$$T = -3.643 \times 10^{11} \left(\frac{\Delta \lambda_b}{\lambda_{bo}} \right)^4 + 4.608 \times 10^9 \left(\frac{\Delta \lambda_b}{\lambda_{bo}} \right)^3$$

$$- 2.404 \times 10^7 \left(\frac{\Delta \lambda_b}{\lambda_{bo}} \right)^2 + 1.671 \times 10^5 \left(\frac{\Delta \lambda_b}{\lambda_{bo}} \right) + 29.42 \qquad [5]$$

In cryogenic conditions, for temperatures down 50 K, the coefficients are no longer constants, and the linearity is lost. In liquid helium (4.2 K) both coefficients are almost zero and the wavelength variation with temperature is negligible under these conditions [GUE 04].

3.3.3. *Response of the FBG to a non-uniform uniaxial strain field*

The reflected spectrum of an unloaded standard Bragg grating is a very narrow (0.1 nm) symmetric Gaussian peak. This spectrum may be acquired by an OSA (Optical Spectrum Analyzer), representing the collected power versus wavelength.

This peak simply moves backwards or forwards when the BG is uniformly loaded, and its pitch, or internal periodic modulation, changes accordingly. If the FBG were submitted to two different strain levels along its length, two different peaks of lower height would be expected. In general terms, a strain gradient along its length will distort the reflected peak. The shape of the reflected peak corresponding to a given strain field can be easily calculated by Transfer Matrix Formalism (TMF), a numerical method for solving the coupled mode equations for aperiodic grating structures.

The coupled mode equations for a single-mode fiber optic with a longitudinal aperiodic diffractive structure written in its core can be written as in [ERD 97]:

$$\frac{dR}{dz} = i\hat{\sigma}R(z) + i\kappa S(z)$$ [6]

$$\frac{dS}{dz} = -i\hat{\sigma}S(z) - i\kappa R(z)$$ [7]

where $R(z)$ and $S(z)$ are respectively:

$$R(z) = A(z)e^{i\left(\delta z - \frac{\phi}{2}\right)}$$ [8]

$$S(z) = B(z)e^{-i\left(\delta z - \frac{\phi}{2}\right)}$$ [9]

$\hat{\sigma}$ is an autocoupling coefficient defined as:

$$\hat{\sigma} = \delta + \sigma - \frac{1}{2}\frac{d\phi}{dz}$$ [10]

where δ is the differential propagation constant associated with the deviation of λ from the Bragg condition, defined as:

$$\delta \equiv \beta - \frac{\pi}{\Lambda} = \beta - \beta_D = 2\pi n_{eo}\left(\frac{1}{\lambda} - \frac{1}{\lambda_D}\right)$$ [11]

where $\lambda_D \equiv 2n_{eo}\Lambda_o$ is the Bragg wavelength for an infinitesimally weak $(\delta n_e \to 0)$ FBG with a period Λ_o, $\phi(z)$ is the term that describes the aperiodicity of the grating, and σ and κ are coupling coefficients for the continuous and periodic components of the propagating wave, respectively:

$$\sigma = \frac{2\pi}{\lambda}\overline{\delta n_e}$$

[12]

$$\kappa = \frac{\pi}{\lambda}v\overline{\delta n_e}$$

[13]

for a uniform grating, that is, a grating with constant modulation period and effective refractive index ($d\phi/dz = 0$ and $\overline{\delta n_e}$ constant). In this case, the equation for the effective refractive index will be:

$$n_e(z) = n_{eo} + \overline{\delta n_e}\left[1 + v\cos\left(\frac{2\pi}{\Lambda_o}z\right)\right]$$

[14]

and the coupled-mode equations will be reduced to a pair of first-order differential equations with constant coefficients, having analytical solutions when boundary conditions are adequately defined.

For an FBG with length L in which the propagating wave is incident from $z = -\infty$, the reflectivity of the grating can be obtained by considering a predetermined value for the amplitude of the field incident on the left of the grating, $R(0) = 1$ for instance, and requiring that the amplitude of the field reflected at the right of the grating be null, that is, $S(L)=0$. The normalized reflectivity ρ, a parameter which contains the complete spectral information of the FBG, will be given by the ratio of amplitudes at the left of the grating:

$$\rho(\lambda) = \frac{S(0)}{R(0)} = \sqrt{r(\lambda)}e^{-i\psi(\lambda)}$$

[15]

where $r(\lambda) = |\rho(\lambda)^2|$ is the relative amplitude or reflectivity of the grating, and $-\psi(\lambda)$ is its phase as a function of the wavelength, that is, its reflective and transitive spectra. As a result of solving the coupled-mode equations, the final equation for the normalized reflectivity is:

$$\rho(\lambda) = \frac{-\kappa\sinh\left(\sqrt{\kappa^2 - \hat{\sigma}}\ L\right)}{\hat{\sigma}\sinh\left(\sqrt{\kappa^2 - \hat{\sigma}}\ L\right) + i\sqrt{\kappa^2 - \hat{\sigma}}\ \cosh\left(\sqrt{\kappa^2 - \hat{\sigma}}\ L\right)}$$

[16]

The reflective spectrum will then be defined by the equation:

$$r(\lambda) = \frac{\sinh^2\left(\sqrt{\kappa^2 - \hat{\sigma}}\ L\right)}{\cosh\left(\sqrt{\kappa^2 - \hat{\sigma}}\ L\right) - \dfrac{\hat{\sigma}^2}{\kappa}} \qquad [17]$$

The TMF offers an elegant and effective solution for estimating the reflectivity of non-uniform FBGs, an alternative to the numerical integration of the coupled-mode equations. Formally, it consists of subdividing the diffraction grating into a discrete number of conjoint sub-gratings with uniform period and constant intensity of modulation, in such a way that the parameters of the defined discrete virtual gratings approximate the continuous parameters of the real aperiodic grating. The solution is calculated by applying the transfer matrix formalism, multiplying the 2 × 2 matrices associated with the solution of the coupled-mode equations for each sub-grating [YAM 87, HUA 94].

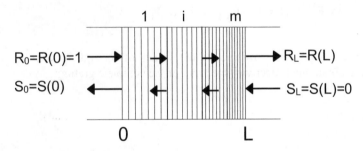

Figure 3.11. *Schematic of the refractive index pattern of a non-uniform FBG*

Consider a grating divided into M uniform sub-gratings, as shown in Figure 3.11, where R_i and S_i are the field amplitudes after passing through sub-grating i, defined for a single uniform grating in equations [8] and [9]. This assumes that the value of the amplitude of field incident on the left of the grating is given by $R_0=R(0)=1$, and that the field amplitude reflected to the left of the grating is null, that is, $S_L=S(L)=0$. $S_0=S(0)$ and $R_L=R(L)$ are unknown.

Propagation through each uniform sub-grating i is defined by a matrix $\mathbf{F_I}$ so that:

$$\begin{bmatrix} R_i \\ S_i \end{bmatrix} = \mathbf{F_I} \begin{bmatrix} R_{i-1} \\ S_{i-1} \end{bmatrix} \qquad [18]$$

where $\mathbf{F_I}$ is given by the equation:

$$\mathbf{F_I} = \begin{bmatrix} \cosh(\gamma l_i) - i\dfrac{\hat{\sigma}}{\gamma}\sinh(\gamma l_i) & -i\dfrac{\kappa}{\gamma}\mathrm{senh}(\gamma l_i) \\ i\dfrac{\kappa}{\gamma}\mathrm{senh}(\gamma l_i) & \cosh(\gamma l_i) + i\dfrac{\hat{\sigma}}{\gamma}\mathrm{senh}(\gamma l_i) \end{bmatrix} \qquad [19]$$

where l_i is the length of the sub-grating with index i, that is, $l_i = z_{i+1} - z_i$, $\hat{\sigma}$ and κ are the values of these parameters in such a section and:

$$\gamma = \sqrt{\kappa^2 - \hat{\sigma}^2} \qquad [20]$$

Finally, the propagation through the complete grating will be given by the expression:

$$\begin{bmatrix} R_L \\ S_L \end{bmatrix} = \mathbf{F}\begin{bmatrix} R_0 \\ S_0 \end{bmatrix}; \mathbf{F} = \mathbf{F_M} \cdot \mathbf{F_{M-1}} \cdot \dots \cdot \mathbf{F_I} \cdot \dots \cdot \mathbf{F_1} \qquad [21]$$

which makes it possible to calculate the values of R_0 and S_0, from which it is possible to obtain the reflectivity spectrum of the complete grating:

$$r = \left| \frac{S_0}{R_0} \right|^2 \qquad [22]$$

Figure 3.12 shows the calculated spectra for a linear gradient, together with the experimentally obtained values for a simply supported beam reproducing this strain gradient (Figure 3.13). The agreement is very acceptable.

For the inverse problem, or to obtain the strain profile applied to the grating by means of integration of the intensity and the phase spectrum of the grating, three approaches have been proposed, the phase based method (PBS), the intensity based method (IBS) and the Fourier transform based method (FTB). Although the Fourier transform method is the most general, being able to measure arbitrary strain profiles, the intensity based spectrum method has the important advantage of requiring only an optical spectrum analyzer for the data acquisition. The most important limitations of the method are those that are derived from the hypothesis of the model and the integration procedure, which reduce its applicability to monotonic strain distributions, the limited spatial resolution of the method (mainly in the case of low-gradient profiles) and the validity of the results at the edges of the grating. The resolution of the OSA affects the strain resolution. However, the procedure is simple and accurate enough to be useful in those applications in which, otherwise, it would be impossible to get information about the strain distribution.

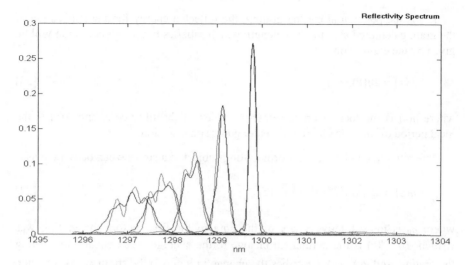

Figure 3.12. *Real and simulated spectra of the grating attached to the beam in several load conditions*

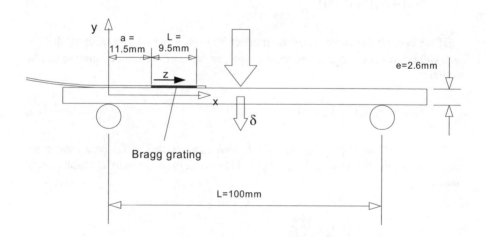

Figure 3.13. *Experimental arrangement for the calibration of the sensor*

Having as input data the parameters of the grating and the experimental intensity spectra acquired by the OSA, a computer program performs the numerical integration of such spectra according to the ISB approach and predicts the strain distribution corresponding to this model.

If z is the position along the grating, the reflected energy for a small section of the grating centered at z, the wavelength which satisfies the Bragg condition will be given by the expression:

$$\lambda(z) = 2n(z)\Lambda(z) \tag{23}$$

where $n(z)$ is the local average refractive index of the fiber core, and $\Lambda(z)$ is the local period of the index modulation along this small section.

The refractive index of a uniform grating subject to a strain ε_f can be written as:

$$n(z) = n_0 + \delta n^{perm}(z) + \delta n^{\varepsilon_f}(z) \tag{24}$$

where n_0 is the refractive index of the core of the fiber before the grating was written, $\delta n^{perm}(z)$ is the permanent change to the average index created by writing the grating, and $\delta n^{\varepsilon_f}(z)$ describes the change in n due to the strain ε_f. As the pitch of the unstrained grating is uniform, the local pitch at the position z of the strained grating can be written as:

$$\Lambda(z) = \Lambda_0\left(1 + \varepsilon_f(z)\right) \tag{25}$$

If we express the variation of the product $n(z)\Lambda(z)$ in terms of a quantity that we shall call optical strain $\varepsilon_{opt}(z)$, the local Bragg condition of the strained grating can be written as:

$$\lambda(z) = 2n_0\Lambda_0\left(1 + \varepsilon_{opt}(z)\right) \tag{26}$$

If $\lambda(z)$ is an invertible function of z (which requires that $\varepsilon_{opt}(z)$ be a monotonic function of z), the following relation will hold between the reflectivity R of the grating and $\lambda(z)$:

$$R(\lambda) = 1 - \exp\left[\frac{-|\kappa(z)|\lambda^2}{2\lambda_0 n(z)} \frac{1}{\left|\frac{d}{dz}\varepsilon_{opt}(z)\right|}\right] \tag{27}$$

where $\chi(z) = \pi\Delta n(z)/\lambda$ is the coupling coefficient of the grating. Substituting for χ and using $d\lambda = \lambda_0\varepsilon_{opt}$, it is possible to integrate the above relation to give the functional relationship between λ and z:

$$\int_{\lambda(z=0)}^{\lambda(z)} \ln(1-R(\lambda))d\lambda = -\frac{\pi^2}{2}\int_0^z \frac{\Delta n^2(z)}{n(z)}dz \qquad [28]$$

A priori, it is not possible to integrate the integral on the right of equation [28], because $n(z)$ depends on the strain, which is unknown. However, the strain-induced variations are very small, and the $\delta n^{\varepsilon_f}(z)$ term can be neglected in the integral. If the refractive index and the amplitude index modulation are not constant over the grating length, the integration cannot be solved analytically either, but the Newton–Raphson method can be employed to solve it numerically.

With $\lambda(z)$ and equation [26], one can obtain ε_{opt} and, using the relation:

$$\varepsilon_{opt}(z) = \frac{\delta n^{perm}(z)}{n_0} + \left[\frac{1}{n_0}\frac{dn}{d\varepsilon_f}+1\right]\varepsilon_f(z), \qquad [29]$$

it is possible to obtain the strain distribution. The coefficient of $\varepsilon_f(z)$, related to the strain sensitivity of the grating, can be obtained experimentally for each grating type.

Figure 3.14 shows a reasonable agreement between the strain profiles calculated from the acquired spectra and the solution from the elasticity theory. The discrepancies grow in low strain gradient conditions, as a result of the reduction in the accuracy and the spatial resolution of the method in this particular situation, but the most severe errors occur near the edges of the grating, due to the requirements imposed on the model by the physical edge of the grating.

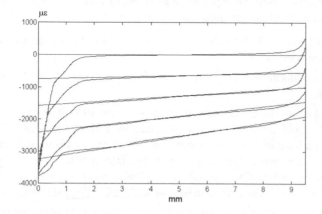

Figure 3.14. *Comparison of the theoretical strain profiles (straight lines for different stress levels) to the results of the integration (curved lines)*

The technique of integration of the intensity spectrum, recently presented in several publications, can be considered as an excellent tool for describing the strain field of specific applications in structural engineering. Although there are some limitations, their strain accuracy and spatial resolution is good enough to yield not only important qualitative information in circumstances in which other techniques cannot be applied, but also reasonably accurate quantitative results.

3.3.4. *Response of the FBG to transverse stresses*

The situation that complicates the suitability of fiber Bragg gratings is their response to a non-unidirectional stress field. The first sign of this effect appeared in fiber Bragg gratings embedded in a thick graphite/epoxy laminate cured at high temperature. Several gratings used to monitor strain and temperature during the cure process produced twin spectra during the cooling of the laminate. Bending of the fiber shows an exchange of optical power between the two sides of the twin spectra, indicating a birefringence of the fiber promoted by the presence of strong residual stresses inside the laminate [LEV 96].

A transverse stress field applied to the grating promotes a birefringence, which modifies the isotropic optical structure of the fiber, generating two principal directions of polarization, with two different refractive indexes. The fiber Bragg grating detects two different Bragg conditions and the optical spectrum of the grating splits in two. If a fiber Bragg grating, written in the core of a standard low-birefringence optical fiber, immersed in a general stress field, were used, it would not be possible to separate the components of the stress field using the spectral information of the grating. These phenomena have been reported by many authors and several methods have been proposed to avoid, or minimize, them: use of special fiber, encapsulation of the gratings, etc.

Figures 3.15 to 3.18 show the spectrum of a single FBG (narrow line) loaded under different conditions:

– Uniaxial uniform stresses (Figure 3.15). This promotes simply a drifting of the reflected peak.

– Uniaxial stress field, with strain gradient along the FBG (Figure 3.16). This will happen, for example, if the FBG is bonded in near the ends of a bonded joint. The peak is distorted and information on the strain distribution can be obtained as explained in section 3.3; another simpler approach to obtain this information about strain gradients is to engrave in the optical fiber several shorter FBGs, making use of the multiplexing capability. FBGs as short as 2 mm can be engraved, still with good reflectivity.

– FBG under transverse stresses (Figure 3.17). The single peak divides into two, owing to the induced birefringence, etc. This behavior can easily be initiated by placing the FBG between two flat surfaces and squeezing it. It is very instructive to see how the heights of the peaks change when the connector is rotated, because it changes the amount of optical power launched in each of the two polarization modes.

– A combination of the previous phenomena (Figure 3.18). This used to happen when the FBG was embedded in a composite made with fabrics instead of tape laminae. In addition to the peak splitting resulting from curing, the texture of the fabric may even bend the sensor, promoting optical losses and distortions. This is a situation to avoid.

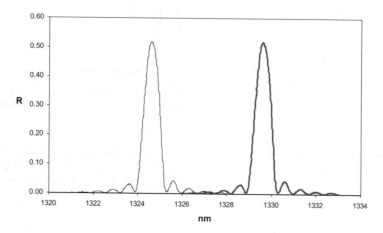

Figure 3.15. *Spectra of a fiber Bragg grating bonded in an aluminium beam: unloaded (left-hand curve) and submitted to a uniform longitudinal strain of 5000 me (right-hand curve). The "ideal" behavior of a grating*

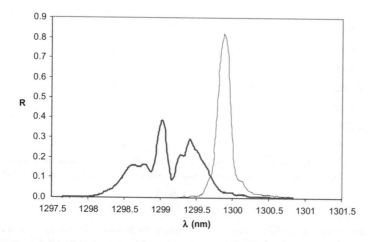

Figure 3.16. *Spectra of a fiber Bragg grating embedded in a unidirectional composite laminate near a bonded joint: unloaded (right-hand curve) and submitted to longitudinal load (left-hand curve). The distortion is due to strain gradients*

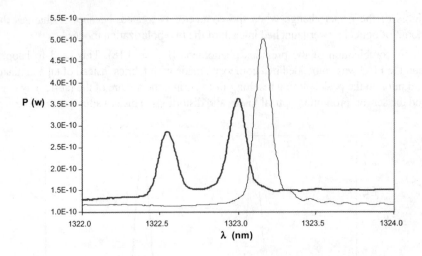

Figure 3.17. *Spectra of a fiber Bragg grating before (right-hand curve) and after being embedded in a thick quasi-isotropic laminate cured in an autoclave (left-hand curve); the split spectrum of the left-hand curve shows the effect of strong transverse stresses over the grating*

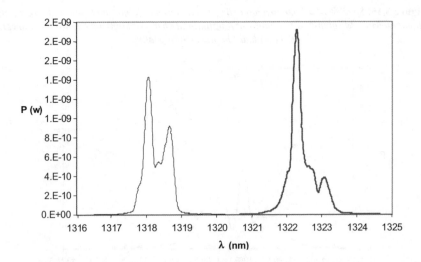

Figure 3.18. *Spectra of an FBG embedded in a composite beam made with fabric and cured in an autoclave: the unloaded (left-hand) curve shows the typical splitting of a biaxial stress state; the loaded (right-hand) curve shows the effect of a complex non-uniform biaxial stress field*

The transverse stresses afford a rich phenomenology of the behavior of FBGs, and this can be exploited. Some authors have tried to solve this problem using a purely optical approach to the solution of the general three-dimensional case by means of fiber Bragg gratings written in a high-birefringence optical fiber. With a precise positioning of the fiber, and solution of a simplified photoelastic model, it is possible to resolve the longitudinal and the two transverse components of the stress applied to the sensors. However, the solution is complex and requires an extremely precise positioning of the fiber, which is almost impossible in the generally rough conditions of a composite laminate, and also requires special fibers and optoelectronic devices for the demodulation.

The plane stress state dominant in laminates makes it possible to simplify the solution of the photoelastic problem. The model considers the longitudinal and transverse stresses applied to fiber Bragg gratings written in low-birefringence fibers and embedded in real manufacturing conditions. It has been validated using simple laminate models and finite element methods, and shows a good agreement between experimental results and the stresses calculated from the lamination theory.

3.3.5. Photoelasticity in a plane stress state

In a highly birefringent optical fiber, the core of the fiber presents an orthotropic symmetry, with two well-defined polarization axes, whose directions are defined by the index ellipsoid of the fiber. A co-ordinate system can be defined with its main axis oriented in the longitudinal direction of the fiber, named as x_1, and two directions normal to it and parallel to the polarization axis, x_2 and x_3 (Figure 3.19). If a Bragg grating is written in the core of a highly birefringent fiber, the intensity spectrum of the grating presents two well-defined peaks corresponding to the refractive indexes in the two polarization axes of the fiber, n_2 and n_3 centered on a wavelength given by the Bragg condition $\lambda_b = 2\overline{n_e}\Lambda_o$, where $\overline{n_e}$ is the average effective refractive index along the grating, and Λ_o is the uniform pitch of its diffractive structure.

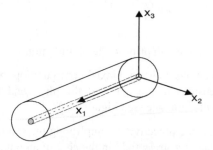

Figure 3.19. *Single-mode optical fiber with co-ordinate axes parallel to the principal polarization axis*

A three-dimensional strain field applied to a regular single-mode optical fiber with an inscribed Bragg grating will promote a birefringent effect. Two different polarization axes with different effective refractive indexes then have to be considered [CAR 93]. The perturbation on the initial refractive index of the grating can be written:

$$\delta\left(\overline{n_{e,2}}\right) = -\frac{\overline{n_e}^3}{2}\left[p_{11}\varepsilon_2 + p_{12}\left(\varepsilon_1 + \varepsilon_3\right)\right] \qquad [30]$$

$$\delta\left(\overline{n_{e,3}}\right) = -\frac{\overline{n_e}^3}{2}\left[p_{11}\varepsilon_3 + p_{12}\left(\varepsilon_1 + \varepsilon_2\right)\right] \qquad [31]$$

where p_{11}, p_{12}, p_{13}, p_{23}, and p_{33} are the elements of the strain–optic tensor of a homogeneous orthotropic material, which define the optical response of a medium submitted to mechanical deformation, and ε_j are the components of the strain field applied to the core of the optical fiber; in contracted notation, $\varepsilon_1 = \varepsilon_{11}$, $\varepsilon_2 = \varepsilon_{22}$, $\varepsilon_3 = \varepsilon_{33}$, $\varepsilon_4 = \gamma_{23}$, $\varepsilon_5 = \gamma_{13}$, $\varepsilon_6 = \gamma_{12}$. The perturbation in the Bragg wavelength due to a strain field applied to the fiber will be given by the equation:

$$\delta\lambda_b = 2\left(\delta\left(\overline{n_e}\right)\Lambda_o + \overline{n_e}\delta\Lambda\right) \qquad [32]$$

where $\delta\Lambda = \Lambda_o \varepsilon_1$ and ε_1 is the strain in the direction of the optical fiber. Therefore, the shift of the Bragg wavelength of a diffraction grating in the principal polarization axis of the fiber will be given by the equations:

$$\frac{\delta\lambda_2}{\lambda_0} = \varepsilon_1 - \frac{\overline{n_e}^2}{2}\left[p_{11}\varepsilon_2 + p_{12}\left(\varepsilon_1 + \varepsilon_3\right)\right] \qquad [33]$$

$$\frac{\delta\lambda_3}{\lambda_0} = \varepsilon_1 - \frac{\overline{n_e}^2}{2}\left[p_{11}\varepsilon_3 + p_{12}\left(\varepsilon_1 + \varepsilon_2\right)\right] \qquad [34]$$

where λ_0 is the Bragg wavelength prior to the perturbation.

A composite laminate has a clearly biaxial configuration, with a thickness much smaller than the dimensions in the plane of the plies, and the loads are always applied in the plane of the laminate (see Figure 3.20).

In this case, and with the objective of simplifying the solution of the problem of calculating the strain due to a stress field in the core of an optical fiber embedded in a laminate, it is reasonable to assume a plane-stress state in plane x_1–x_2, which implies that $\sigma_1, \sigma_2, \sigma_6 \neq 0$, $\sigma_3 = \sigma_4 = \sigma_5 = 0$ (for simplicity, we assume shear stress

$\sigma_6 = 0$). The components of the strain fields in the principal directions of the fiber, ε_1, ε_2, ε_3, are then:

$$\varepsilon_1 = \frac{1}{E_f}\left(\sigma_1 - v_f\sigma_2\right) \tag{35}$$

$$\varepsilon_2 = \frac{1}{E_f}\left(-v_f\sigma_1 + \sigma_2\right) \tag{36}$$

$$\varepsilon_3 = -\frac{v_f}{E_f}\left(\sigma_1 + \sigma_2\right) \tag{37}$$

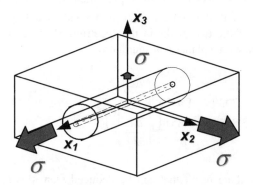

Figure 3.20. *Element of laminate with an embedded optical fiber and a common coordinate system*

Substituting these equalities in equations [33] and [34], it is possible to obtain the value of the relative shifts of the Bragg wavelength of the split peaks corresponding to the two principal axis of polarization x_2 and x_3 as a function of the longitudinal stress σ_2 and the transverse stress in the plane of the laminate σ_2:

$$\begin{pmatrix} \dfrac{\delta\lambda_2}{\lambda_0} \\ \dfrac{\delta\lambda_3}{\lambda_0} \end{pmatrix} = \begin{bmatrix} U_1 & U_2 \\ U_1 & U_3 \end{bmatrix}\begin{pmatrix} \sigma_1 \\ \sigma_2 \end{pmatrix} \tag{38}$$

where:

$$U_1 = \frac{1}{E_f}\left\{1 - \frac{\overline{n}_e^2}{2}\left[\left(1 - v_f\right)p_{12} - v_f p_{11}\right]\right\} \tag{39}$$

$$U_2 = \frac{1}{E_f}\left\{ -v_f - \frac{\overline{n_e}^2}{2}\left[p_{11} - 2v_f p_{12} \right] \right\}$$ [40]

$$U_3 = \frac{1}{E_f}\left\{ -v_f - \frac{\overline{n_e}^2}{2}\left[\left(1-v_f\right)p_{12} - v_f p_{11} \right] \right\}$$ [41]

It is possible to estimate the array of coefficients U using the same coefficients: 70 GPa and 0.17 for the elastic modulus and the Poisson coefficient, $p_{11} = 0.113$ and $p_{12} = 0.252$, and $\overline{n_e} = 1.4496$. Equation [42] then allows the field of transverse stresses in the core of the optical fiber to be calculated, given the measured wavelength shifts of the two peaks in the spectrum of the grating embedded in a composite laminate with a planar configuration:

$$\begin{pmatrix} \sigma_1 \\ \sigma_2 \end{pmatrix} = 70 \begin{bmatrix} 2.7020 & -1.4526 \\ 5.8514 & -5.8514 \end{bmatrix} \begin{pmatrix} \dfrac{\delta\lambda_2}{\lambda_0} \\ \dfrac{\delta\lambda_3}{\lambda_0} \end{pmatrix} GPa$$ [42]

The most immediate result that can be deduced from this equation is that the distance between the two peaks is only affected by the transverse stress applied to the fiber, and is proportional to that stress (Figure 3.21):

$$\sigma_2 \cong k \left(\frac{\delta\lambda_2}{\lambda_o} - \frac{\delta\lambda_3}{\lambda_o} \right)$$ [43]

where $k \cong 70 \times 5.8514$ GPa is the coefficient of proportionality, λ_o is the Bragg wavelength of the non-perturbed grating, and $\delta\lambda_2$ and $\delta\lambda_3$ are the wavelength shifts of the split peaks.

It is also possible to deduce an equation for the longitudinal strain of the grating as a function of the displacement of the wavelength of one of the peaks and the distance between the two peaks, $\delta\lambda = \delta\lambda_2 - \delta\lambda_3$, that is:

$$\varepsilon_1 = 1.25\frac{\delta\lambda_2}{\lambda_o} + 0.46\frac{\delta\lambda}{\lambda_o}$$ [44]

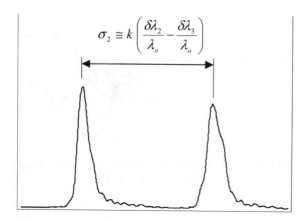

$$\sigma_2 \cong k\left(\frac{\delta\lambda_2}{\lambda_o} - \frac{\delta\lambda_3}{\lambda_o}\right)$$

Figure 3.21. *The distance between the two peaks is only affected by the transverse stress applied to the fiber, and is proportional to it*

In order to prove the validity of these results, two different experimental tests have been implemented. The first, very simple, one uses the well-known thermal expansion of aluminium to prove the importance of the effect of transverse stresses over fiber Bragg gratings in apparently very simple applications. The second test, with four gratings embedded in different positions in a graphite/epoxy laminate, makes it possible to evaluate the effect of the residual stresses generated in a laminate during curing.

The use of fiber Bragg gratings to monitor strain distributions in bonded joints in aluminium elements using high temperature adhesive film was one of the first applications in which the importance of transverse stress in their spectral response became clear. The following test has been configured as an illustrative example of this behavior, and shows the necessary precautions that have to be taken when using this technology in applications in which temperature or external loads can promote transverse strain fields over the gratings.

Figure 3.22 shows a bare fiber Bragg grating 10 mm long, written in photosensitive fiber and submitted to thermal stabilization at 300°C for two hours; it has been bonded in a beam of AA5083 aluminium using an epoxy adhesive film. The adhesive has been cured using a standard cycle shown in Figure 3.23a). Figure 3.23b) shows the evolution of the Bragg wavelength of the grating with temperature during the curing cycle. Initially, during the first part of the heating stage, the wavelength of the grating increases according to the wavelength–temperature curve for a free grating.

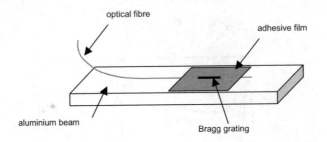

Figure 3.22. *Schematic of an aluminium beam with a fiber Bragg grating bonded using adhesive film*

The thermal characterization curve of the grating adjusted at UPM for the series of gratings is:

$$T = T_0 + 1.671 \times 10^5 \left(\frac{\Delta \lambda_b}{\lambda_{bo}} \right) - 2.404 \times 10^7 \left(\frac{\Delta \lambda_b}{\lambda_{bo}} \right)^2$$
$$+ 4.608 \times 10^9 \left(\frac{\Delta \lambda_b}{\lambda_{bo}} \right)^3 - 3.643 \times 10^{11} \left(\frac{\Delta \lambda_b}{\lambda_{bo}} \right)^4 \qquad [45]$$

valid from 0 to 600°C, with a deviation of ±1.6°C. This curve includes both effects: free expansion of the optical fiber and changes in the refractive index.

At 170°C, there is a discontinuity in the slope promoted by the fast gelification of the resin, which stretches the grating following the expansion rate of the specimen. During cooling, the wavelength decreases corresponding to this last slope, but the grating splits in two at 70°C (the discrimination of the two different peaks depends on the resolution of the spectral analyzer used) due to the transverse thermal strain in the beam.

To prove the validity of the results, a simple calculation can be done: the Bragg wavelength of the grating is given by the well-known Bragg condition:

$$\lambda_b = 2 \overline{n_e} \Lambda_o$$

The wavelength shift of a grating bonded to an aluminium specimen with a linear expansion coefficient α_{Al} is given by:

$$\delta \lambda_b = \delta \lambda_{bT} + \lambda_{bo} \left(\alpha_{Al} - \alpha_f \right) \delta T \qquad [47]$$

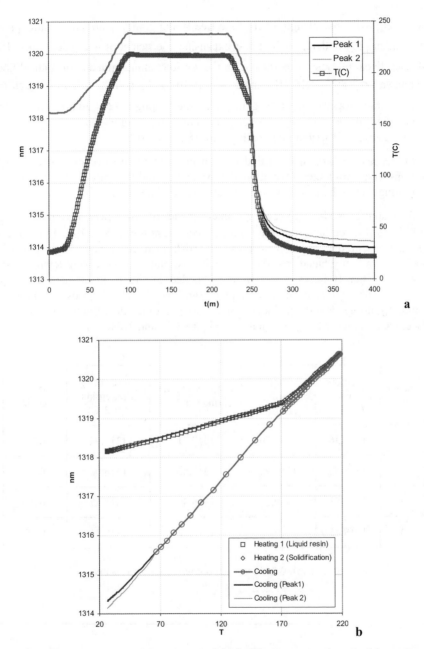

Figure 3.23. a) *Evolution of the temperature and of the Bragg wavelength of the grating during the curing cycle. The spectrum of the grating split in two during the cooling of the specimen;* **b)** *Evolution of the Bragg wavelength with temperature during the curing cycle*

where $\delta\lambda_{bT}$ is the wavelength shift of the free Bragg grating with temperature, given by equation [45], and α_f is the linear expansion coefficient of the optical fiber. The values of linear expansion coefficient for the AA5083 aluminium and the optical fiber are known ($\alpha_{Al} = 23.5\times10^{-6}\,K^{-1}$ and $\alpha_f = 5.5\times10^{-7}\,K^{-1}$), and the value estimated for the final Bragg wavelength of the grating after the cooling is 1314.20 nm, whereas the empirical result obtained by extending the curve representing the cooling step in Figure 3.23b) is 1314.19 nm for cooling from 180°C to 20°C.

In this case, it is difficult to find an immediate quantitative relationship between the transverse shrinkage of the aluminium specimen and the shift of the spectrum of the grating due to the complex host/adhesive/sensor interaction.

A graphite–epoxy laminate specimen made of 24 plies of unidirectional T300/F155 prepreg tape with the following configuration: $[[90]_8[0]_4]_s$, $150 \times 125 \times 8$ mm, was cured in a hot-plate press at 180°C in a quasi-isothermal cycle (the specimen was introduced into the press with the plates at the curing temperature). Four bare fiber Bragg gratings were embedded in the laminate parallel to the reinforcing fibers, two in the middle plane (gratings 1 and 2) and two in the outer laminae (gratings 3 and 4). Spectra of the four gratings at different stages in the process and numerical data are shown in Figure 3.24 and Table 3.1.

Grating no.	λ_b before curing (24°C)	λ_b after curing (180°C)	λ_3 (peak 1) after curing (21.2°C)	λ_2 (peak 2) after curing (21.2°C)	$\delta\lambda = \delta\lambda_2\,\delta\lambda_3$
1	1318.36	1319.70	1317.71	1317.94	0.23
2	1318.50	1319.92	1317.83	1318.08	0.25
3	1318.53	1319.96	1318.30	1318.57	0.27
4	1318.45	1319.78	1318.20	1318.36	0.16

Table 3.1. *Numerical data from the experiment*

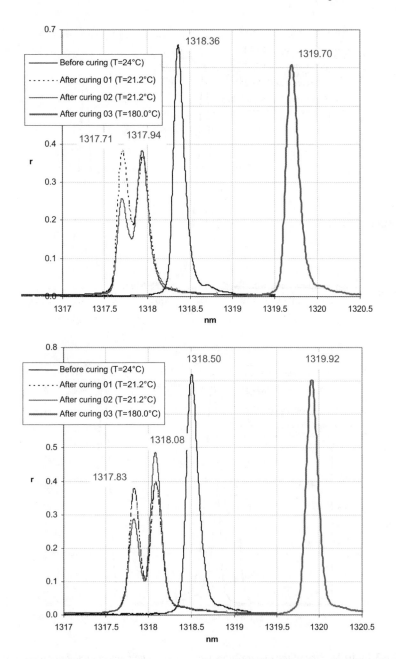

Figure 3.24. a) *Spectra of the two gratings embedded in the middle plane at different stages of the process: grating at 24°C before curing process, grating at 180°C after curing process, and grating at 21.2°C after curing process*

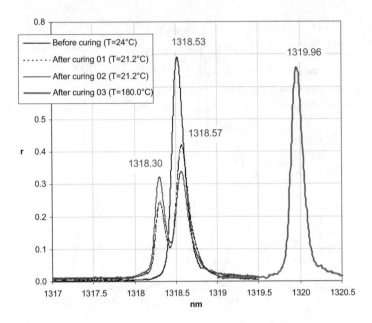

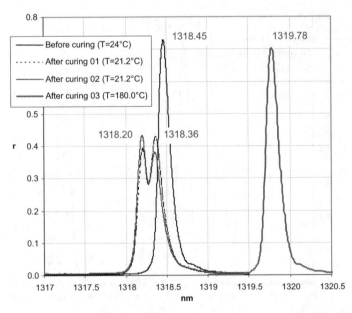

Figure 3.24. b) *Spectra of the two gratings embedded in the outer laminae at different stages of the process: grating at 24°C before curing process, grating at 180°C after curing process, and grating at 21.2°C after curing process*

To use these results to calculate the residual strains and stresses generated during the curing process, it is necessary to compensate for the effect of the temperature in the optical properties of the optical sensors. The wavelength shift of a grating due to the effect of the temperature on its refractive index is:

$$\delta\lambda_{bT} = \lambda_{bo}\varsigma_f\delta T \qquad\qquad [48]$$

where ς_f is the thermo-optic coefficient of the fiber. This equation can be also written:

$$\delta\lambda_{bT} = \delta\lambda_{bTf} - \lambda_{bo}\alpha_f\delta T \qquad\qquad [49]$$

where $\delta\lambda_{bTf}$ is the wavelength shift of the grating given by the wavelength–temperature curve [45] for the temperature increment considered. As a first result, this value, 1.27 nm for a temperature increment from 24°C to 180°C is roughly the same as the shift of the four gratings during heating, as might be expected after a quasi-isothermal curing process.

With these data, and using the equations deduced from the proposed photoelastic model, the values of strains and stresses can be calculated (see Table 3.2 where an f has been added as a subscript to those results that are the strains and stresses in the core of the grating). Stresses and longitudinal strain in the laminate and the fiber have to be roughly the same, but the difference of one order of magnitude in the stiffness of the sensor/host pair implies important differences between the transverse strains given by the lamination theory and those calculated with the grating (the transverse strain obtained from lamination theory is an averaged value and there are large differences in the resin and fiber values at the micromechanical level). Data obtained by applying the lamination theory are given in Table 3.3.

Grating n°	σ_{1f} (MPa)	σ_{2f} (MPa)	ε_{1f} ($\mu\varepsilon$)	ε_{2f} ($\mu\varepsilon$)
1	−15	71	−385	1057
2	−19	78	−453	1155
3	13	84	−20	1167
4	24	50	−87	705

Table 3.2. *Residual strains and stresses calculated using the photoelastic model*

Laminae	σ_1 (MPa)	σ_2 (MPa)	ε_1 ($\mu\varepsilon$)	ε_2 ($\mu\varepsilon$)
Middle plane	−58	30	−530	4410
Outer laminae	−15	29	−190	4070

Table 3.3. *Residual strains and stresses calculated using the lamination theory*

3.4. Structures with embedded fiber Bragg gratings

The possibilities of structural integration do not only depend on the sensor characteristics. The structural material, which from now on will be described as the host, also has to fulfill a set of requirements to allow the integration of fiber Bragg gratings, without disturbing the manufacturing process. The host has to be immune to the presence of the optical fiber, so that neither its mechanical properties nor its structural integrity are modified.

Although the fiber optic, and therefore the sensors, can be bonded externally to any material on any structure, the concept of Smart Structures inherently implies the concept of integration. Some of the advantages associated with embedded fiber-optic sensors are:

– increased survival possibilities of the optical fiber and integrated sensors, which are intrinsically brittle, but are protected by the external structure;

– the existence of an intimate sensor/host union;

– the absence of external wiring, important not only from the aesthetic point of view, but also allowing the use of this method for applications that require clean surfaces, as is the case for aeronautical structures;

– the possibility of internal monitoring the structure, in places that cannot be reached by conventional sensors, or where conventional sensors would be too intrusive because of their size and the large number of wires that are usually necessary to connect to them.

Full integration of sensors into the host material is limited by the nature of the material and by the manufacturing processes. For example, the maximum temperature for silica fiber (without external shell) is about 800°C, limiting its possibility for embedding in metals and metal matrix composites. Concrete structures, and advanced composite materials with a polymeric matrix, are more amenable to integration purposes.

The manufacturing process of structural elements also limits the use of embedded fiber optics, because it is necessary to avoid potentially aggressive operations, such as high-pressure applications, shear stresses and very small chamfers. Another critical factor is the survival of fiber terminals needed to connect structural elements to each other, for the output monitoring systems.

To use fiber Bragg gratings as strain sensors it is necessary to have the maximum information possible about the surroundings in order to establish basic criteria that make it possible to guarantee the accuracy, repeatability and stability of the measurements.

The main purpose of the coating is mechanical and environmental protection. Commonly used materials for coating optical fiber are polymers, acrylate or polyimide. The characteristics of the coating are also very important because the transmission of the stress and strain from the structural element to the fiber is achieved through it, and its characteristics will determine the response of the sensor/structural element. Standard optical fiber has an acrylate coating, with a low elastic modulus, low intrinsic strength, oxidizing at temperatures used for curing composite materials (around 180°C). Polyimides have greater rigidity and intrinsic strength, and do not degrade at these temperatures, which is why they are preferred for embedding in composite materials.

Some factors influencing the response are:

– the coating thickness, which has to be sufficient to isolate the sensor from cross-sectional stresses, but not excessive, since it could be intrusive and could delay the transmission of longitudinal deformations [SIR 90, MAD 93, TAN 99];

– its tolerance to the adhesive curing process or the matrix in which the fiber is embedded, decreasing its degradation as a result of high pressures and temperatures;

– its chemical compatibility with the host material, and degradation resistance in humid conditions;

– its rigidity, which should be as close as possible to that of the resin matrix.

3.4.1. *Orientation of the optical fiber optic with respect to the reinforcement fibers*

One of the main factors in the fiber/host behavior is, in the case of composite materials with long fiber reinforcement, the optical fiber orientation relative to this reinforcement. Many articles have analyzed the different behavior of a laminate with embedded fibers depending on this orientation [CAR 93, DAV 91]. The single mode fiber used in Bragg grating sensors has a diameter of 125 µm, approximately the same as the thickness of the lamina of a composite material. In the surroundings of the optical fiber, a resin-rich zone appears, called a resin pocket (Figure 3.25). In weave laminates, the determination of the resin pocket geometry is very difficult, because of the complexity of the phenomena involved by its formation. In the case of tape laminates, the resin pocket size is reduced when the optical fiber/reinforcement fiber angle is reduced, and disappears when the fiber is embedded longitudinally to them.

The resin pocket geometry and size depend on the orientation of the reinforcement fibers in sheets adjacent to the optical fiber. The best configuration is to have the optical fiber embedded parallel to the reinforcing fibers. In this way, the resin pocket size is zero, and the optical fiber interference with the laminate is practically zero. It will be necessary to set specific points at which the optical fiber/reinforcement angle is not null and identify these as possible sensitive zones.

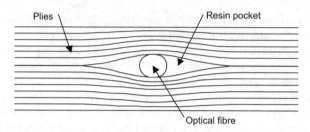

Figure 3.25. *Sketch of the laminate cross-section, showing the "resin eye"*

It is evident that the resin pocket, as well as creating a laminate zone without any reinforcement material, causes a distortion of the stress field that may initiate a delamination process.

The existence of a free end in a laminate also causes a resin accumulation and a local discontinuity in the stress field of the laminate. However, the small fiber size will not produce a very significant effect in the material, and it will only be necessary to avoid locating the sensor near the fiber end, because of the possible complexity in the stress field (Figure 3.26). The failure of the fiber/host union is very likely in this case and, in any case, the output from the sensor in these conditions is not what is required, which is why this configuration must be avoided. Leblanc *et al.* deal with the details of this problem, and use Bragg gratings as a tool to analyze it [LEB 93, LEB 95].

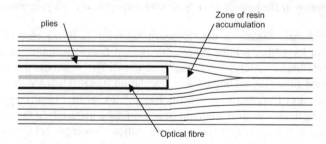

Figure 3.26. *Cross-sectional view of the free end of the optical fiber*

In laminates made with cloths, the fiber weft added to the pressure applied during the curing process and the differential contraction that appears during cooling may cause a fiber breakage or, if the fiber has survived the application of pressure, the build-up of a complex stress field and a residual strain around the Bragg grating, making analysis of its response very difficult. After many tests on the integration of Bragg gratings in materials, we concluded that the best way to use Bragg gratings in this type of laminate is surface bonding or embedding them between unidirectional tapes that protect the optical fiber.

3.4.2. *Ingress/egress from the laminate*

The input/output ends of the fiber are considered critical points for mechanical interference of the optical fiber with the structural host element. Standard procedures for input/output of the optical fiber from composite materials still do not exist. This is because of the difficulty of edge machining composite material parts with optical fiber terminals or connectors leaving the edges of the laminates. The developments carried out during the last decade have been aimed at two possibilities:

– Use of composite material manufacturing techniques that avoid the need for later edge trimming (RTM, SMC, RIM, etc.) and which would allow semi-embedded connectors on any surface of a laminate. Their effect on the mechanical behavior of the part, and therefore the precautions to take from a point of view of fiber survival, will be similar to those for conventional rigid inserts.

– Output by means of pigtails through the laminate flat surface, which obviously do not required any later manipulation. The effect on the laminate characteristics will depend on several factors: laminate configuration in the surroundings of the output, size and rigidity of the cable, etc. The existence of a low module inclusion in a contour piece zone is not an optimal solution for applications with high structural requirements, mainly in aeronautical applications, because the technique would have to be restricted to lightly loaded elements or laboratory applications. In the case of outputs by rigid inserts, the precautions to be taken will be adapted to these types of elements.

3.5. Fiber Bragg gratings as damage sensors for composites

The multiple advantages of fiber Bragg gratings over conventional electrical gauges as strain sensors allow a progressive introduction of "sense by light" technologies into load monitoring for aerospace structures.

However, load monitoring is only part of the problem. Major efforts have been made in recent years to go a step further: damage detection. Many technologies have been investigated in metallic and composite airframes to prove their applicability and reliability in this context: acousto-ultrasonics, acoustic emission, Lamb waves, cut wire sensors, eddy-current foil sensors, comparative vacuum monitoring, etc.

3.5.1. *Measurement of strain and stress variations*

Fiber Bragg gratings are passive, point-strain sensors, and this is a limitation on their use in damage sensing applications. The most straightforward technique would be an array of FBGs to monitor the strain distribution in a known load situation. A small crack would only change the strain field in its surrounding area, so the system would be ineffective. A large amount of structural damage is needed to modify the internal load distribution, and consequently to change the strain profile.

A further step could to locate sensors in very close but different load paths. In this case, the requirement for a known load status could be avoided, as local damage in one of the load paths would promote a sensible differential response. A typical application would be a monolithic structure composed of a thin composite skin with co-cured or bonded stringers. In the case of a stringer debonding, the change in the stress and strain distributions would be significant.

Fiber Bragg gratings have an advantage over conventional sensors that is due to their unique characteristics: low profile, which even allows them to be embedded into the structure, and the high capacity for multiplexing. The applicability of this technique is clearly determined by the configuration of the structure, and by its probable mode of failure. A good knowledge of strain and stress distributions and possible critical points is necessary, or even a dense array of sensors could be insufficient.

Figure 3.27 shows a schematic of a compression test panel with 16 fiber Bragg gratings bonded in different positions. Optical sensors can be bonded, for example, at the head of the stringers' web, a place not accessible for conventional sensors.

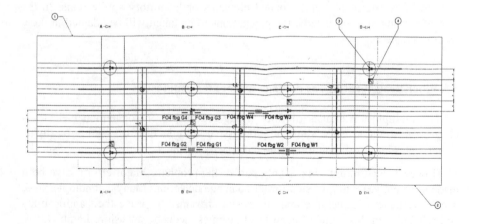

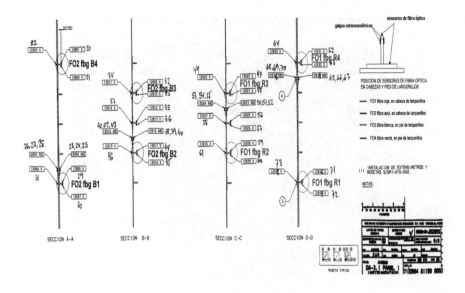

Figure 3.27. *Schematic of a compression test panel with 16 FBGs*

The results of the test, summarized in Figure 3.28, show the good correspondence between strain data given by conventional strain gauges and that from fiber-optic sensors up to the initiation of buckling.

However, the most interesting results of this kind of test are not reflected in these graphics. A careful analysis of the data acquired by the FBGs produces interesting results that make it possible to foresee their use as damage sensors.

Figure 3.29 shows premature buckling of the webs of two stiffeners of a panel, still at a low level of strain. The decrease in the load absorbed by these stringers is compensated by the increase in the load, and subsequent compressive strain, in other stingers. Figure 3.30 shows the progressive failure of the stringers, with different severity.

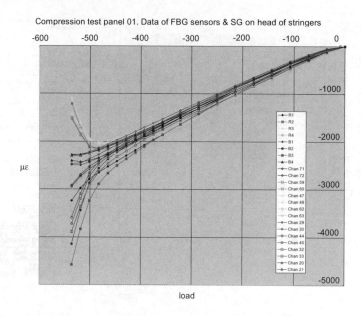

Figure 3.28. *Comparison of results obtained by conventional resistive gauges and FBGs. Differences are due to the position of the FBG, at the head of the stringer*

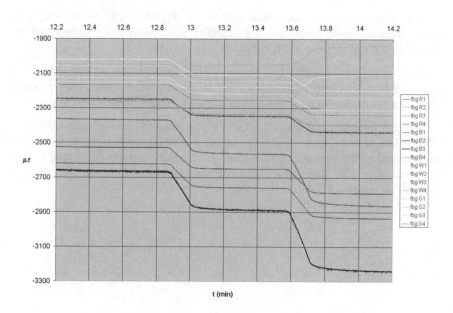

Figure 3.29. *Buckling of the webs of two stiffeners of the panel*

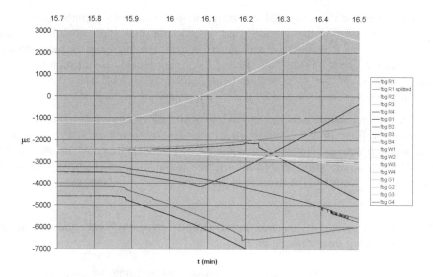

Figure 3.30. *Progressive failure of the stringers*

The behavior shown by the FBG R1 in Figure 3.30 (magnified in Figure 3.31) is very significant. The signal of this grating splits into two as a result of spectral distortion, revealing the complex character of these sensors. It is very difficult to identify whether this distortion is promoted by a non-uniform longitudinal strain or by a spectral splitting due to transverse stresses.

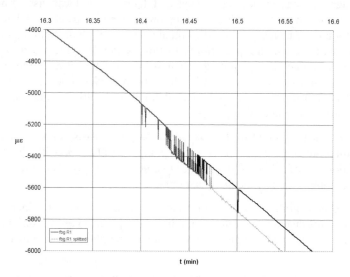

Figure 3.31. *Splitting of the signal of an FBG, promoted by a spectral distortion*

This phenomenon, observed by many authors, results in perturbations in the spectral response of the grating, and affects not only the reliability of the measurement, but even the technique and the device used to make the measurement (conventional "peak locators" are useless for giving reliable information about the load condition of the specimen, it being necessary to obtain the full spectral response of the grating and to have a clear understanding of the problem).

These effects add a new issue to the difficult task of demodulating the spectral response of fiber Bragg gratings when they are employed as strain sensors in composite structures. However, the photoelastic model presented in section 3.3 allows us to obtain considerable information about these phenomena, and relate them to local changes in the stress field promoted by structural damage.

3.5.2. Measurement of spectral perturbations associated with internal stress release resulting from damage spread

Internal residual stresses cause undesirable peak splitting, which in turn causes reading difficulties as explained in the previous section. However, this phenomenon holds an important amount of information that can be exploited: structural damage can release part of the residual stresses, and the effect can be seen as a distortion in the spectrum of the grating, opening up a new line of research in the use of fiber-optic sensors as embedded damage sensors in composite structures.

This is still an open research field, and important efforts are being made to analyze the behavior of fiber Bragg gratings when submitted to these phenomena (fiber Bragg gratings embedded in tape and fabric, near local discontinuities as holes, bonded and bolted joints, bonding lines in repair patches, delaminations, impact promoted damage, etc.), and to obtain the actual strain field applied to the grating by demodulating its spectral information. A simple test is presented to show these effects: 4 fiber Bragg gratings 10 mm long written in a 25 mm long bare portion of a photosensitive optical fiber have been used in this test. A process of annealing at 250°C has been applied to them in order to stabilize their thermal behavior during the embedding process.

The specimen proposed for the test was a 44-ply quasi-isotropic laminate made with unidirectional graphite epoxy prepreg tape T300/F593, with the following configuration: $[[+45/-45/0/90/0/-45/+45/90/0/90/+45/-45]_s]_2$, and cured following the cycle recommended by the manufacturer.

Fiber Bragg gratings were embedded without coating, to increase their sensitivity to residual stresses, in the middle of the laminate ($\pm45°$ configuration of the two central laminae facilitate the embedding of the fibers), as is shown in Figure 3.32.

After the curing process, the fiber Bragg gratings show the expected peak splitting and a drift of the peak wavelength of both peaks, as a result of the differential expansion coefficient of the laminate and the optical fiber. Residual stresses and strain fields in the middle plane of the laminate can be estimated using lamination theory and, for a quasi-isotropic laminate, residual strain and stresses are roughly independent of direction: a residual stress of −16.35 MPa, and a residual strain of −235 $\mu\varepsilon$ (both for a thermal load of 150°C).

Using these data and equation [43], a peak splitting of about 500 pm can be expected. Real peak splitting observed in the embedded gratings ranges from 200 to 500 pm. This result shows, on the one hand, considerable unexpected irregularities in the internal stress state of the laminate, but, on the other, not too bad an approximation to the theoretical estimated results, bearing in mind the simplifications considered in the solution of the photoelastic problem.

Five impacts of 20 joules with a projectile having a semi-spherical head of 10 mm diameter were applied to the panel (approximate positions are shown in Figure 3.32). Indentation produced is in the range of barely visible impact damage. Slight delaminations can be observed with ultrasonic testing, but the energy of the impact is not enough to produce visible delaminations on the back of the panel. There was no effect over the spectral response of the gratings.

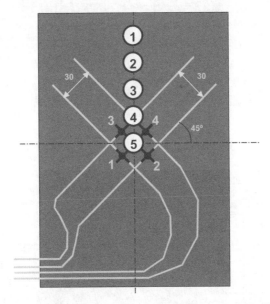

Figure 3.32. *Schematic of the panel and relative position of the gratings (fiber optics are light in color, and gratings are marked with an X). The schematic also shows the position of the impacts received by the panel*

An impact of 40 joules (impact 6) is applied at the same point as impact 5, just in the middle of the four gratings. Superficial damage is only barely visible, and a clear delamination is detected using ultrasonic testing, but no visible delaminations in the back side of the panel can be detected. In this case, only two of the gratings show visible effects in their spectral behavior.

A final impact of 80 joules (impact 7) is applied at the same point as impacts 5 and 6. Superficial damage is clearly visible, and a big internal delamination is detected using ultrasonics, but no visible delaminations on the back of the panel can be detected.

The spectral response of the gratings submitted to a superimposed stress field promoted by a hole drilled in the same position as the last impacts can be used as an easy way to compare the seriousness of the damage caused by the impacts. Figures 3.33 to 3.36 show the intensity spectrum of fiber Bragg gratings after 20, 40 and 80 joule impacts, and after the drilling of a 15, 20 and 25 mm diameter hole.

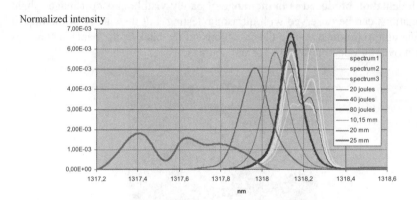

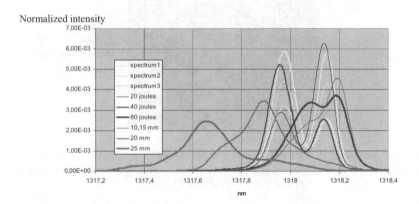

Figures 3.33. and 3.34. *Intensity spectrum of FBG #1 and #2, before and after impact, and after drilling holes of different diameters*

Normalized intensity

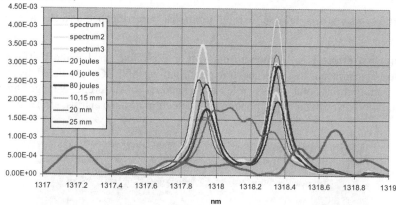

Normalized intensity

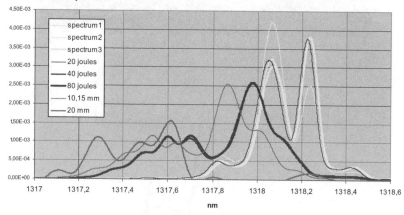

Figures 3.35 and 3.36. *Intensity spectrum of FBG #3 and #4, before and after impacts, and after drilling holes of different diameters. Note the poor reproducibility of this technique for damage assessment*

The direct application of this effect is complex: it is impossible to predict where impact damage will occur. However, similar effects can be observed in FBGs embedded near artificial defects. Damage progression promotes spectral distortion that can warn about debonding in co-cured/co-bonded joints, or repair patches, critical points where the low range of this technique is not a drawback (Figure 3.37).

The experience summarized above makes it possible to establish the basis for SHM based on a fiber Bragg grating and the following phenomena:

– global strain perturbations in a loaded structure;

– perturbations in local strain fields, which promote spectral distortions close to FBGs;

– transverse stress release, which promotes birefringence effects in nearby FBGs.

The first effect is well-known and widely exploited, so that most efforts are directed at characterizing the spectral perturbations of FBGs promoted by strain field perturbations and stress release (response to different damage events, repeatability, discrimination of phenomena, detection of false alarms, etc.).

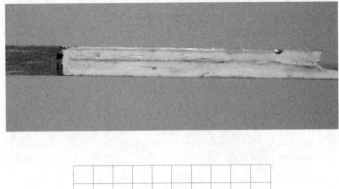

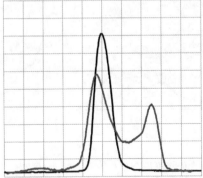

Figure 3.37. *Damage progression in a bonded joint with an artificial defect, and spectral distortion of an embedded FBG*

3.6. Examples of applications in aeronautics and civil engineering

Although the technique is still not incorporated in on-going production, nor accepted for certification procedures, a large number of demonstrations can be found in the open literature. The practical details of the implementation of two structural elements instrumented with fiber Bragg gratings are described next, illustrating the special precautions that need to be taken, and the results obtained.

3.6.1. *Stiffened panels with embedded fiber Bragg gratings*

Three stiffened panels with co-cured stiffener webs were manufactured by means of manual lay-up and autoclave curing. This structure could simulate a portion of the skin panel of the tailplane or the wing of a small aircraft. Two of the panels (panels 1 and 2) were instrumented with optical fiber Bragg grating sensors bonded onto the stiffeners and back-to-back on the skin surface. The other panel (panel 3) was provided with two embedded optical fiber sensors in each stiffener. In additional, FBGs and strain gauges were bonded on the stiffener webs and on the skin back to back.

The panels have dimensions of 560 × 280 mm with three blade stiffeners of cross section 20 × 20 mm situated along the centre line and on the left and right at 105 mm from the centre line. Skin thickness is 1 mm and stiffener thickness 1.6 mm. Fiber lay up of the panels has been optimized, being ($\pm45°_{(f)}$, $\pm45°_{(f)}$, $0°$, $\pm45°_{(f)}$, $\pm45°_{(f)}$) in the panel 3 skin and ($\pm45°_{(f)}$, $0°$, $0/90°_{(f)}$, $0°$) in the stiffener, using a combination of unidirectional and fabric (f) carbon fibers.

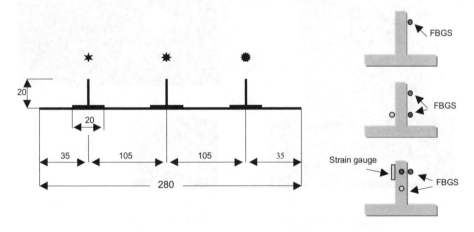

Figure 3.38. *Dimensions of the panels with details of the sensor positions in panels 1, 2 and 3*

To manufacture the stiffened panels, a toughened epoxy prepreg material HEXCEL A-193-P/8552S carbon fabric and AS4/8552 unidirectional tape with AS4 high-strength, high-strain carbon fibers were employed.

The panels were manually laid-up over a reinforced elastomeric mould made of AIRPAD polyacrylic rubber (AIRTECH). The elastomeric mould turned out to be very suitable for embedding the optical fibers in the laminate. The elastic rubber surface of the mould enables the optical fibers, covered by their protection tubes, to be guided between the laminate and the mould and protects them from high-pressure concentrations. The applied curing cycle was the standard one recommended by the prepreg manufacturer (180°C, 6 bars, 2 hours). For panel 3, a lower pressure of 4 bars was applied. Panel edges were machined with a diamond disc saw. After edge machining, panel edges were potted into an epoxy/gypsum mixture inside a normalized U-transverse section steel profile (80 mm width, 45 mm high) to ensure uniform loading (Figure 3.39).

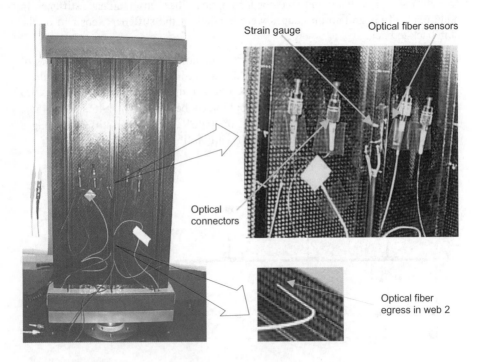

Figure 3.39. *Integration of optical fibers in the stiffener webs*

The optical fiber sensor used was a 230 mm long portion of photosensitive optical fiber with a 10 mm long Bragg grating sensor zone located at 65 mm from the end of the photosensitive fiber. For panels 1 and 2, sensors were bonded by means of a cyanate adhesive to the slightly ground and degreased surface of the panel. The photosensitive fiber was spliced with a conventional single-mode fiber provided with an optical connector.

In panel 3, a 165 mm long photosensitive fiber was spliced with an 800 mm long single-mode conventional optical fiber. Two FBGs were embedded in the web of each of the three stiffeners of the panel. The optical fibers were placed between the two central 0° layers and parallel to the reinforcement fibers. Their sensing zone was located at mid-span of the stiffener. Embedding the optical fibers between 0° prepreg layers ensured that they would not be destroyed during the consolidation process and they would provide a reliable sensor signal, which would not be spoilt by the lobed curved surface of the fabric layers. A tiny hole was made through the laminate to enable the optical fiber to exit from it. A 0.8 mm diameter PTFE tube was used to protect and guide the two optical fibers in each web out of the laminate. This tube was sealed with silicone to prevent resin contamination. A lead of about 25 mm of this PTFE tube was left inside the laminate to guarantee good protection of the fiber at the point where it came out of the laminate. The rest of this PTFE tube was guided between the laminate and the mould coming out of the mould at the front. A long portion of the protected optical fiber was left outside the laminate and a connector was fixed to it after curing.

Additional FBGs and conventional strain gauges were surface-bonded onto panel 3 to verify the strain measurements of the embedded FBGs. Conventional strain gauges were also bonded onto the skin to prevent the skin buckling. The strain gauges were bonded onto the left-hand side of each web at the same location as the upper embedded FBGs, while FBGs were bonded onto the right-hand side of each web. Additional gauges were also bonded onto both sides of the middle web in front of the lower FBGs. All the sensors survived manufacture of the part, showing clearly defined spectral peaks with no significant distortion due to complex local strain fields caused by residual stresses.

To verify the exact geometrical position of the embedded optical fibers inside the stiffener webs, several X-ray images of panel 3 were made. The images show 300 mm of the stiffener length including the fiber egress. A microfocus X-ray system was used, operating at a low voltage level to obtain sufficiently high resolution so as to make possible the detection of the small optical fiber inside the composite. Figure 3.40 shows the X-ray image of stiffener web 3. The two optical fibers can be clearly identified within the composite. On the left-hand side, the fiber egress with the protective Teflon tube can be seen. The resolution is high enough to detect not only the fiber path but also possible optical fiber cracking within the composite.

Compression tests at room temperature were conducted for the three panels using for, panels 1 and 3, a mechanical testing machine with fixed compression plates and, for panel 2, a hydraulic testing machine with bearings allowing longitudinal rotation of the upper plate and transverse rotation of the lower one.

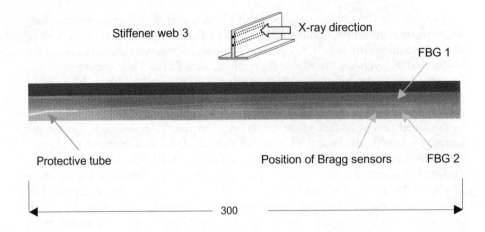

Figure 3.40. *X-ray image of stiffener web 3 in panel 3 indicating the position of the optical fibers and the protective tube at fiber ingress*

Tests were conducted up to rupture of panels 1 and 2. Panel 3 was not loaded up to rupture and several compression tests were performed. In panels 1 and 2 the sensors worked correctly up to the panels' failure. In the explosive rupture, the splice of the photosensitive and the conventional single-mode fiber broke at almost every sensor. Fixing an optical connector directly to the photosensitive fiber would leave the sensors operative even after rupture of the panel.

In the case of panel 3, the six FBGs were embedded on the centre line of the stiffener web so that they were able to measure tensile/compression strains but not transverse bending or twisting of the web. At the beginning of the compression test all the sensors detected the same compressive strain up to the moment when buckling occurred at a strain level of 800 $\mu\varepsilon$. A large bending strain in the skin can be identified from the bifurcation of the sensor signals from the bonded strain gauges. Skin buckling increases the load, which has to be carried by the stiffeners and also induces a twisting and transverse bending of the stiffeners. With increasing load, longitudinal bending of the web occurs. Figure 3.41 shows the strain measured with the upper embedded FBGs, the bonded FBGs and the bonded strain gauge of the stiffener webs 1, 2 and 3. The measurement of the skin-bonded strain gauges is also included in the graph to indicate the skin buckling.

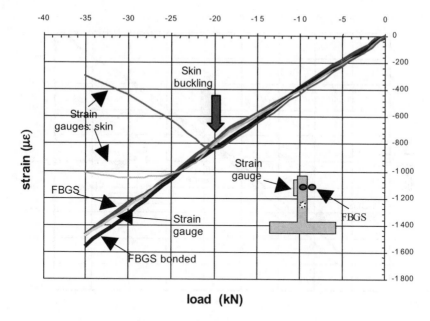

Figure 3.41. a) *Strain versus load values from the FBGs and strain gauges of stiffener web 1 and strain gauges in the skin of panel 3*

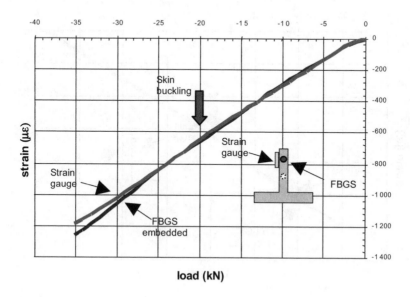

Figure 3.41. b) *Strain versus load values from the FBGs and strain gauges of stiffener web 3 in panel 3*

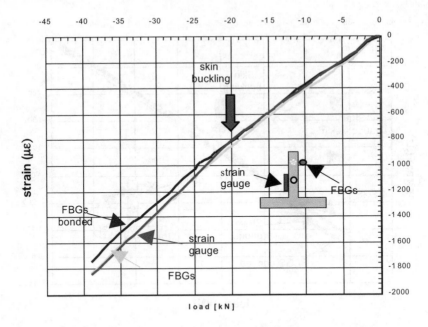

Figure 3.41. c) *Strain versus load values from the FBGs and strain gauges of stiffener web 1 in panel 3*

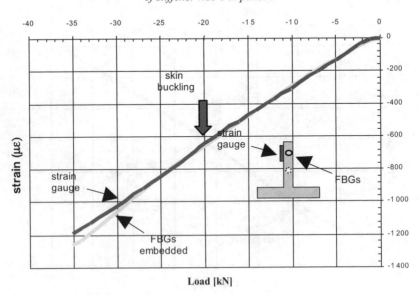

Figure 3.41. d) *Strain versus load values from the FBGs and strain gauges of stiffener web 3 in panel 3*

3.6.2. *Concrete beam repair*

Advanced composites offer some advantages over traditional procedures for repairing concrete structures, due to their optimal corrosion properties, low weight and decreasing costs. Thin cured laminates may be externally bonded or dry fabrics can be applied *in situ*.

On the other hand, Bragg gratings show the advantages discussed above when compared to conventional strain gauges: absolute measurements, spectrally encoded output, no EMI, no drift (long-term stability), small size, multiplexing capability and their ability to be embedded into laminates without degradation or detrimental effects.

The combination of the two techniques is easy and offers important advantages. The long-term mechanical behavior of the repair can be checked and information on environmental degradation obtained. In the short term, information on the stress transfer from the concrete to the laminate is obtained, and the validity of models is verified.

3.6.2.1. *Tests*

A reinforced concrete continuous beam with two spans (7.2 m + 7.2 m), representative of current in-service civil engineering structures, was loaded up to failure and repaired. After removing loose concrete near the cracks, new concrete was poured. Composite patches, made of two layers of carbon fabric, *in situ* impregnated with room-temperature epoxy resin, were extended over the crack locations, to transmit the load from the steel bars, which are now not continuous. Bragg gratings were placed with the fabrics, both in the internal and external surfaces. External pre-stressing was added to increase the failure load. Arrangement is shown in Figure 3.42.

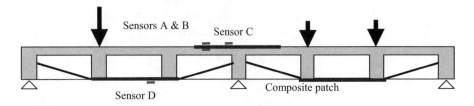

Figure 3.42. *Test arrangement and Bragg grating locations*

3.6.2.2. *Results*

Strain versus load is plotted in Figure 3.43, obtained with strain gauges and fiber-optic sensors. Agreement is very satisfactory, even in the non-linear region of the concrete beam. Damage starts at around 80 kN, still with considerable residual strength. It is worth mentioning that the ultimate load on the repaired beam (156.6 kN) was higher than that for undamaged beam (100 kN).

Results for Bragg grating sensors A and B are shown in Figure 3.44. They were at the same position, one on each side of the composite patch. A shear lag due to the finite thickness of the adhesive layer can be seen. The strain measured by sensor A, located in the interface, is higher than that measured by sensor B, at the same location but on top of the composite patch. This behavior should be expected, because of the adhesive joint, but had never been previously experimentally demonstrated.

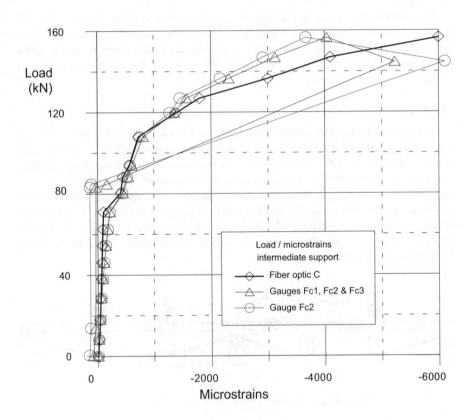

Figure 3.43. *Bragg grating measurements*

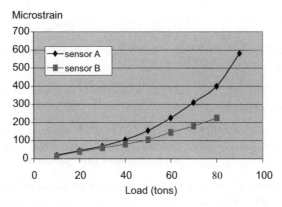

Figure 3.44. *Strain measurements from Bragg grating sensors*

3.7. Conclusions

The technology of fiber-optic sensors, and particularly of the fiber Bragg gratings, is well-matured for strain monitoring at a number of points, and can be used for load monitoring in conventional and advanced structures. It offers several important advantages over conventional strain sensing, namely:

– low size and weight, embeddable capability, single-ended cabling;

– long-term stability; it can be used for load monitoring throughout the structural life;

– inherent multiplexability, typically 10 sensors/fiber without decreasing reading speed;

– immunity to electromagnetic noise and the ability to work in harsh or explosive environments.

Current efforts are addressing the development and demonstration of large sensor arrays, affording a detailed map of the strain field in a complex structure such as a ship, an aircraft or a bridge. This will require new optoelectronics and signal processing systems, able to handle at high speed the information coming from several hundred sensing points and to reduce this information to the significant events. This is needed if damage detection is the goal, because in continuous structures local damage will produce only a very small change in the global strain field. Only by considering the dynamic response or the transient strains can information be derived about damage occurrence.

Only the strain and temperature response of FBGs has been discussed in this chapter. In addition, fiber-optic sensors can provide very valuable information on chemical processes, such as corrosion in metals, degradation phenomena in concrete structures, or concerning resin curing in composite materials. This is a very active research area and, combined with process modeling, is paving the way towards intelligent materials processing.

3.8. References

[BUT 78] BUTTER C.S., HOCKER G.B., "Fiber-optic strain gauge", *Applied Optics*, vol. 17, no. 18, pp. 2867-2869, 1978.

[BYO 03] BYOUNGHO LEE, "Review of the present status of optical fiber sensors", *Optical Fiber Technology*, 9 (2003).

[CAR 93] CARMAN G.P., PAUL C.A., SENDECKYJ G.P., "Transverse strength of composites containing optical fibers", *Smart Structures and Materials* 1993, Smart Structures and Intelligent Systems; Proceedings of the Meeting, SPIE-1917, pp. 307-316, 1993.

[DAV 91] DAVIDSON R., ROBERTS S., "Mechanical properties of composite materials containing embedded fiber-optic sensors", Fiber-optic Smart Structures and Skins IV, Proceedings of the Meeting, SPIE-1588, pp. 326-341, 1991.

[ERD 97] ERDOGAN T., "Fiber grating spectra", *Journal of Lightwave Technology*, vol. 15, no. 8, pp. 1277-1294, 1997.

[GRA 00] GRATAN K.T.V., SUN T., "Fiber optic sensor technology: an overview", *Sensors and Actuators, A: Physical*, vol. 82, no. 1, May 2000, pp. 40-61.

[GUE 04] GUEMES A., FROVEL M., PINTADO, J.M., BARAIBAR I., OLMO E., Fiber-optic sensors for hydrogen cryogenic tanks. Proceedings of the 2nd European Workshop on Structural Health Monitoring, Munich 2004.

[HOC 79] HOCKER G.B., "Fiber-optic sensing of pressure and temperature", *Applied Optics*, vol. 18, no. 9, pp. 1445-1448, 1979.

[HUA 94] HUANG S.Y., LEBLANC M., OHN M.M., MEASURES M., "Bragg intra-grating structural sensing", *Applied Optics*, vol. 34, no. 22, pp. 5003-5009, 1994.

[KER 97] KERSEY *et al.*, "Fiber grating sensors", *Journal of Lightwave Technology*, 15, 1442-1463, 1997.

[LEB 93] LEBLANC M., MEASURES R.M., "Micromechanical considerations for embedded single ended sensors", Smart Sensing, Processing and Instrumentation; Proceedings of the Meeting, SPIE-1918, pp. 215-217, 1993.

[LEB 95] LEBLANC M., HUANG S.Y., MEASURES R.M., "Fiber-optic Bragg intra-grating strain gradient sensing". Smart Structures and Materials 1995; Proceedings of the Meeting, SPIE-2444, pp. 136-147, 1995.

[LEV 96] LEVIN K., NILSSON S., "Examination of reliability of fiber-optic sensors embedded in carbon/epoxy composites", 3rd International Conference on Intelligent Materials and 3rd European Conference on Smart Structures and Materials; Proceedings of the Meeting, SPIE-2779, pp. 222-229, 1996.

[LOP 02] LOPEZ-HIGUERA J.M. (ed.), *Handbook of Optical Fiber Sensing Technology*, John Wiley and Sons Ltd. (Chichester, UK), 2002.

[MAD 93] MADSEN J.S., JARDINE A.P., MEILUNAS R.J., TOBIN A., PAK Y.R., "Effect of coating characteristics on strain transfer in embedded fiber-optic sensors", SPIE-1918, pp. 228-236, 1993.

[MEA 89] MEASURES R.M., "Smart structures with nerves of glass", *Progress on Aerospace Science*, vol. 26, no. 4, pp. 289-351, 1989.

[MEA 01] MEASURES R.M., *Structural Monitoring with Fiber-optic Technology*, Academic Press, Ontario, 2001.

[ROD 99] RODRIGUEZ-LENCE F., MUNOZ-ESQUER P., MENEDEZ J.M., PARDO DE VERA D.S., GÜEMES J.A., "Smart sensors for resin flow and composite cure monitoring", Proceedings of the 12th International Conference on Composite Materials, ICCM-12, Paris, France, 1999.

[SAL 95] SALEH B.E.A., TEICH M.C., *Fundamentals of Photonics*, Wiley Interscience, 1991.

[SIR 90] SIRKIS J.S., DASGUPTA A., "Optimal coating for intelligent structure fiber-optical sensors", Fiber-optic Smart Structures and Smart Skins III; Proceedings of the Meeting, SPIE-1370, pp. 129-134, 1990.

[TAN 99] TANG L., TAO X., CHOY C.-I., "Effectiveness and optimization of fiber Bragg grating sensor as embedded strain sensor", *Smart Materials and Structures*, vol. 8, no. 1, pp. 154-160, 1999.

[TOD 05] TODD M., Chapter 17 of the book: *Damage Prognosis For Aerospace, Civil and Mechanical Systems*, Wiley, 2005.

[UDD 94] UDD E., *Fiber-optic Smart Structures*, Wiley Interscience, New York, 1994.

[YAM 87] YAMADA M., SAKUDA K., "Analysis of almost-periodic distributed feedback slab waveguides via a fundamental matrix approach", *Applied Optics*, vol. 26, pp. 3474-3478, 1987.

Chapter 4

Structural Health Monitoring with Piezoelectric Sensors

4.1. Background and context

Damage detection and characterization of complex industrial structures are conventionally performed through a wide variety of classical Non Destructive Evaluation (NDE) techniques such as tap testing, visual inspection, ultrasonics, eddy currents, flux leakage, thermography, liquid penetrant, acoustic emission, X-ray inspection, and so on. Some of these techniques are carried out during periodic scheduled inspections or after a casual event such as, for example, in the aeronautical context, bird or object impact, thermal shock, violent landing, over-stress, thunder impact. Nowadays, NDE procedures take place during manufacturing, transformation or heat treatment of structural material and during its whole life cycle. Practically, to assess the integrity of a structure, most NDE methods require a reliable knowledge of the damaging process of the material prior to the detection and characterization of hidden internal defects. On this point, it is helpful to distinguish the case of metallic structures (relatively homogeneous and isotropic materials), for which the damaging modes are rather well-known, from that of composite materials, which present a huge variety of possible defects.

In the latter field of application, one can observe the growing use of composite materials over the two last decades. This was permitted because of many advantages offered by composites compared to conventional metallic materials for their use as structural materials in aeronautics, the automotive industry, ground transportation, ship-building or in civil engineering. Among these advantages, one initially quotes

---

Chapter written by Philippe GUY and Thomas MONNIER.

their great mass/stiffness ratio, which allows a consequent mass reduction of the structures, which is particularly required in the above-mentioned fields. For example, in civil air transportation, the use of reinforced plastic fiber composites (of glass, carbon or aramide fibers) allows a weight saving of up to 20% for the structural parts, keeping the structure's stiffness equal to or even greater than those made of metallic parts. This produces a weight reduction estimated as 6% of the total mass of the plane; in an A340 Airbus, a reduction of 1% in the structural mass represents an economy estimated to be equivalent to at least 600 tonnes of fuel for an average service of between 20 and 25 years [SCH 99].

Composite materials also offer good resistance to fatigue and corrosion and a good adaptability of their mechanical properties to the specific requirements during their use. However, the intensive use of this kind of material as a primary structural element, in particular in the field of the civil aviation, remains limited by the requirement of proving its reliability, given that the presence of defects can severely degrade its mechanical properties. The second limitation to the more widespread use of composite materials is their cost. For example, manufacturing composite parts made of Carbon Fiber Reinforced Plastic (CFRP) are significantly more expensive than manufacturing their equivalents in metals. In addition, the cost of inspection procedures accounts for approximately 30% of the total investment for the implementation of a composite structure [BAR 99]. Indeed, many NDE methods are already available for composite structures. Thus, visual inspection continues to be the most widespread method of damage monitoring for in-service aircraft [SCH 99], but it does not offer a real quantification of the severity of the detected defects. The vast majority of the other methods usually employed require the immobilization of the equipment in the ground workshops. This results in a trading loss, which further increases the cost of this kind of passive detection technique [MOU 99].

Thus, the development of more effective non-destructive testing techniques able to quickly evaluate the health of a piece of equipment, without the need for it to partially disassembled to give access to the internal structures, became an economic issue of paramount importance. One reason for this was the increase of the operational safety of the structures (avoiding the risk of human error during maintenance), and the other was the reduction in the effective cost of maintenance in terms of the direct hardware and manpower cost and through a reduction in the time for which the plane is out of service.

Consequently, there appeared to be a need for fully integrated SHM techniques, which assess the integrity of the structure more easily and more rapidly. The future *in-situ* and continuous monitoring systems should be small in size in order to minimize the risk of their initiating damage. They must also have good sensitivity and robustness, at least as good as those of conventional techniques. This approach is part of the more general concept called "Smart Materials and Structures". Actually, if sensors are embedded in a composite structure before curing, in order to monitor and optimize its processing parameters, they are able, since they remain within the structure, to carry out the health monitoring function during the second stage of the lifetime of the structure. Furthermore, in slightly damaged systems, it becomes reasonable to use the data collected for on-line predictive estimation of the residual lifetime of the structure (prognosis). The final goal of SHM can thus be regarded as locating the tested structure at a point on its lifetime scale. In this context, some authors [GIU 03a] have recently underlined the increasing need for reliable on-line devices for SHM.

Traditional ultrasonic non-destructive evaluation techniques suffer from problems such as the reproducibility of the acoustic coupling, accessibility to the structure and poor signal-to-noise ratio in highly attenuating materials. The use of embedded or bonded piezoelectric sensors overcomes some of these difficulties. As they remain permanently attached to the structure, these sensors can be used to monitor the state of health of composites, from their curing process to the end of their life cycle. Up to now, the vast majority of work dealing with acousto-ultrasonics has used piezoelectric transducers.

The recent literature in the field of SHM with piezoelectric elements is constant expanding and reflects the huge interest of both industrial and scientific communities. Giurgiutiu describes quite comprehensively the various techniques of NDE that can be implemented with piezoelectric inserts, which he calls PWAS (Piezoelectric Wafers Active Sensors) [GIU 03a]. Even though such a technique may seem very simple, it can actually be implemented in many different ways, each one giving different information about the state of health of the structure.

In this chapter, we shall first present a review of the most significant sensing techniques based on piezoelectric transducers. We will group these in three classes, insofar as their behavior is passive, active or mixed. These main classes are acoustic emission, acousto-ultrasonics using piezoelectric transducers and electromechanical impedance. We shall outline the various kinds of sensor, the signal and data reduction methods, and the inverse techniques used to identify the influence of homogenous or localized damage on the physical parameters for each method. Then, we shall consider the particular problem of the life-long health monitoring of composite materials using piezoelectric implants, because the work undertaken by the Group of Metallurgy and Materials Science (GEMPPM) of INSA de Lyon allows us to illustrate the three aforementioned possible approaches to SHM with piezoelectric sensors.

4.2. The use of embedded sensors as acoustic emission (AE) detectors

This section aims to describe the use of acoustic emission technique to contribute to the general problem of SHM and, more generally, to the prediction of the remaining lifetime of industrial materials and structures.

Acoustic emission is primarily used to study the physical parameters and the damage mechanisms of a material, but it is also used as an on-line non-destructive testing technique. The AE phenomenon is based on the release of energy in the form of transitory elastic waves within a material having dynamic deformation processes. The waves, of various types and frequencies, propagate in the material and undergo possible modifications before reaching the surface of the studied sample. The surface vibration is collected by a piezoelectric sensor. Amplified, it provides the so-called acoustic emission signal.

A typical source of an AE wave within a material is the appearance of a crack from a defect when the material is put under constraint, or when a pre-existing crack grows, which causes a transitory mechanical wave to emerge from the latter. Thus, this technique makes it possible to detect in real time the existence of evolutionary defects, whereas passive defects are not detected [HUG 02].

It is helpful to distinguish continuous acoustic emission from discrete AE, which involves acoustic bursts. For the latter, the AE burst has the shape of a damped sinusoid. When AE bursts are so frequent that they overlap, the AE signal results in an apparent increase of the background noise. This is what is called continuous acoustic emission. It is mainly observed during plastic deformation in metallic materials. In a composite material, continuous AE is generally not observed.

Modal acoustic emission is largely used for composite structures. Nevertheless, some reports which are concerned with metallic structures can be found. For instance, in [FIN 00], Finlayson et al. reported the application of an Acoustic Emission Usage Monitoring System for an on-the-ground fatigue test on an F-15 fighter aircraft. This study focused on several structures within the aircraft (see Figure 4.1), such as the connecting lugs between the wings and the main fuselage. These critical structural parts are machined from 2124 aluminum alloy, 7075-aluminium alloy and 6Al-4V-titanium forging. This reference demonstrates the ability of embedded AE sensors to detect fatigue cracks that are conventionally difficult to locate in sub-structures that have previously been very difficult or impossible to inspect.

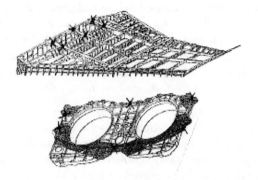

Figure 4.1. *Typical sensor locations on a bulkhead and wing of an F-15 fighter aircraft*

The real challenge of the application of AE in aviation industry is the in-flight monitoring of aircraft structure. Much effort has been spent on this respect, although little success has been achieved so far. The main challenge still comes from the background noise, such as vibration-induced noise, airflow noise, electromagnetic interference and transient noise. The fretting of fasteners and the rubbing noise between bolt and bolt hole are other sources of loud noise. Studies have shown that the rubbing noise between bolt and bolt hole has almost the same characteristics as fatigue-related AE, which makes it difficult to differentiate the two [GEN 04]. It appears that the acoustic emission method could be used on an aircraft in flight, since a difference of at least 30 dB is found between the impact signal and flight background noise. Several flight tests of acoustic emission systems have been achieved, demonstrating the feasibility of classical AE even in the very noisy environment of rocket motors [HUA 97], [FIN 00] and of modal AE on a fighter aircraft [DIT 03]. In reference [FIN 00], in-flight AE has been successfully demonstrated on board the DC-XA Delta Clipper Technology Demonstrator (see Figure 4.2). The system was a commercially available unit that was modified for autonomous control and was designated AEFIS, which stands for Acoustic Emission Flight Instrumentation System. The system is used to monitor and feed back information to the on-board vehicle computers about the condition of the spaceship, keeping in mind that the main threat in space travel is micrometeorite impacts. This becomes especially relevant for structures, such as fuel tanks and fuel systems for which composites are becoming mainstream. The first in-flight testing was prototyped to feed back information about the composite tank structure, as shown by AE sensor locations in Figure 4.2, and aspects of the operating environment within a rocket, such as temperature limits, vibration, and background noise. Once the AE system has detected and located the occurrence of an impact, the system performs an Acousto-Ultrasonic test across the impact in order to rate the damage severity. This is achieved by actively pulsing the AE sensors and capturing the received digitized waveform for comparison with waveforms captured during calibration on the ground. The results of this test are then fed into an algorithm that gives a go/no-go command for re-entry (if the damage occurs on the ceramic heat shield, thus saving the structure from breakup caused by a burn-through condition says Finlayson). One can only regret that such a system was not operational when Space Shuttle Columbia STS-107 was launched in 2003, because it experienced impact damage on the heat shield, which presumably caused its destruction and the death of its crew.

In [DIT 03], a comparison of impact signals registered during ground tests with the noise level registered during the aircraft flight shows that a difference of 30 to 50 dB exists, which ensures a satisfactory use of the localization algorithms. In this experiment, the signals were issued from four piezoelectric patches (10 mm in diameter and 1 mm thick) surface-mounted on the CFRP main landing gear door and registered on a flight test tape recorder. The bandwidth of the signal was 30–200 kHz.

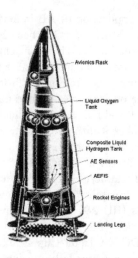

Figure 4.2. *Location of AEFIS and AE sensors*

The other difficulty in developing an in-flight monitoring system lies in the limitation of space available for mounting the AE facility. The requirement for the facility to adapt itself to changing environmental conditions (temperature, humidity, vibration and shock) is another tricky challenge.

Again, for the detection and localization of damage or/and damaging events (impacts in particular), AE is mainly used in composite structures. For impact detection in metallic structures, it is useful to refer to the work done in NASA Langley Research Center by Prosser *et al.* [PRO 99]. Piezoelectric sensors were mainly used, although there were some attempts to use optical fiber sensors. For short- and mid-term developments of systems, people prefer to use commercial off-the-shelf AE sensors [HUA 97, SAN 01, SAN 02]. In [PRO 99], heavily damped, 3.5 MHz ultrasonic transducers for thickness gauging (Panametrics model V182) are used to detect the signals produced by low-velocity impacts on aluminum plates. Operated far away from resonance, in the low-frequency AE range (less than 1 MHz), these sensors provide a flat frequency response, with high-fidelity displacement. This was previously demonstrated by the same author [PRO 91] by comparison of the response with that of a calibrated laser interferometer. As the large aperture (1.27 cm) of this sensor leads to problems with phase cancellation across the sensor face at higher frequencies, a smaller diameter sensor (Digital Wave Corp., model B1025) was used for studying high-velocity impacts.

The current tendency is to use piezoelectric patches specially designed for modal AE, which offer more versatility since they allow networks of sensors to be built that can be used with AE, Lamb wave propagation and electromechanical impedance techniques. For optimizing the number and position of piezoelectric sensors, a genetic algorithm has been used at Sheffield University [STA 01, COV 03, HAY 05].

4.2.1. *Experimental results and conventional analysis of acoustic emission signals*

Acoustic emission has been developed over the last two decades as a non-destructive evaluation technique and as a useful tool for materials research. AE is an efficient method for monitoring, in real time, damage growth in both structural components and laboratory specimens [GOD 98, JAC 99]. In loaded fiber reinforced composite materials, the strain energy release due to microstructural changes results in stress-wave propagation. In the case of composite materials, many envisaged source mechanisms have been confirmed as AE sources [BAR 94, ELY 95], including matrix cracking, fiber–matrix interface debonding, fiber fracture and delamination. However, many other physical phenomena can be at the origin of AE events and are referred to in [BEA 83, EIT 84].

In the work of Huguet *et al.* [HUG 00], AE is used to discriminate in real time the different types of damage occurring in a fiber-reinforced composite under mechanical load. These authors worked on polyester and glass/polyester unidirectional specimens, subjected to tensile loading within different configurations, causing preferential damage modes in the material. Three types of sample were used in this study: pure resin, 45° and 90° off-axis unidirectional samples. Each of these samples was chosen to preferentially generate a particular damage mode during the tests: matrix fracture for resin samples; mainly matrix fracture with some decohesion for 90° off-axis; and mainly decohesion with some matrix fracture for 45° off-axis. Acoustic emission measurements were achieved by the use of two resonant micro80 PAC sensors (Figure 4.3), applied to the faces of the samples during testing. The composite used was made of E-type Vetrotex glass fibers and a Scott Bader polyester matrix with Peroximon K1 hardener. Polymerization was achieved at room temperature within approximately 12 hours. Unidirectional laminates were made of 12 prepregs and were about 2.5 mm thick with the fibers being 75% in weight.

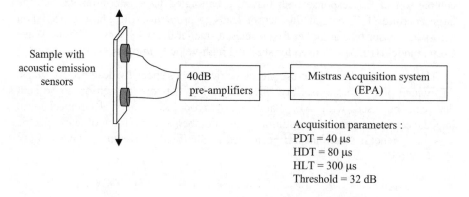

Figure 4.3. *Tensile test arrangement with acoustic emission*

Tensile tests on pure resin samples generated little AE activity. All detected AE waves can be considered as burst type signals, which imply that the AE waves were generated during dynamic and discontinuous micro-fracture processes inside the sample. Signals appearing during the main part of the tests were associated with the nucleation and growth of vacuoles inside the resin, in other words, the formation of matrix micro-cracks. From tests on pure resin samples, the acoustic signature of matrix fracture was determined. The associated waveform presents slow rising and low amplitude.

The amplitudes of the collected signals recorded during tests on unidirectional samples loaded in the direction transverse to the fibers (90° off-axis) are mainly distributed in two zones exhibiting a bimodal behavior (Figure 4.4). About 70% have amplitudes in the range of 50 to 70 dB, with waveforms similar to those observed in the tests on pure resin. These signals are referred as "A-type" (Figure 4.4, lower left). The remaining 30% of the signals have amplitudes in the range of 70 to 90 dB and waveforms quite different from the waveforms of A-type signals, with a shorter decay time and higher energies. These signals will be referred to as "B-type" (Figure 4.4, lower right). The definition of the signal amplitude ranges is somewhat arbitrary. There was overlap in the parameter data for various signals. Nevertheless, such simple arbitrary guidelines appear to be useful.

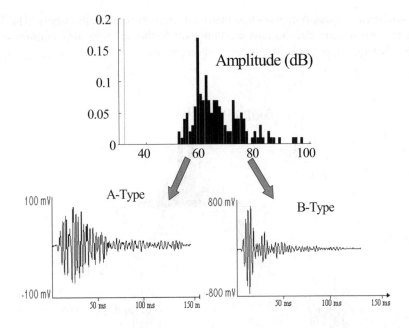

Figure 4.4. *Amplitude distribution and typical waveforms of signals from tensile tests on 90° samples*

The similarities in waveforms found between A-type signals in 90° off-axis tests and signals from pure resin tests makes one think that the source mechanism is the same in both cases, i.e. matrix fracture, in agreement with the fact that mode I cracking is the main fracture mode for this kind of sample. B-type signals (of amplitude between 70 and 90 dB) are detected only during the second half of the test, whereas the activity of A-type signals seems to be almost continuous. Thus, A-type and B-type signals seem to correspond to different mechanisms inside the material. B-type signals must have a mechanical origin that occurs after a certain matrix damage level has been attained.

Figure 4.5 represents the amplitude distribution and typical waveforms for the tests on samples at 45° to the fibers. A-type and B-type signals can be observed in the same amplitude zones as previously and with very similar parameters and waveforms. However, B-type signals are much more numerous than A-type signals in these tests (80%). The main expected damage mode in this type of tests, considering that 45° is the direction in which the shear stress is maximum along the direction of the fibers [GAY 91], is fiber/matrix decohesion. So the source of B-type signals, predominant in this configuration, corresponds to decohesion. This is also consistent with the presence of B-type signals in the 90° tests, appearing chronologically after A-type, when matrix fracture surrounds the fibers. The fact that the fiber/matrix decohesions are dominant in the 45° tests was confirmed by SEM (Scanning Electronic Microscopy) observations.

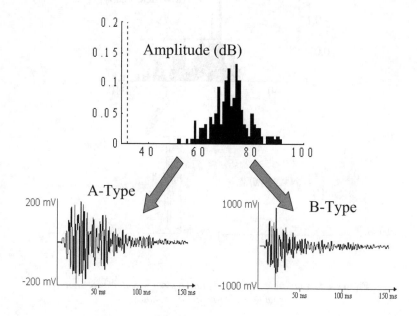

Figure 4.5. *Amplitude distribution and typical waveforms of signals from tensile tests on 45° samples*

4.2.2. *Algorithms for damage localization*

Whatever kind of damage we are looking at, the location and severity of damage have to be identified from the signal generated by the various piezoelectric transducers. The vast majority of the literature about SHM with AE sensors deals with low-velocity impacts and the associated damage. To do this, several techniques exist, and these are presented in the following sections.

4.2.2.1. *Time-of-flight based triangulation*

This is *a priori* a simple approach but its applicability is questionable with composite structures, which often present high anisotropy of wave propagation and structural complexity. For instance, the Tobias algorithm, which assumes a circular shape for the wave propagation front, is used by the BALRUE Crack Growth Monitoring System [OBR 02] for metallic structures (and is described in section 4.2.4.2). Even if the same approach can still be used for quasi-isotropic composites, it is not adequate for other composite lay-ups. Reference [PAG 03a] proposes using an "elliptical" or "quasi-elliptical" algorithm covering most of the composites used in aeronautical structures (see section 4.2.2.3).

4.2.2.2. *Energy-based triangulation*

ONERA [DUP 99, OSM 00a, and OSM 00b] uses the High-Frequency Root Mean Square (HF-RMS) value $\langle s \rangle$ of the electric signal s received by four surface-mounted piezoelectric sensors laid out in an array around the tested area. This quantity represents the signal energy:

$$\langle s \rangle = \sqrt{\frac{1}{T} \int_0^T s^2 \, dt} \tag{1}$$

The cut-off frequency, below which the signal is not considered for the calculation of $\langle s \rangle$, was fixed at 400 kHz for a C/epoxy plate of 2 to 4 mm thickness. In a first order of approximation, $\langle s \rangle$ is independent of the anisotropy of the wave propagation. It is then used both for triangulation, which leads to the impact localization, and for the evaluation of the delamination area (after calibration for the given material; see section 4.2.3.3). If it is assumed that the acoustic emission is generated by an acoustic point-source, the HF-RMS of the displacement $\langle u \rangle$ at the location of a detector distance D from the source can be estimated, if the plate is assumed to be transversely isotropic, by the following:

$$<u>= \frac{A \cdot e^{-\xi \cdot D / \lambda}}{\sqrt{D}} \tag{2}$$

where A is the characteristic amplitude of the source, λ is the wavelength of the acoustic wave and ξ the attenuation for a propagation distance equal to the wavelength. Parameters λ and ξ are *a priori* defined from Lamb wave propagation analysis. Discussion of these assumptions can be found in [DUP 99]. It is then possible to identify the actual location and the amplitude of a source A by means of an optimization process using the different $\langle s \rangle$ values measured by each sensor.

The technique works well for plate-like structures, but needs improvements to take into account the complexity of real structures, stringers in particular. Furthermore, it is worth noting that the technique is able to discriminate between damaging impact and non-damaging impact. This is an important aspect for a future system, as it is not necessary to store data concerning non-damaging impacts, which statistically could be the more frequent events.

4.2.2.3. *Time-of-flight based triangulation with respect to the angular dependence of the propagation velocity*

In [COV 03], the angular dependence of the propagation velocity is experimentally determined by making non-damaging impacts on a plate equipped with three sensors only (no more than in a classical triangulation procedure in isotropic materials). The triangulation is achieved thanks to an iterative optimization procedure based on a genetic algorithm. The article shows that the method has potential for effective impact damage location although the authors agree that the method cannot be guaranteed to give the true optimum. The use of the genetic algorithm avoids the learning and modeling difficulties associated with other techniques. The resultant speed and accuracy with which the algorithm performs the damage localization favors the implementation of the technique in real-time structural health monitoring systems. Nevertheless, it must be pointed out that the calibration (here the determination of the influence of the propagation direction) is obtained from non-damaging impact experiments, whereas the technique must be applied to the localization of damaging impacts. Such a procedure needs to be verified by application to a real damaging impact, which is not the case at present. Furthermore, the influence of the structure complexity is not taken into account in this study and it is not obvious how the technique could overcome this difficulty.

The same Sheffield research team recently proposed an automatic impact-monitored composite panel, inside which is embedded the commercially available SMART layer technology (see section 4.3.6.1). This study investigates two approaches to locating impacts (non-damaging impact introduced by a hammer): a triangulation procedure including a genetic algorithm and a procedure based on an artificial neural network.

Reference [PAG 03a] should also be mentioned; here the wave propagation front is approximated by an elliptical shape, depending on the lay-up, which has the advantage that it requires only two velocity measurements (longitudinal and transverse directions). The device measures the group velocities in the 0° and 90° directions using artificial AE sources such as a pencil break, and the times of flight. Validation experiments with surface-mounted piezoelectric sensors and pencil lead breaks were conducted, showing good accuracy for the location predicted by the elliptical approximation. In this case, one can make the same observation that no experiment has been made to study real damaging impact.

4.2.2.4. *Structure modeling*

This consists of modeling the dynamic response of the structure to a localized impact. From the comparison of the calculated and real responses, the impact location is revised in the model and the process is iterated (see the illustration of the process in Figure 4.6). In the Seydel and Chang model from Stanford University

[SEY 99, SEY 01a, SEY 01b], stiffeners are taken into account through adjustable and possibly discontinuous properties (mass and stiffness) given to the structure, which makes it possible to model a complex panel as an equivalent plate.

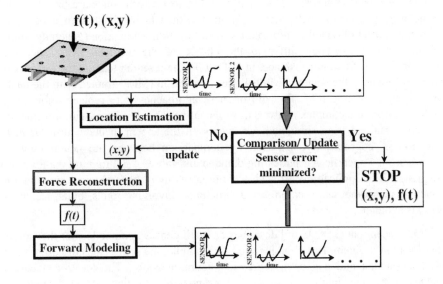

Figure 4.6. *Overview of the model-based impact identification scheme [SEY 01b]*

The two last papers are important for practical situations. They state that the critical parameter is the spacing between sensors, which depends on the maximum load during impact. The spacing scales roughly with the square root of the load and, for a given load, the error increases linearly with the spacing. This means that the number of sensors required for successful impact detection inside a given structure can be estimated by dividing its dimensions by the maximum sensor spacing that will provide adequate impact identification. For instance, it was estimated that approximately 250 sensors would be required to cover a Boeing 777 tail. Furthermore, it is stated that the exact placement of sensors is not so important once the structural geometry (i.e. the presence of the stiffeners) is considered. Thus, this affirmation means that there is less interest in work on optimization like those in reference [STA 00].

Again, one must regret that the model, produced by the Stanford team, only considers elastic phenomena and, consequently, only non-damaging impacts. Although the model works well in this case, its performance when compared to damaging impacts, which produce irreversible phenomena, have not been investigated yet.

4.2.2.5. *Learning from examples using neural networks*

Neural networks have been tested by several groups, such as EADS-France [SAN 00, SAN 01, SAN 02], the University of Sheffield [MAS 98, STA 00] and Korea Aerospace Industries (KAI) [SUN 00]. Various techniques are based on the same principle: neural networks are trained with data obtained from experiment (generally, a set of pencil lead breaks) or resulting from simulation (generally finite-element analyses). They differ in the choice of the artificial neural network paradigm to be used. For instance, in [STA 00], the research team at the University of Sheffield used the standard Multi-Layer Perceptron (MLP) trained with the back-propagation learning rule. The aim of these approaches is to replace information obtained from the complex analysis of propagating AE waveforms with information directly obtained from simulated AE events on the specimen under test. Damage location is, consequently, achieved by comparison between the neural network output from a specific AE indicating damage and those of the training (calibration) sequence. The applicability of such technology is more general, since propagation analysis in geometrically complicated structures (involving joints, bolts, fasteners, etc.) is avoided.

One recent paper [KOS 05] describes an intelligent AE source locator that solves the location problem. Based on learning by example, the source locator comprises a sensor antenna and a general regression neural network. The location accuracy achieved by the intelligent locator is comparable to that obtained with the conventional triangulation method. Moreover, blind source separation (BSS) by independent component analysis (ICA) has been used to solve the problem of the separation and location of two independent simultaneously active AE sources on an aluminum beam specimen. The conclusion is that the method is promising, though it is still mainly applied to one-dimensional cases.

4.2.3. *Algorithms for damage characterization*

The AE sources are related to irreversible phenomena. In composite materials, the collected AE bursts can be attributed to various damage mechanisms or to friction phenomena. Here, the damage terminology designates all the phenomena related to a material loss of cohesion, the accumulation of which leads to the deterioration of the structure. When decohesion occurs, only part of the released energy can be transformed into an acoustic wave. In the case of a composite material, conventional AE analysis (the investigation of diagrams of cumulative hits, or counts versus test parameters, and of the amplitude or duration histogram) is an inadequate approach for discriminating the AE signatures, due to different damage mechanisms.

A wide variety of techniques can be considered for identifying the nature and the severity of impact-generated damage. Reference [PED 01] presents a general overview of the different techniques that can be used: time analysis of the signal; various statistical moments and in particular kurtosis; frequency analysis using the moments of the power spectrum, in particular, the positive zero crossing and the number of peaks; time-frequency analysis using wavelet transforms; etc. The following two questions can be asked:

– What are the best features, extracted from the impact strain data, to map the different impact energy levels?

– Is it possible to detect damage directly from features without using a learning process?

It is concluded in this reference that the question whether information about different types of damage in composite materials can be directly obtained from the impact strain data remains open.

Improvement can also be brought about by the use of classifiers to group into clusters signals having similar shapes. The following sections will describe the two kinds of damage characterization: discrimination of the nature of the damage using supervised or unsupervised neural networks, and discrimination of the severity of the damage through energy-based techniques.

4.2.3.1. *AE-based supervised classification of failure modes in a glass fiber reinforced composite*

Here, a combination of an unsupervised pattern recognition scheme (*k*-means method [LIK 03]) with the supervised *K*-Nearest Neighbors method (KNN method [GOD 04]) is proposed for taking into consideration several AE descriptors. The *k*-means method is used for data clustering and the KNN method for classification. The flow chart of Figure 4.7 shows the procedure developed for the analysis of the AE data that have been described in section 4.2.1, showing its main steps. The *k*-means algorithm is used to split the AE data into two classes and the data segmentation is validated by the use of the experimental results based on the validated results of the previous phase. Supervised pattern recognition is then used to classify AE data by means of the KNN method. In this approach, a training data set from a representative sample is used. Thus, an unknown signal is classified according to the majority of the votes of its nearest neighbors in multi-dimensional descriptor space.

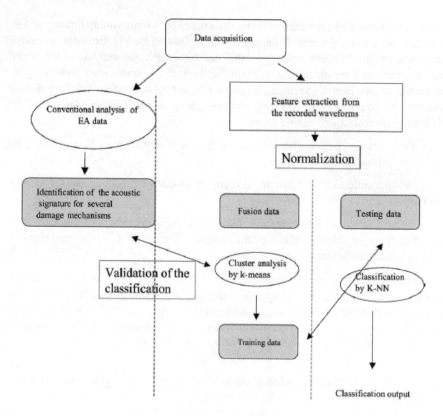

Figure 4.7. *Flow chart representation of the pattern recognition method proposed by Huguet [HUG 02] for the analysis of AE data*

4.2.3.2. *Unsupervised clustering: Kohonen self-organizing feature map*

In order to separate numerically the two types of signals presented in section 4.2.1 without resorting to supervised learning, a self-organizing neural network approach can be implemented. The capabilities of such a technique have been demonstrated in functionally modeling processes involving many variables [FIS 98]. More precisely, a Kohonen map [HUG 00] was used to organize feature vectors into clusters, so that vectors within one cluster are more similar to one another than to vectors belonging to other clusters. The inputs were six AE features taken from the signals. The network response clearly exhibits two main clusters (Figure 4.8), corresponding to the two types of signals previously observed.

The Kohonen map could then be used to compute the chronology of appearance of A-type or B-type signals during tests, as an indicator of the kinetics of each damaging mechanism. Interpretation of clustering results on a macroscopic level is performed by plotting the cumulative events of each class versus load. An example is given in Figure 4.9 for two tests at 90° and 45° from the orientation of the fibers. It shows a typical evolution of an AE event for the discriminated source mechanisms. For the 90° sample, it is observed that the two classes exhibit similar activity trends. The fraction of A-type signals for the 90° sample evolves more prominently than for the 45° sample, from the beginning to the end of deformation. In contrast, a larger fraction of B-type signals was detected for the 45° sample than for the 90° sample, for which decohesion is the dominant damage mechanism. Let us also note that the number of B-type signals associated with decohesion greatly increases at the end of tests, causing the final failure of the material.

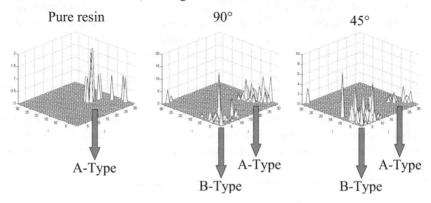

Figure 4.8. *Kohonen maps for the three types of tensile test*

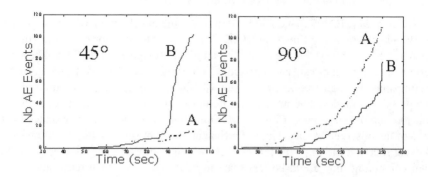

Figure 4.9. *AE activity evolutions of A-type and B-typesignals during tests*

Recently, Godin *et al.* presented an original combination of the use of the Kohonen Self-Organizing Map (SOM) and the *k*-means methods. To classify AE signals, collected during tensile tests on cross-ply glass/epoxy composites in order to monitor the chronology of the damaging process, the Kohonen SOM is applied as an unsupervised clustering method. The input vectors of the signal descriptors, used in the clustering procedure, are calculated from the signal waveforms. Then, the k-means method is applied to the neurons of the map, in order to delimit the clusters and visualize the topology of the map [GOD 05].

4.2.3.3. *Energy-based techniques*

References [DUP 99, OSM 00a, OSM 00b] state that this energy-based technique is able to give the extent of the damage, considered as a delamination because the HF-RMS severity factor (equation [1] in section 4.2.2.2) has been shown to be linearly linked to the delaminated area. Moreover, the technique makes it possible to discriminate between damaging and non-damaging impacts, which can be considered as the first level in the characterization of the damage. For the same aim of discrimination, reference [STA 99] presents another technique, which relies on the same physical basis (damaging impacts generate high frequencies and these are not present in non-damaging impacts): high frequency components of the signal are isolated by wavelet transform decomposition. The spikiness of a high-rank wavelet level is evaluated by calculation of its kurtosis (4th spectral moment). This parameter is almost null when impacts are not damaging and strongly increases when damage appears.

4.2.4. *Available industrial AE systems*

4.2.4.1. *Broadband acoustic emission system (Boeing/Digital Wave Corp.)*

References [SEA 97a] and [IKE 99] describe the development of a system based on Broadband Acoustic Emission (BAE), which aims to monitor crack growth with good accuracy. The system was tested on an F16 Fuselage Station 479 bulkhead and results during ground-tests are reported. Flight tests were planned. Crack growth is accompanied by high-frequency emission (0.7 to 1.5 MHz) and propagation of bulk compressional modes. The problem is the noise due to various mechanical sources and the number of events. Careful discrimination is needed and eliminates 90 to 95% of the observed events. Time of arrival at the eight sensors allows automatic 3-D event source location. With 1300 AE events, and using a C-language program, the total computing time could be less than an hour. Times of occurrence and source locations are registered. The depth and size of the crack measured in this way compared favorably to post-test examination. Reference [SEA 97b] reports the crack detection ability of Broadband Acoustic Emission Monitoring from fatigue testing on the 777 full-scale aircraft.

4.2.4.2. *Other piezoelectric-based systems*

A similar new crack monitoring system has been developed by British Aerospace (BA), in cooperation with Lloyd's Register of Shipping (LR) and Ultra Electronics Ltd., Ocean Systems (UE). This is called the BALRUE Crack Growth Monitoring System [OBR 02]. First demonstrated in late 2001 at Shiprepair & Conversion (London), it has been recently applied by Paget *et al.* to a full-scale Airbus metal wing loaded in fatigue [PAG 04]. For this study, the in-house sensors of the BALRUE system were surface-bonded to the structure and used to continuously monitor a large area of the wing for about 30,000 simulated flight cycles, corresponding to a testing period of about 14 months. The BALRUE system located all simulated and artificial damage after the first couple of months. In addition, the system located early damage, which only became detectable by conventional NDT after a period of fatigue loading ranging between 8 to 11 months. In other words, the BALRUE system was able to foresee future structural repairs. Paget *et al.* further demonstrate the great accuracy of damage (fretting and cracks) localization in such complex wing structures.

Another similar system has been developed by BAE Systems and tested as part of the AHMOS Program [MAR 03]. The system was evaluated on a Eurofighter design wing-box with a composite center section incorporating a pylon fitting and on a GLARE fuselage panel containing repair patches. Little information and few results are given, and these only for the composite panel. The system seems less advanced than the one described by Boeing (see above), especially for signal de-noising prior to waveform identification and accuracy of localization.

4.2.5. *New concepts in acoustic emission*

This section outlines mid-term and long-term solutions which are considered by the acoustic emission community insofar as they provide better integration of this technique in future health monitoring systems. The first solution is the concept of the continuous sensor or artificial nerve, which has mainly been developed by a group composed of Sundaresan, Ghoshal, Schulz and others (see references: [GHO 01, GHO 02, GHO 03, SCH 02, MAR 03, KIR 03a]). It essentially consists of electrical linking of several sensors in a single channel and subsequent data processing of the sum of the individual sensor signals traveling in this channel in order to drastically decrease the hardware complexity of such systems.

The second kind of long-term solution under development is the concept of Residual Lifetime Prediction through the statistical analysis of AE events. This is being carried out by collaborating groups of the Université de Nice–Sophia Antipolis, Nizhny Novgorod State University and INSA de Lyon (see references [NEC 04, NEC 05] and [SAI 05]).

4.2.5.1. *Continuous sensors and artificial neural systems*

All previously mentioned methods present difficulties when applied to structures having complex geometry (discontinuities, joints, stiffeners, etc.) and/or containing hybrid or absorbing materials, such as honeycomb sandwich structures. Furthermore, when considering the monitoring of a large structure, the number of sensors needed to monitor each area becomes high and the associated wiring, amplification, multiplexing and computational devices make the systems more complex.

Back in 1998, Egusay and Iwasawaz of the Japan Atomic Energy Research Institute, first, proposed the concept of a continuous AE sensor in the form of a piezoelectric paint. Such a paint is prepared using PZT ceramic powder as a pigment and epoxy resin as a binder. Reference [EGU 98] describes the application of the piezoelectric paint as an integrated SHM sensor for vibration modal sensing, when it is used on aluminum cantilever beams and for AE sensing, when used on glass/epoxy cylindrical test blocks. The paint film thickness varies from 25 to 300 μm. For AE purposes, the paint film sensor is evaluated in the frequency range from 0.3 to 1.0 MHz, where it exhibits an almost flat frequency response to AE waves, which is an indication of its promising sensitivity.

Piezoelectric polyvinylidene fluoride (PVDF) film should also be mentioned as a continuous sensor. PVDF polymer is known to show a relatively strong piezoelectric effect; it is not as sensitive as PZT sensors, although it is sometimes preferred to piezoceramics because of its flexibility, low weight, low profile (a few tens of microns) and very low cost. Nevertheless, few papers demonstrate the effective use of PVDF film as an AE sensor in the context of structural integrity monitoring. However, [KWO 04] has reported the practical application of commercially available PVDF film for the detection of AE, due to fatigue crack growth in lap joints of graphite/epoxy laminates. This technology still suffers from having a signal-to-noise ratio, which is not as high as those measured by PZT and a low Curie temperature, which currently limit its integration as an embedded sensor.

A bio-inspired concept proposes to mimic a biological nervous system. Several techniques are under development, but all are based on arrays of distributed sensors, linked to form artificial nerves, like neurons in biological systems. If the summation of multiple sensors in a reduced number of channels does not degrade the diagnostic ability too much, compared to classical AE systems, the result of this activity will be the design of an interconnected sensory system with tens to hundreds of units distributed in the structure [GAR 02]. Some important research on this has been carried out by various collaborating groups at the University of Cincinnati, North Carolina A&T State University, Arizona State University, and NASA Langley with the aim of developing Artificial Neural Systems [MAR 01, GHO 02, MAR 02, KIR 03a, KIR 03b] and continuous sensors [SUN 02, DAT 03]. In [SUN 02], Sundaresan *et al.* propose a continuous sensor for monitoring stress waves and AE signals. This sensor is made of a long tape with a number of sensing nodes, which are electrically connected together to form a single sensor with one data output channel. The performance of the sensor on a bar-like structure is evaluated through numerical simulations and experiments. It is shown that there is a possibility of optimizing the parameters of sensor connectivity (series or parallel) and sensor polarity (positive or negative) to obtain the maximum amplitude of the sensor response for a particular structure. The main advantages of such a sensor are:

– the ability to capture the leading edge of the AE signal before it undergoes significant loss of amplitude, due to attenuation and dispersion, during its propagation in the tested material;

– the ability to monitor relatively large and complex areas with a single AE channel with a higher probability of detecting critical damage;

– the reduction in the cost, complexity (number of channels) and weight of the AE monitoring system to levels where it becomes practical to perform Structural Health Monitoring.

4.2.5.2. *Acoustic emission as a tool for residual life time prediction*

This consists of early-stage research, which is aimed to predict the remaining lifetime of a macroscopic structure (mainly composites) from the modeling rupture phenomena at a microstructural scale. These theoretical studies are supported by the experimental data from AE tests, and will undoubtedly find an application in future AE-based SHM systems.

In [NEC 04], it is shown that strain rates and acoustic emission (AE) recorded during creep (with constant stress) experiments on fiber composite materials exhibit both a power-law relaxation in the primary creep regime and a power-law acceleration before global failure. In particular, it is observed that the time-to-failure follows a power law in the tertiary regime. A very important correlation has been discovered between some characteristics of the primary creep regime (the exponent of the so-called Andrade law and duration) and the time-to-failure of the samples. This result indicates that careful monitoring of the primary regime strain rate or AE rate makes it possible to predict the time-to-failure, and thus the remaining lifetime of the structure under test. In [NEC 05], fiber composite materials with clear controlled heterogeneity have been tested in the same way as the materials in the previous reference. These experimental results are rationalized with a mean-field model of representative elements, which have nonlinear viscoelastic rheology and a large heterogeneity of strength. Saichev and Sornette [SAI 05] introduce thermal fluctuations in the rupture model, which allows them to present analytical theories explaining three empirical observations widely made in creep experiments: the initial Andrade-like $1/t$ decay of the deformation rate, the Omori-like $1/t$ decay of the fiber cracking rate and the $1/(t_c - t)$ critical time-to-failure behavior of acoustic emissions just prior to the macroscopic rupture.

4.2.6. *Conclusion*

AE techniques are significant in SHM for many kinds of structure, and particularly in air safety. In the latter, however, the application of SHM is limited to very narrow areas, because of the lack of a quantitative appraisal of defects and a shortage of standards and programs [GEN 04]. Traditional AE parameter analysis allows a simple, rapid and cost-effective inspection or monitoring of a structure, but a fully integrated SHM system demands more understanding of the actual link between AE source mechanisms and the damage phenomena. Much effort has been made towards achieving this. Modal acoustic emission is close to the source mechanism, and research towards the implementation of this technique should focus on the optimization of data processing, which still takes up much time and computer memory.

New concepts, such as continuous sensors and bio-mimetic information acquisition and processing, would also be fundamentally helpful to realize real-time AE instrumentation for the health monitoring of in-service (in-flight in airspace industry) structures.

4.3. State-of-the-art and main trends in piezoelectric transducer-based acousto-ultrasonic SHM research

The acoustic emission technique described above makes use of piezoelectric sensors bonded on or embedded in a structure and implemented in a passive way. The same kind of attached sensors can also be used in an active way to produce and detect high-frequency vibrations. The resultant waves propagate and interact with defects. Since, in aircraft, a large part of the structure corresponds to the plate type, most laboratory work is oriented towards plate waves (Lamb waves). Experimenters use a transmitter to send a diagnostic stress wave along the structure and a receiver to measure the changes in the received signal caused by the presence of damage in the structure. This wave propagation approach is a natural extension of traditional NDE techniques, and is very effective in detecting damage in the form of geometrical discontinuities.

For more 3-D parts, bulk waves can be used (see the SWISS system in section 4.3.6.2). Also to be mentioned is the possibility of using surface acoustic waves (SAWs), although only exploratory work in laboratories has been carried out so far, such as in reference [TIT 03].

Many publications are devoted to the issue of sensor attachment. One can summarize the situation by saying that two different strands are developed. The first strand does not really concerned the way the vibrations are introduced into the structure to be tested, and thus concentrates on the post-processing of the acoustic signatures (sometimes complicated), which are collected. The second strand tends to optimize the generation of particular modes, and thus requires an understanding of the physical phenomena involved in the intimate connection between the actuator and the host structure. This is studied in several references: [GIU 01, ZAG 02, GIU 02a]. Reference [GIU 02a] also suggests the possibility to self-diagnose all piezoelectric patches with a simple electromechanical impedance measurement: if the wafer is not perfectly bonded or embedded, resonance peaks appear in the impedance versus frequency curve. This technique is also used at ONERA to validate all piezoelectric implants, before starting tests. Giurgiutiu mentions the possibility of degradation of the adhesion, due to cycling of humidity and temperature, which show the importance of using self-diagnosis with piezoelectric acousto-ultrasonic sensors. This problem was first identified by Saint-Pierre at INSA de Lyon, who proposed different ways of changing electromechanical impedance-based methods when the bonding of the piezoelectric element becomes defective [SAI 96b].

The implementation of SHM technologies also causes the problem of transmitting and processing a huge amount of data, which requires expensive and cumbersome hardware. For a "real world" setting, the latter is added to the concern of including actuators, sensors, connecting, processors, etc, which require complex wiring configurations. Thus, the need to develop stands alone or networked wireless SHM systems is emerging (see section 4.3.2.2).

4.3.1. *Lamb wave structure interrogation*

Lamb waves are basically two-dimensional propagating waves in plate-like structures. The mathematical formulation of these guided modes was originally proposed by Lord Horace Lamb [LAM 17]. Lamb waves are dispersive, that is to say that their velocities depend on the frequency and the thickness of the plate. The usual approach is to plot the so-called dispersion curves, which represent the phase velocities of all existing modes plotted versus the product of the frequency and the thickness [VIK 67]. Lamb waves have the advantage of propagating over quite long distances without appreciable attenuation, so the tested volume in a single acquisition is increased in comparison with most bulk wave-based techniques. Furthermore, the various Lamb modes offer specific sensitivities to different kinds of defects [GUO 93, MAS 97]. Two modes of particular importance, called fundamental modes, do not have cut-off frequencies. Hence, they are present even at very low frequency. The fundamental symmetric mode S_0 is often used to detect surface cracks in metallic structures, while the fundamental anti-symmetric mode A_0 is widely used for damage detection in composites because of its sensitivity to delamination. This does not always apply because the polarization of a given mode changes with frequency, so that the induced stress and strain distributions in the through-thickness direction also vary [DIE 75, NAY 95].

Vibration testing has been the subject of research for several decades. The most attractive parameters to measure are natural frequencies and damping, and this can be done through a limited number of point-to-point measurements on the structure. The reproducibility of the first parameter is better, even if its sensitivity to damage is lower. With several measurements, mode shape can be recovered and used to detect damage. To locate damage, various parameters deduced from the measurement can be used: flexibility matrix, stiffness error matrix, modeling updating methods, neural networks, etc. The main drawback of the vibration-based methods is essentially a low sensitivity to localized damage and sensitivity to other changes in the structure and in the surroundings. In an attempt to avoid the influence of the boundary conditions, the idea of using propagating waves with a short time window emerged. Rayleigh waves can be used to detect sub-surface flaws in substrates or thick plates. Since they can explore the entire thickness of a plate, Lamb waves are particularly well-suited to the aerospace industry, ground transportation and civil engineering, where composite material plates are becoming increasingly common.

For instance, Chimenti and Martin [CHI 91] used Lamb waves to detect various defects such as delamination, porosity, ply gap, presence of foreign material and changes in fiber volume fraction of carbon/epoxy laminates. Tan *et al.* [TAN 95] compared Lamb waves with the normal incidence pulse-echo approach for the detection of near-surface delaminations. Changes in the Lamb wave amplitude are used to determine both the size and the depth of the delamination. Another comparison between Lamb wave tomography and conventional C-scan was carried out by Jansen *et al.* [JAN 94] for the localization of delamination and matrix and fiber cracking in composite plates. Seale *et al.* used Lamb waves for the characterization of composites submitted to mechanical and thermal fatigue [SEA 98a, SEA 99] and for the non-destructive determination of their fiber volume fraction [SEA 98b]. Lamb waves were also largely used for the characterization of the quality of bonds [GUY 92, DAL 00], of welds and joints [SUN 94], and more generally of interfaces [XU 95]. These plate waves were also successfully used for the long-range non-destructive evaluation of corrosion-induced damage in pipes, in particular by Rose [ROS 96] and Alleyne and Cawley [ALL 97]. Kessler *et al.* [KES 02] used piezoelectric actuators to provide Lamb wave scans in damage detection in various composite structures (laminates, sandwich beams, pipes and stiffened plates). It is shown that for localized flaws, as stated before, Lamb waves make it possible to retrieve richer information about damage type, location and extent than do frequency response techniques. The high sensitivity of pulse-echo signals to the presence of cracks in aluminum plates is demonstrated by the simple subtraction of the baseline obtained in the pristine state. This type of result has been obtained with both piezoelectric transducers [GIU 02a] and magnetostrictive transducers [KWU 02a, KWU 02b]. The detection of such damage by measuring the amplitude of the signature of the Lamb waves is here assisted by the use of a simple material and the high signal-to-noise ratio of in-the-laboratory experiment. Outdoor tests on real structures cannot satisfy these requirements and thus demand the development of more complex data analysis procedures (multivariate data processing, Fourier and Wavelet Transforms, etc.). More recently, Diamanti *et al.* [DIA 04] attached a linear array of piezoceramic patches onto the surface of a composite structure in order to investigate the interaction of Lamb waves with impact damage over large areas in CFRP laminates.

Very soon, the question of the integration of guided waves in a fully on-board system arose. The potential of such propagating waves and the practical problems linked to the monitoring of large areas of metallic aircraft fuselage structure are studied in reference [DAL 01]. The complex features of the aircraft are supposed to fall into only six types (free skin, tapered skin, lap joint, stringer joint, skin with sealant overlay and double skin, as illustrated in Figure 4.10). Each feature presents a different guided wave system for which useful rules are given. This simplification neglects the fact that more than two skins can be encountered and also ignores the scatter due to fasteners. Nevertheless, this paper is fundamental for designing a system, which will be relevant to a large part of a modern aircraft. Its final conclusion is that an integrated, single mode, active system to monitor an entire fuselage is not feasible.

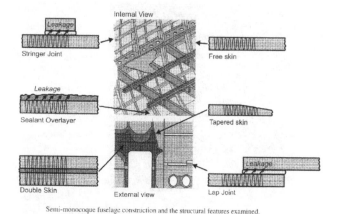

Figure 4.10. *Various structural features of a complex aircraft fuselage examined by guided waves (from [DAL00])*

Two main drawbacks of Lamb wave-based testing are the co-existence of numerous modes (at least two modes at any given frequency) and their strongly dispersive nature. Thus, many attempts have been made to develop sensors, which could somehow preferentially generate one particular Lamb mode. The selection of the most interesting mode, from the viewpoint of the sensitivity to damage, is the subject of references [WIL 99] and [WIL 01]. Furthermore, even if a single and pure Lamb mode can be generated, it could produce a variety of other modes, either by interacting with an internal defect or by reflection at structure boundaries: this is called mode conversion. Consequently, the output signal becomes more complex and its interpretation requires a detailed physical analysis of the propagation mechanism.

Many important attempts have been made to provide a better understanding of the physics of the interaction between Lamb waves and localized defects. Some of them are based on numerical simulations, like the work of B.C. Lee from the University of Sheffield [LEE 02, LEE 03a, LEE 03b], which simulates the interaction of Lamb waves with various kinds of defects using the Local Interaction Simulation Approach (LISA) developed by the P.P. Delsanto Group in Torino [AGO 00]. Others use Finite Element Modeling (FEM); for instance, Su and Ye from the University of Sydney simulated the interaction of Lamb waves with a delamination [SU 02, SU 04]. Figure 4.11 shows the FEM mesh in the vicinity of the delamination, which is simulated by modification of the upper and lower contact surfaces of the delaminated area.

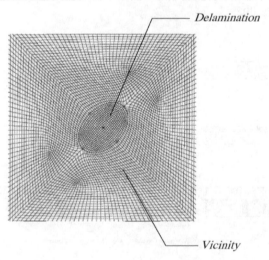

Figure 4.11. *FEM mesh near a delamination area, from [SU 04]*

In Figure 4.12, the disturbance of the propagating wavefront due to the interaction of Lamb waves with the delamination is clearly visible; one can see that this FEM image is remarkably similar to those from shearography experimentally obtained by Taillade [TAI 00] and shown in Figure 1.16.

Some other papers present basic experiments on beams or plates with real or artificial cracks, but the one-dimensional character of experiments in beams makes their conclusions of no practical use for real application on a complex aircraft structure. With the same aim of getting better knowledge about propagation and interaction with defects, wavelet transforms can also help to process the Lamb wave response data. This topic will be discussed later (see section 4.3.4.2).

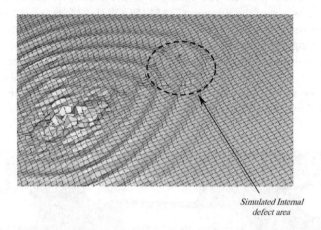

Simulated Internal
defect area

Figure 4.12. *Defect-imposed disturbance on Lamb wave propagation
in composite laminates*

4.3.2. *Sensor technology*

Most of the time, simple shapes are used for piezoelectric transducers. They are mainly disc-shaped, rectangular or square. Sometimes, Inter-Digited Transducers (IDT) are implemented for specific purposes. PZT piezoceramics are often chosen as actuators because of their high force output at relatively low voltages and their good response at low frequencies. The shape of the actuator is generally chosen on the basis of the desired propagation or reception direction. Simply speaking, ultrasonic waves propagate parallel to the edges of the actuator, that is to say longitudinally and transversely for a rectangular patch and circumferentially for a disc-shaped actuator. Though the width of the actuator in the propagation direction is not critical, the wider it is, the more uniform is the created waveform. In the direction of propagation, the actuator length for optimal signal is given by a sinusoidal relationship between actuating wavelength λ (or frequency *f*) and actuator length *L* [VIK 67]:

$$L = \lambda \left(n + \frac{1}{2} \right) = \frac{c_p}{f} \left(n + \frac{1}{2} \right) \qquad n = 0, 1, 2, \ldots \qquad [3]$$

C_p is the phase velocity of the generated wave. The value of L could either be the length of a rectangular patch or the diameter of a disc-shaped patch. This equation could also be used to determine actuator minimum dimensions, in order to inhibit waves from propagating in undesired directions [KES 03a].

A few years ago a number of attempts were made to optimize the transduction by developing matched patches. For example, S.G. Pierce in [PIE 00] presents such systems made as part of the European DAMASCOS (DAMage Assessment in Smart COmposite Structures) project (see Figure 4.13, pictures b and c). Among them, the solution of using a pair of transducers working in phase or with opposite phases and situated at two different depth levels to enhance the symmetric or the anti-symmetric modes (see [BLA 97]) remains an efficient way to improve the emission, which can be useful when applying acousto-ultrasonics to relatively highly absorbing materials such as GFRP or sandwich composite structures [OSM 02].

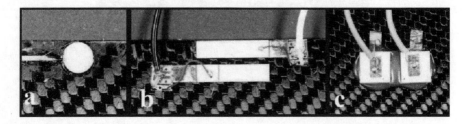

Figure 4.13. *Three different surface mounted Lamb wave sources, from [PIE 00]*

Using numerical simulations, reference [HAY 01] compares the efficiency of two types of piezoelectric transducer. The first one is a 23 × 23 mm plate made of modified lead titanate. The second one is an IDT made of piezoelectric ceramic (modified lead titanate) platelets held together by a polymer and placed between two thin, flexible printed circuit boards, one of them having a finger-type electrode pattern, the other acting as a ground electrode. Both transducers are embedded inside a plate of pure epoxy resin. The plate transducers appear to be more efficient, with smaller size, and more versatile (bi-directional instead of unidirectional).

In [KES 03a, KES 03b], Kessler *et al.* present an original matched patch made of a PZT-5A disc working as receiver, surrounded by a PZT-5A ring working as emitter. Experimental tests of all four combinations (disc–disc, ring–ring, ring–disc and disc–ring) showed that the best signal was obtained with ring as actuator and disc as sensor. The material and the geometry of this actuator/sensor were selected as an overall optimum from analytically derived figures of merit for Lamb wave-based health monitoring of composite structures.

4.3.2.1. *Beam steering and focusing*

A recent trend in research on acousto-ultrasonics is an increase in the number of works oriented towards the enhancement of localization of the emitted ultrasonic energy by beam steering and focusing. To achieve this, several piezoelectric patches are arranged to form a Lamb-wave beam steering phased array. Phased arrays have been extensively used during recent decades in ultrasonic medical imaging and therapy because they advantageously allow both beam steering and focusing. The principle of the superposition of individual beams produced by distinct piezoelectric patches also applies in the SHM context. For instance, Sundararaman *et al.* illustrated beam steering on a steel plate with an array of five piezoelectric patches [SUN 03]. Most work is concerned with metallic structures, but Moulin *et al.* [MOU 03] present results obtained with a three-element array mounted on the surface of a glass/epoxy composite plate. An experimental demonstration of the angular localization of delaminations generated by impacts was made, but only in transmission mode, which does not provide a complete proof of the method.

Giurgiutiu [GIU 03a] presents a more sophisticated linear array of nine elements bonded onto an aluminum plate of large size. The system, called Embedded Ultrasonic Structural Radar (EUSR), is capable of both steering and focusing. Detection, in pulse–echo mode, of cracks, 19 mm long with two different directions was achieved (see the experimental setup in Figure 4.14). In the direction orthogonal to the linear array (the "axis" of the transducer), the damage signature is well identified as the ultrasonic beam is reflected back to the source. Detection of off-axis cracks is relatively more challenging because the ultrasonic beam is deflected sideways to some extent and only the back-scattered signal from the tip of the crack, which acts as a diffracting source, is received. It appears that, even at a 46° direction off the transducer axis, at a distance of 409 mm, detection was possible.

A Graphical User Interface (GUI) displaying the signals has been developed. Figure 4.15 shows a snapshot of the GUI for the experiment shown in Figure 4.14. The sweep is performed either automatically or manually thanks to the EUSR-GUI. The reconstructed signal is presented to the user (maximum amplitude of the echo as a function of the angle of the direction of propagation of the steered beam). The reconstructed signal is shown in the lower pane. In Figure 4.15, the lower plot shows the reconstructed signal at the beam angle of 46° corresponding to the direction at which the damage lay. The distance of the damage is determined from time-of-flight (TOF) estimation and, consequently, the exact location of the damage is fully determined. In [PUR 04], Purekar *et al.* used a similar array made up of a set of 11 PZT-5H piezoceramic sensors bonded onto a 24 × 34 inch (60.96 × 86.36 cm) aluminum plate. Damage, in the form of increased stiffness, was simulated at various locations by clamping a 1 inch (2.54 cm) diameter circular section of the plate. The array was able to determine the angle and distance of damage correctly for most of the cases, with an average distance accuracy of 1 inch.

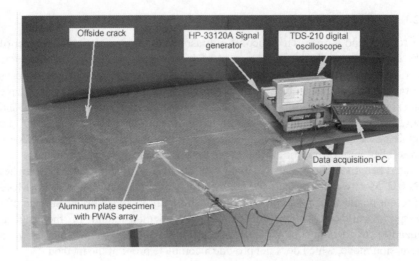

Figure 4.14. *Thin aluminum plate with an array of nine piezoelectric elements and a 19 mm offside crack, from [GIU 03a]*

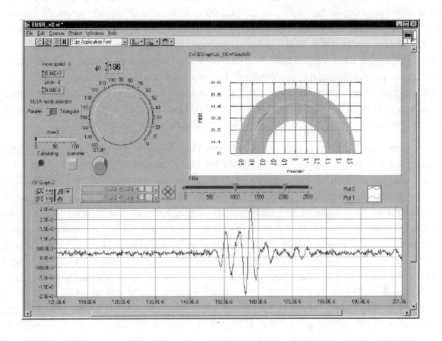

Figure 4.15. *Graphical User Interface of the EUSR showing the image of a 19 mm crack at an angle of 46° on the left of transducer axis, from [GIU 03a]*

These references are interesting because, even though they are just laboratory systems, they are representative of what could be a fully integrated system capable of Lamb wave beam steering and focusing for an aircraft structure. The advantage of systems using detector arrays is not so critical in the particular case of metallic structures, because of the low attenuation. For more absorbant media such as composite materials, this will be an important improvement.

Recently, Fromme *et al.* designed a self-contained, permanently attached guided ultrasonic wave array for the constant long-term monitoring of structural integrity. The array consists of two concentric rings of piezoelectric transducer elements, one for excitation and the other for reception of guided waves (see Figure 4.16). The circular design of the array is intended to give the system the same performance in all directions. Multiplexing units are used to switch between the different excitation and receiving transducers. A phased addition algorithm is used to synthesize a guided wave beam that can be steered in any direction from the array [FR0 04]. Preliminary measurements were made on a 5 mm thick aluminum plate of dimensions 2.45 × 1.25 m. Based on observations in a previous study [WIL 99], the A_0 Lamb mode below the cut-off frequencies of the higher order Lamb modes was used. Figure 4.17 shows the resulting omni-directional B-scan for the introduction of a through-hole ($r = 15$ mm), with the position of the array and the plate edges marked. This image shows the reflections of the guided wave at the four sides and the four corners of the plate. The plate edges are only seen in the direction where they are normal to the waves propagating from the array. The data processing algorithm is designed to pass signals transmitted and received along the same radial line and to reject signals from other directions. The presence and location of the defect are clearly in evidence. [FR0 04] also presents a second prototype, which uses a single ring of elements operating as both transmitters and receivers in order to reduce inter-element reflection problems.

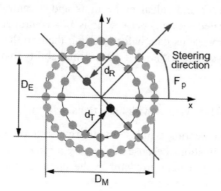

Figure 4.16. *Array layout of a beam-steering prototype with 16 excitation elements (inner circle DE = 50 mm) and 32 receiving elements (outer circle DR = 70 mm)*

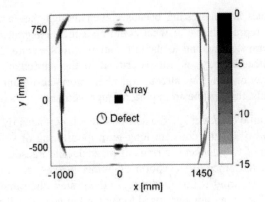

Figure 4.17. *Omni-directional B-scan of plate (edges marked) from the indicated array position. Defect is a through hole r = 15 mm (15 dB scale)*

4.3.2.2. Moving to wireless systems

The main idea of the wireless sensing concept is to develop systems with local transmitting or processing capabilities. SHM based on wireless data transfer should improve the fidelity of subjective visual inspections and decrease inspection costs by reducing inspection periods and focusing them on critical areas only. Wireless SHM can be envisaged from two significantly different viewpoints. The first one consists of acquiring data on a structure and wirelessly transmitting the data to a computer for post-processing. Different transmission methods and their advantages have been reviewed by Sohn *et al.* [SOH 03b] and some of them will be presented in the next section. The second way to design a wireless-based system consists of building a cluster of wirelessly connected probes and constructing the architecture for communication between individual probe units and a central processing unit. An even more sophisticated system, because it is not centralized, would allow each cluster node to preprocess its data before passing them to the host PC. These kinds of wireless SHM systems with local processing ability have been reviewed by Grisso [GRI 04].

4.3.2.2.1. Wireless SHM systems with local transmission

The idea of implementing wireless monitoring systems naturally came first in civil engineering applications because of the very large size of the infrastructures to be monitored. Indeed, practical problems in the implementation of advanced SHM system in real-world structures are:

– the assembly of massive and complex wiring from the sensors to the processors;

– the need to transmit and process large amount of data.

The severity of these two issues is linearly linked to the size of the structure, or to the number of sensing units that have to be handled, which in a sense is the same. To deal with these problems, Straser *et al.* [STR 98] stated that the deployment of long-term SHM systems for civil infrastructures is not possible because of the high cost per channel, the extensive cabling, signal deterioration over long transmission distances and due to environmental exposure and maintenance costs. Wireless SHM should overcome these difficulties but it needs the introduction of key design factors such as ease of installation, low cost per unit and broad functionality. In [STR 98], the original proposal was to implement an AM type radio to perform time synchronization between sensors, which has been a major obstacle for real-time SHM systems until then. Embedded micro-processors, radio modems, batteries, data acquisition equipment, Micro-Electromechanical Systems (MEMS) accelerometers and analog/digital converters (ADC) are also specified.

Mitchell *et al.* [MIT 99] discussed the application of smart sensors and radio frequency (RF) wireless data transmission for health monitoring, putting the emphasis on removing computing equipment from a harsh environment to protect it from possible damage. The authors developed a sensing unit, which integrates a microcontroller, a sensor and an RF transmitter/receiver. The microcontroller can multiplex signals sampled from several different sensors, which reduces the total number of data channels needed. The feasibility of the wireless transmission was demonstrated through a simple experiment, which consisted of monitoring the natural frequency of a cantilever beam. Excitation and sensor data were transmitted over the wireless channel at a rate of 19,200 bps, which corresponds to a 400 Hz sample rate. The wireless system proved to give similar results when compared with wired results using the same test setup.

In [IHL 00], Ihler *et al.* identified the main steps in planning and implementing a modular wireless SHM system: choosing a power supply, selecting signal frequency and modulation, and embedding sensors. They designed a simplified wireless sensor system based on a smart "crackwire", which is composed of four conducting wires in an epoxy-based substrate associated with an RF module. To monitor the structural integrity of aging aircraft, the sensor is placed around a rivet where cracks are expected. Crack growth is monitored as the conducting wires break. The RF module consists of a power divider, a rectifier and a voltage-controlled oscillator. To determine the optimum carrier frequency, numerical computations of the propagation of RF waves must be conducted.

Recently, Zhao *et al.* developed a prototype wireless guided wave system for the inspection of layered structures such as an aircraft wing. The system includes a stationary antenna and an active antenna as transceivers, an on-board antenna as transponder, and PVDF comb transducers for Lamb wave generation and detection [ZHA 04]. A series of preliminary experiments on a 0.8 mm thick aluminum plate with a 12 mm long, 50% through-the-wall crack clearly showed the feasibility of flaw detection, in comparison with conventional wired or semi-wired approaches. Tests were conducted using a commercial off-the-shelf Vectronics® active antenna (designed for the reception of amateur radio waves from 300 kHz to 40 MHz) and hand-made passive monopole antennae made of aluminum tubes 580 mm in length and 6.5 mm in diameter, specially designed as leave-on-board flat panel antennae for aircraft wings. The originality of this contribution resides in the fact that the on-board antenna is simply wired to the PVDF sensor, the energy of the received signal being directly converted into ultrasonic guided waves through the transducer.

4.3.2.2.2. Wireless SHM systems with local processing ability

Wirelessly transmitting all of the data is much more inefficient than local computing followed by the transmission of the computed results only, that is to say waving the flag to say that the structure is damaged or not. A review of smart sensor technologies for civil infrastructure health monitoring is presented by Spencer *et al.* [SPE 04]. Smart sensors, as opposed to standard integrated sensors, are defined as sensors having intelligence capabilities, or hosting an on-board micro-processor. The micro-processor should enable data processing, analog/digital conversions, control of data flow, and interfacing with other systems. The four features which must be present in a smart sensor are an on-board central processing unit, small size, wireless communications, and the promise of being low cost [GRI 04].

In 2000, Mitchell *et al.* propose the use of distributed computing and sensing to detect damage in critical locations [MIT 00]. The system consists of several units, which are distributed over the structure and communicate with a central processing unit and amongst themselves through wireless technologies. Radio Frequency (RF) communication or commercial cellular phone networks are used. Each individual unit has a microcontroller, a wireless transmitter, actuators, sensors and data acquisition/processing capability. Each unit can either deal with information locally or pass it to the central processing unit. The laboratory setup is designed around a Motorola 8051 microcontroller development board. The proposed three-layer network architecture allows effective distributed computing and sensing and supports the detection and correction of data coding/retransmission errors.

Basheer *et al.* propose a new wireless SHM system architecture called the Redundant Tree Network (RTN) to address the issue of fault-tolerant self-organizing wireless sensor networks. In fact, wireless networks imply a number of issues such as interference, fault-tolerant self-organizing, multi-hop communication, energy efficiency, routing and finally reliable operation in spite of massive complexity [BAS 03]. RTN is a hierarchical network that exploits redundant links between nodes to provide reliability. Here, the sensor network is used for strain monitoring, as each sensor has the ability to process and transmit the raw strain data using neighboring sensors. The developed smart sensors, called ISC-iBlue, have the following hardware specifications:

ISC-iBlue v1.0	ISC-iBlue v2.0
25 MIPS, 8 bit Microcontroller from Cygnal (C8051F022)	ARM7TDMI , High performance, low power, 32-bit RISC processor
Wireless connection through Bluetooth (ROK 101007)	Wireless connection through Bluetooth (Bluebird from Inventel)
	eCOS-Real Time Operating System
8 channel ADC	3 channel ADC with 2KHz sampling rate
128 KB on-board RAM	512 KB on-board RAM

The focus of this work is to assess what communications are necessary for the network to self-organize with master and slave nodes. To this end, an algorithm is introduced to achieve a tree network of sensors with self-routing capability. The resultant SHM system should be able to perform complex feature extraction, with wavelet analysis or other methods, for accurate fault detection.

[TAN 03] demonstrates local processing capabilities with off-the-shelf components in a wireless SHM system. A Mote, an open hardware and open software platform for smart sensing developed at the University of California, Berkeley, is used as the wireless sensing system. Now distributed by Crossbow Inc. (www.xbow.com), Mote hardware platforms consist of two circuit boards. The first board is the data-processing/transmitting board, which holds a micro-processor, an ADC and a wireless transmitter (see Figure 4.18). The other is a sensor board containing two MEMS accelerometers. The Mote does not produce its own actuation signal and relies on input from an external shaker. The bandwidth of the accelerometers is 1 kHz and the micro-processor can sample eight channels at 4 kHz, but only by sequential multiplexing (the sampling rate is divided by the number of channels being used). The Mote is powered by standard AA batteries and has its own damage detection scheme, programmed in C and downloaded to the board, which simply consists of the computation of the cross-correlation coefficient of the two accelerometers. Tests carried out on a structure made of two aluminum plates connected with aluminum pillars, steel angles and steel bolts assess the severity of simulated damage very intelligibly: a yellow flashing LED indicates that a bolt is loosened, a red LED indicates the failure of the joint and a green LED flashes if the system is restored to its original state.

Figure 4.18. *Wireless data-processing/transmitting board of the Mote (dimensions 5.70 × 3.18 × 0.64 cm)*

Limitations pointed out by Tanner *et al.* concern the sensing resolution and range of the Mote (limited by the 10 bit ADC and low frequency bandwidth respectively), the difficult multi-channel feature extraction due to the low sampling rate, and the too small amount of ROM and RAM of the system, which limit computational and data storage capability respectively. The major limitation is the inability of the present system to resolve high frequencies, where much of the damage would, however, be detected.

To monitor civil engineering structures, Lynch *et al.* proposed in 2004 a wireless active sensing unit [LYN 04a], constructed with off-the-shelf components and having the ability to control both actuation and sensing. The sensing circuits wirelessly communicate with the computational core, which uses algorithms to process the acquired data and to broadcast the structural condition. It is equipped with a Motorola PowerPC MPC555 microcontroller, the 32-bit architecture which allows floating point calculations, and with 512 Kbytes of external RAM for sensor data storage. The MPC555 can process 32 channels at sampling rates up to 40 kHz. An actuation signal (10 V peak-to-peak at up to 40 kHz) is provided by the combination of a Texas Instruments DAC 7624 digital-to-analog converter and an Analog Devices AD620 amplifier. Wireless radio communications are integrated into the system with a MaxStream XCite radio. All of the components and software can be powered by a lithium battery pack. Feasibility tests of damage detection have been done on an aluminum plate with two SAW piezoelectric patches surface mounted in a pitch–catch arrangement.

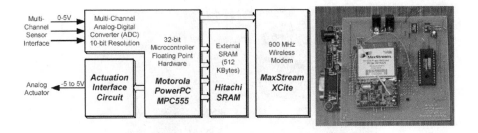

Figure 4.19. *Architectural design of the wireless active sensing unit. Dimensions of the fully assembled unit are 9 × 9 cm in area and 4 cm in height. [LYN 04a]*

The damage detection methodology proposed by the same author is included in the core of the wireless active sensing unit. It is capable of detecting damage as well as providing an estimate of its severity using the location of system identification model poles. The poles are determined from an auto-regressive with exogenous inputs (ARX) time-series model calculated using the input–output behavior of the structural element. The ARX model is used to detect damage in the structure by observing the shifting of poles. This is somewhat outside the scope of this chapter and a more detailed description of this novel methodology can be found in [LYN 04b].

The original prototype of [LYN 04a] and the evolution of the Mote system used by [TAN 03] are the two systems that have proved their ability to perform localized processing. Both are based on a series of three thin cards (sensors, processor, radio) sandwiched together, but are still cumbersome. Although the original Mote distributed by Crossbow Inc. has been upgraded to the Mica series (Mica, Mica2, and Mica2dot), the lack of programming memory and processing power noted by Tanner *et al.* still remains. To address these issues, Intel has started development of the next-generation of Mote platforms. The Intel project focuses on additional hardware and software improvements and increased levels of integration. The aim of this research is to develop an Intel Mote in the form of a single microchip with layered components, including sensor, nonvolatile storage, digital/analog silicon, battery and RF MEMS. By late 2003, the Intel Mote® prototype had reached the size of a Lego® brick, as shown in Figure 4.20.

Figure 4.20. *Intel mote® prototype (3 × 3 cm)*

Nowadays, the most miniaturized hardware platform in a single-chip CMOS device which integrates the processing, storage and communication capabilities to form a complete system node is the Spec chip, developed by JHL Labs (founded by Jason L. Hill, www.jlhlabs.com). This single chip only measures 2.5 × 2.5 mm (see 04.21) and contains: an AVR-like RISC microcontroller, an 8 bit On-chip ADC, an FSK radio transmitter, general purpose I/O ports, an RS232 universal asynchronous receiver–transmitter, 3Kb of memory and an encryption engine. This kind of tiny (and low cost when manufactured in quantity) wireless MEMS sensor shows the way for the development of future SHM systems based on networks of hundreds of distributed sensors.

Figure 4.21. *Single chip Smart Sensor Spec (dimension 2.5 × 2.5 mm)*

4.3.3. *Tested structures (mainly metallic or composite parts)*

Most SHM-related work deals with plate-like metallic or composite structures. Complexity can come from geometry (the presence of stiffeners in metallic parts for instance) or from the anisotropy of materials (composites), knowing that many current structural parts already combine both features. Strong anisotropy and complex geometry would unquestionably need specific data processing.

For instance, as a result of the specific fiber orientation, filament wound composite tubes are highly anisotropic (ultrasonic wave velocities vary with the angular direction). In [BEA 97], delamination detection in a graphite–epoxy wound tube using a time-of-flight algorithm thus needed to take this anisotropy into consideration. Estimation of delamination size seems feasible. More recently, Lin *et al.* proposed monitoring the integrity of filament wound structures (such as solid rocket motors and liquid fuel bottles) using built-in sensor networks (embedded SMART layer®) [LIN 03]. In this reference, however, the anisotropy of composite wound bottles does not need to be fully characterized for subsequent damage localization as this is undertaken through direct path approaches (see section 4.3.4.6).

Sandwich composite structures are perhaps the most difficult to monitor, because of the high damping of such structures. ONERA (France) is working on this problem in the particular case of radome materials and has had some success in detecting, localizing and sizing impact damage in a radome sandwich composite made up of GFRP skins and a low density foam core [OSM 01, OSM 02]. Kessler also investigated the possibility of applying acousto-ultrasonic SHM techniques to sandwich beams with four different cores: low and high-density aluminum honeycomb, Nomex™, and Rohacell™ [KES 03a]. To allow comparison, for each type of core material damage consists of partial skin-to-core debonding. For high-density aluminum specimens, impacted facesheets and core gaps are also introduced. Finally, a few samples with multiple core types are prepared for advanced testing. All kinds of damage are clearly detected and located through wavelet analysis of the collected Lamb wave signatures, though they have different effects on the wavelet spectrum (introducing phase shift of the central peak, widening the peak bandwidth, or producing a secondary peak).

4.3.4. *Acousto-ultrasonic signal and data reduction methods*

In the SHM context, the decision to repair or remove a structural part from service is the last stage of what K. Worden calls in [WOR 04] the waterfall model, which comprises 1) Sensing, 2) Signal processing, 3) Feature extraction, 4) Pattern processing, 5) Situation assessment and finally 6) Decision making. From the often extremely large amount of information collected with SHM sensors, the signal and data reduction (also called data fusion) methods described in this section encompass more or less all the tasks of stages 2 to 5 cited above in order to ease human decisions. In addition to the presence of a defect, any damage parameter that is extracted can contribute to improving the precision of the description of the current or future situation. The majority of studies focus on the recovery of the defect position and size, although some also evoke the criticality, the probable evolution, the orientation and the direction of propagation of the defect.

4.3.4.1. *Short-time Fourier transform*

The damage identification, localization and sizing procedure developed by the Department of Aeronautics and Astronautics of Stanford University is based on the use of a time–frequency analysis. This consists of calculating the spectral components of time-domain signals over sliding time windows using the short-time Fourier transform [WAN 99, WAN 01]. For each sensor, the scatter due to the presence of the damage is given by subtracting the baseline spectrogram of the sensor (measured in the original state) to the sensor spectrogram obtained after the occurrence of damage. The time of flight (TOF) of the scatter is deduced from the comparison of its peak amplitude occurrence with the actuator peak amplitude occurrence (see Figure 4.22). Using the TOF corresponding to the six possible actuator–sensor path configurations for a set of four transducers, and assuming an elliptical shape for the uniformly damaged area, the location, size and orientation of the damage are identified. Experimental validations of the technique with CFRP plate are presented in these references.

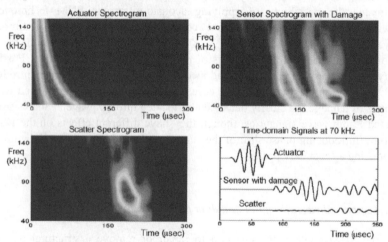

Figure 4.22. *Actuator and sensor signal spectrograms and time histories*

4.3.4.2. *Wavelet transform*

Wavelet transform-based data processing is increasingly used in acousto-ultrasonic techniques to detect, localize and size delaminations in composite materials.

ONERA has proposed the use of wavelet transform analysis to locate delaminations [LEM 99, LEM 00, LEM 01, LEM 03]. The signals delivered by disc-shaped PZT transducers, before and after the damage occurrence, are used. The multi-resolution processing by discrete wavelet transform (DWT) allows the isolation of various propagating modes in order to measure the time delay between arrivals of the direct burst and of the outgoing modes generated by the interaction with the delamination. Everything happens as if the delamination were a secondary acoustic source. Two comparative experimental validations are presented:

– the first one is made on a quasi-isotropic [452, 02, 452, 902]s carbon/epoxy plate of dimensions 700 × 700 mm (see Figure 4.23),

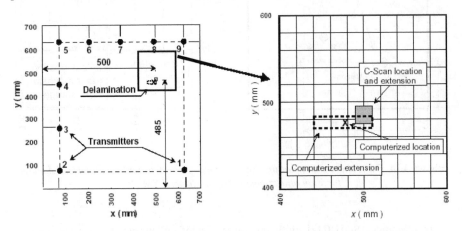

Figure 4.23. *Experimental setup of the 700 × 700mm plate (left) and computerized damage location and extension (right), from [LEM 00]*

– the second one uses data registered by the aforementioned Stanford University group on an orthotropic [02, 902]2s carbon/epoxy plate of dimensions 300 × 300 mm (see Figure 4.24).

In both cases, from the interrogation of various propagation paths, the localization of the delamination is quite good, although the estimate of the size of the damage is not very precise. The second case is close to the estimation provided by Stanford University using a different algorithm and given in reference [WAN 99].

Paget *et al.* have also used wavelet transforms [PAG 02a, PAG 02b, PAG 03b]. The decomposition of the signal into wavelet coefficients is improved compared to classical Morlet or Debauchies wavelets (used in [LEM 01]) by utilizing specially adapted wavelets based on the recurrent waveforms of Lamb waves. Changes of the wavelet coefficient magnitudes when the Lamb wave interacts with the damage allows the size of the delamination to be identified. The wavelet technique also shows great sensitivity in detecting damage of small size, but suffers from its inability to locate the damage. All things considered, this wavelet-decomposition method is essentially complementary to that of ONERA presented above.

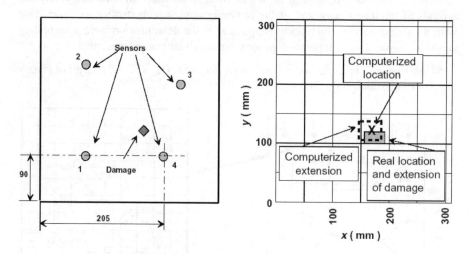

Figure 4.24. *Experimental setup of the 700 × 700mm plate (left) and computerized damage location and extension (right), from [LEM 00]*

In reference [SOH 03a], with the same aim of enhancing the wavelet method sensitivity by making the physical Lamb signal and the mother wavelet very similar, Sohn *et al.* input into the emitter the electrical duplicate of the specific wavelet waveform. The receiver signal is then processed by the matched wavelet transform to extract the flaw features.

Reference [GIU 03b] gives a comparative study of the Short-Time Fourier Transform (STFT) and discrete or continuous Wavelet Transforms (WT) for data processing of acousto-ultrasonic signals. The authors have integrated the two methods in their GUI (see Figure 4.15), the WT using the functions of the MatLab Wavelet toolbox. Reference [KEH 02] compares simple time-of-flight analysis and wavelet transform analysis, assuming the existence of mode conversion processes.

Mode conversion causes distinct wave packets to overlap. In this case, the resolution of STFT is found to be insufficient to extract the exact TOF of each packet. The WT technique is efficient, but becomes difficult to apply with low signal-to-noise ratios and/or strong dispersion of the wave.

Other groups working with wavelet transform analysis can also be mentioned:

– the department of Aeronautics and Astronautics at MIT ([KES 02, KES 03a, KES 03b]), which uses wavelet decomposition with a modified Morlet mother wavelet,

– the Laboratory of Smart Materials and Structures (LSMS) of the Center for Advanced Materials Technology (CAMT) at the University of Sydney ([SU 02]), which processes the signal with an algorithm based on DWT to identify the various activated modes and with a continuous wavelet transform (CWT) to make a spectrographic analysis of the signals,

– the group at the University of Nanjing ([XIA 99]).

4.3.4.3. *Time-reversal techniques*

In the work discussed above, the detection of a defect at an unknown location requires a systematic exploration of the full 2-D space by varying the angle/distance parameters. Even if the process can be totally automated using phased arrays, the majority of the acquired data are of no use. This problem can be solved by automatically adjusting the focus of the phased array to point at the reflection/diffraction sources responsible for the main echoes. This is possible since Matthias Fink demonstrated the time-reversal principle for ultrasound [FIN 92, FIN 97]: the wave motion equation only possess a second-order time-derivative operator and is thus invariant to the time reversal transform ($t \Rightarrow -t$). Ultrasound autofocusing is, in practice, achieved in three steps, which are as follows.

First, the sample of interest is illuminated with a widely diverging wave, using a small number of emitting elements. In response to this incident insonification, the defect behaves as an active source and generates an echo in the direction of the array.

Second, the array behaves as a receiver and measures the ultrasonic field coming from the target. Temporal signals on each element of the array are digitized with an Analog/Digital converter and stored in register memories (at this point, the signals can be amplified and filtered).

Lastly, using a Digital/Analog converter, the registered signals are time-reversed, that is to say they are simultaneously re-emitted by all the elements of the array.

It emerges from the theory and experimental observations that this procedure ensures that the acoustic field is focused at the target position.

Depending on the control software, the procedure could be carried out on the source which produces the first echo to arrive, or on the one that produces the strongest echo. However, in neither case is it certain that these sources are the damage which should be monitored. Thus, definite knowledge of the origin of the echoes, in particular those issued from sources that are not of interest (interfaces, edges, etc.), is needed, and could be obtained from the data registered before any damage occurrence (baseline). A widely used strategy in case of several interesting sources is to sequentially focus on the sources in descending order of the amplitude of their echoes.

Such a pulse–echo system will theoretically work correctly only if no dispersion phenomena occur in the transmission. While this condition is fulfilled in the case of bulk waves or Rayleigh waves, it is not fulfilled for Lamb waves used in many SHM systems. For instance, the time-reversal technique would work with the SWISS System, but not with the laboratory reference system [GIU 03a] (see section 4.3.6). However, some experimental work tends to use the time reversibility of Lamb waves and there is some evidence that the dispersion of Lamb waves can be compensated through the time reversal process. Indeed, dispersive Lamb waves have wave packets traveling at higher speeds, which thus arrive at the sensor earlier than those traveling at lower speeds. During the time-reversal process, the slower wave packets are re-emitted first. Therefore, wave packets traveling at different speeds all concurrently converge at the source point during the time-reversal process, compensating the dispersion. Figure 4.25 illustrates this principle in a typical experimental setup for a Lamb-wave plate investigation [SOH 04]. In contrast to the case of bulk waves, the time reversibility of Lamb waves has not yet been fully investigated. [PAR 05] reports an investigation of the time reversibility of Lamb waves exerted by PZT patches on a composite plate by introducing the time-reversal operator into the Lamb wave equation based on the Mindlin plate theory.

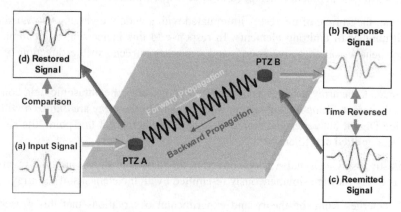

Figure 4.25. *Schematic concepts of damage identification in a composite plate through time-reversal processes, from [PAR 04]*

In references [PAR 04, SOH 04], an enhanced time reversal method is proposed. It consists of a combination of input waveform design (Morlet mother wavelet) and a multi-resolution signal processing based on Wavelet Transforms. The authors show that the time reversibility of Lamb waves could be preserved within an acceptable tolerance in the presence of background noise. The detection of defects is achieved when the time reversibility of Lamb waves is violated by the wave distortion due to wave scattering by a defect along a direct wave path (for more details about the direct path approach, see section 4.3.4.6).

4.3.4.4. *Synthetic time-reversal method*

Wang *et al.* try to define a methodology which would work with distributed transducer networks functioning in transmission, as well as pulse–echo modes and which could take into account the effect of dispersion in plate-like structures. A plate theory based on Mindlin flexural theory and on Kane–Mindlin extensional theory has been developed to describe the dispersive behavior at frequencies below the cut-off frequency of the second flexural wave. The experimental validation described in [WAN 03a] is based on a series of two experiments.

In the first experiment, four piezoelectric patches are bonded on an undamaged aluminum plate. Sensor A emits a Gaussian pulse, which excites both flexural waves, for which dispersion is important, and extensional waves. Sensor B receives a signal showing dispersion. When transducer B is subjected to the time-reversed version of the Gaussian signal, the signal received by A is compared to calculation and to the initial Gaussian signal. There is a relatively satisfactory agreement between calculation and experiment, and the recompressed signal received by transducer A is not very different from the initial Gaussian excitation, the difference being due to dispersion. This experiment demonstrated to a certain degree that temporal and spatial focusing result from the time-reversed waves. Nevertheless, the waveform of the time-reversal response cannot be identical to the original excitation, as demonstrated theoretically and experimentally by the authors.

In the second experiment, on the damaged plate (a mass bonded onto the plate simulates a delamination), a tone burst excitation is used instead of a Gaussian pulse. This defines a method that is not precisely time reversal, which is impossible in such 2D configuration with this type of transducer, but strongly increases the signal-to-noise ratio for the damage diffracted wave. The proposed method, called the synthetic time-reversal method, consists of emitting a signal from one particular transducer, registering the diffracted signals at the other transducers of the network, calculating their relative time delays and finally reconstructing the diffracted signal that would be received by the chosen transducer by superimposing the responses of the other transducers after time-shifting each signal with the appropriate delays. This method will only work with tone burst excitation since the time-reversed tone burst is itself a tone burst. When applied to a network of four transducers, the technique provides an amplification of the diffracted signal by a factor higher than 2, with a better signal-to-noise ratio.

In conclusion, the authors of [WAN 03a] have demonstrated that the shape of the time-reversal wave is not exactly the same as the original excitation, this difference being mainly due to the frequency dependence of the mechanical transduction efficiency of the circular piezoelectric transducers used as sensors and actuators, and that both spatial and temporal focusing can be achieved by time reversal.

4.3.4.5. *Migration technique*

References [LIN 99, LIN 01a] propose the use of the "migration" technique, which is derived from geophysics, to recover the location and shape of reflecting/refracting/diffracting defects. The technique is based on the same idea as the previous one. The reconstruction in this case is made via numerical finite-difference calculations. The signals recovered by the receivers (positioned along a line including the emitter) are time-reversed, back-propagated and "stacked" to create image snapshots of the displacement field, in particular at the moment at which all back-propagated waves precisely converge on the defect. An image of the defect is thus obtained. More detailed explanation of the procedure is given in [LIN 01b].

The technique described in these references was limited by the assumption of isotropy of the material, which has been verified for metals. Recently, the technique has been applied to anisotropic composite materials [WAN 03b], the narrow-band signal generated being considered as quasi-non-dispersive and the group velocities taken as a function of the propagation direction. The method is now capable of imaging several defects present in a composite plate, and the quality of the defect image has been improved thanks to the use of several excitations from distinct transducers.

4.3.4.6. *Direct path approach*

In parallel to the very sophisticated techniques described above, real-world applications sometimes use simpler algorithms. Reference [DUG 03] presents an application of the SMART Layer® to a large aeronautical composite structure. Here, the direct path approach is preferred to the time-of-flight approach because the wavelength is relatively short compared to the target dimensions (delaminations 4 cm in diameter).

The method consists of acquiring the baseline signals corresponding to all possible actuator–sensor paths in the structure in its undamaged condition, and repeating the same operation after damaging the structure. The difference between the new set of data and the reference set gives the set of scattered signals. The extent of the damage is deduced from the amplitude of the scattered signal.

Two techniques are possible:

a) Signal time-windowing: a time-window, no larger than one bending-mode wavelength and corresponding to the first arrival of the signal, is selected, ensuring that only scatter from defects near the direct emitter–receiver path is analyzed. Only paths showing significant scatter will be visualized in the final image of the structure, with a fixed width and an intensity proportional to the amplitude ratio of the scatter to the baseline signal (weight factor). Both location and extent of the damage are assessed by visually identifying the points of intersection of relevant paths. By summing the contributions of all paths, the damaged region is visually highlighted. After all contributions have been summed, the highest peak is re-scaled to match the maximum weight found for the region. In reference [LIN 03], Lin *et al.* implement this technique on a filament-wound tank with embedded PZT patches. The scatter signals from all possible paths on the tank are displayed together to produce a comprehensive image that shows their relative magnitudes. Since the amplitude of each path is a function of its propagation distance, this factor has to omit in order to make the signals directly comparable. The result in Figure 4.26 indicates the presence of damage. This display technique can be used as a fast imaging method to help visualize the approximate location and extent of damage due to an impact. Reference [SOH 04] presents an interesting procedure that iterates the process in order to detect multiple damage locations. To achieve this, once the more critical damage is localized, any "damaged" paths that are not intersecting the first location that is obtained are processed again for the identification of an additional damage location on the virtual grid that pictures the plate.

b) Intensity weight factoring: this alternative approach to computing the weight factor, leading to analogous results and not requiring time-windowing, consists of taking this factor as equal to the ratio of the cross correlation of the scatter signal with the baseline signal to the baseline signal auto-correlation. Although intensity weight factoring does not lead to results that are as accurate as signal time-windowing, it has the advantage of not requiring any information about the structure since it does not use TOF. This topic is discussed in the recent work of Prasad *et al.* from the Center for Non-Destructive Evaluation of the Indian Institute of Technology of Madras, who presents an SHM technique of Lamb wave tomography on anisotropic composite plates with PZT surface-mounted sensors [PRA 04]. The accuracy of the method depends on the total number of sensors and their spacing. Both of these parameters are influenced by the ultrasonic attenuation and the anisotropy of the host material, as well as by the potential presence of stringers. The latter may cause a strong attenuation along the through-stringer paths.

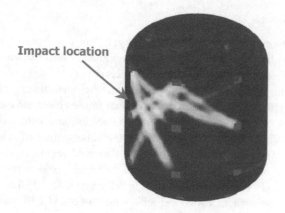

Figure 4.26. *Damage identification result using signals collected by a network of built-in sensors, from [LIN 03]*

4.3.5. *The full implementation of SHM of localized damage with guided waves in composite materials*

This section highlights the chronological steps of a comprehensive implementation of a Lamb wave-based technique using piezoelectric sensors surface mounted onto anisotropic composite plates.

If it is operated in the frequency range of its in-plane vibration mode, a piezoelectric disc-shaped sensor generates guided waves able to propagate across the tested structure. A particular Lamb mode for a given structure can be selected by tuning the driving frequency in relation to the radial dimensions of the sensor [MON 99, MON 00]. As mentioned above, Lamb waves provide the advantage of propagating over quite long distances, so the tested volume in a single acquisition is increased in comparison with most techniques based on bulk waves.

Giurgiutiu [GIU 03a] proposes an interesting simple model describing the actual shear layer interaction between the sensor and the structure. He also explains how the selection of a particular mode works. The point is that bonded piezoelectric sensors have high efficiency because they are strain-coupled to the structure. As a consequence, the selection results from a wavelength mode tuning phenomenon.

The increasing development of techniques based on Lamb-like modes results from their interesting properties:

– the dependence of the stress and strain distributions on frequency, which gives them a multi-mode character, allowing the selection of the mode best fitted to a given type of damage;

– their long-range propagation, which results in a reduction of the inspection time and makes their use cost-effective;

– their ability to follow complex shaped structures (curvatures, stringers, rivets, corners, etc.), which is useful when accessibility problems have to be overcome.

In the following section, the ability to apply piezoelectric patches in SHM will be illustrated through experimental results obtained on two different composite materials.

4.3.5.1. *Materials characterization*

Two different types of composite samples were studied. The first one was a Glass fiber Reinforced Plastic (GRP) plate (of thickness 3 mm), where the short glass fibers (volume fraction 33%) are randomly oriented in a polyester matrix. The second type of sample was a Carbon Fiber Reinforced Plastic (CFRP) plate (of thickness 3 mm), made up of 22 carbon/epoxy unidirectional plies stacked according to the sequence $[45°,[135°_2,45°_2]_{10},135°]$.

In dealing with Lamb-like modes, the first step consists of measuring the stiffness tensor and the density of the host material. Then the guided-mode dispersion curves can be calculated together with their stress and strain distributions. This theoretical approach is crucial in order to be able:

– to identify the various echoes in the transmitted waveforms;

– to simplify the data interpretation;

– to assess the different mode sensitivities to different kinds of defect;

– to locate damage.

The ultrasonic evaluation of the elastic properties of materials is based on the relationship between the elastic constants and the phase velocities of ultrasonic bulk waves. The elastic constants of the GRP sample were measured using an immersion device, which allows the determination of most of the elastic constants of the material. This technique is based on the measurement of ultrasonic bulk wave phase velocities under variable incidence. An ultrasonic wave is launched in a water tank in which the sample is immersed. According to Snell's law, at the surface of the sample this wave is converted into quasi-longitudinal and quasi-shear bulk waves propagating through the sample. These modes are then converted again into bulk longitudinal waves at the back face of the sample. The signals are transmitted to a receiver. The emitter and receiver make the same angle with respect to the surface of the sample. The shear and longitudinal wave velocities in the sample are computed from the cross-correlation with a reference signal obtained between the emitter and receiver when the sample is removed. Finally, from the whole set of quasi-longitudinal and quasi-transversal velocities, and using a non-linear optimization process, the elastic constants of the material and their uncertainties can be deduced.

For the GRP sample, whose density was $\rho_{GRP} = 1926 \pm 6$ kg.m^{-3}, the following stiffness tensor was measured:

$$C_{GRP} = \begin{pmatrix} 20.3\pm1.2 & 6.0\pm0.7 & 6.5\pm0.5 & 0 & 0 & 0 \\ 6.0\pm0.7 & 20.3\pm1.2 & 6.5\pm0.5 & 0 & 0 & 0 \\ 6.5\pm0.5 & 6.5\pm0.5 & 16.2\pm0.2 & 0 & 0 & 0 \\ 0 & 0 & 0 & 4.3\pm0.3 & 0 & 0 \\ 0 & 0 & 0 & 0 & 4.3\pm0.3 & 0 \\ 0 & 0 & 0 & 0 & 0 & 7.1\pm0.4 \end{pmatrix} \text{GPa}$$

In the case of the CFRP samples, the ultrasonic quasi-longitudinal and quasi-shear bulk waves overlap. This prevents the use of the method described earlier for the characterization of this material. Hence, the elastic constants on the diagonal of the stiffness tensor were determined from contact measurements using classical bulk wave transducers. The remaining elastic constants were deduced through the inversion of S_0 Lamb wave velocity measurement with respect to the direction of propagation. For the CFRP sample, whose density was $\rho_{CFRP} = 1578 \pm 8$ kg.m^{-3}, the following stiffness tensor was obtained:

$$C_{CFRP} = \begin{pmatrix} 72.5\pm6 & 3.9 & 4.5 & 0 & 0 & 0 \\ 3.9 & 72.5\pm7 & 4.5 & 0 & 0 & 0 \\ 4.5 & 4.5 & 15.0\pm0.8 & 0 & 0 & 0 \\ 0 & 0 & 0 & 5.1\pm0.2 & 0 & 0 \\ 0 & 0 & 0 & 0 & 5.2\pm0.2 & 0 \\ 0 & 0 & 0 & 0 & 0 & 7.1\pm0.3 \end{pmatrix} \text{GPa}$$

4.3.5.2. Damage introduction and characterization

Whatever the material, the sample is instrumented with pairs of identical sensors, one acting as source and the other as detector. Instead of being embedded, they are bonded onto the surface for practical reasons, but without any loss of efficiency in the generation and detection of Lamb waves.

To introduce realistic damage, such as that observed during the service life of real aeronautic or automotive structures, it was decided to use low-speed impacts, that is to say, impacts with a typical velocity below 20–40 m s^{-1}. For example, these impacts can be representative of the accidental fall of a tool onto an aircraft composite part during maintenance.

The drop-weight calibrated device that is used can record the time history of the load and the displacement of a 20 mm diameter hemispherical impactor. As the impacts are controlled in a very reproducible way, the damaging threshold below which no critical damage occurs can easily be determined. Whatever the damaging mechanism inside the material, the energy corresponding to the creation of new surfaces is released and results in abrupt variations in the load history. In the literature, it is widely agreed that low-speed impacts produce matrix cracks, fiber breakage and delamination. Many parameters, such as the mechanical properties of the composite material, the plate thickness and the stacking sequence of the plies, determine the response of the structure to an impact. Figure 4.27 represents the load history obtained for a series of impacts with increasing energy carried out on several identical samples. It appears that the first damaging impact corresponds to impact number four of energy 6.38 J. One can therefore conclude that in the 6 mm thick CFRP plate, the damage threshold is between 6.13 J and 6.38 J.

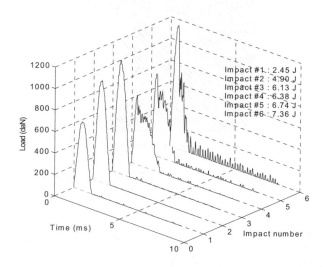

Figure 4.27. *Determination of the damage threshold in CFRP*

In the case of the GRP samples, the damage threshold was found to be about 0.7 J, which is much lower than in CFRP. Preliminary visual observations also reveal a quite different damage mechanism: large cracked areas are visible around the impact point. Local heterogeneity in fiber density inside the plate is probably responsible for the observed dispersion in the shape and size of the cracked areas as the impact location changes.

4.3.5.3. *Lamb wave measurement results*

4.3.5.3.1. Damage detection through Lamb wave interrogation: CFRP case

In this experiment, four successive impacts of energy 2, 4, 8 and 8 J were performed at the same location, keeping in mind that the damage threshold was around 6 joules for this plate. The location of the impacts was in the direct path between the emitter and the receiver, this path being parallel to the direction of the fiber's first ply (Figure 4.28). In Figure 4.29, it can clearly be seen that the two lightest impacts are non-damaging, as there is no abrupt decrease in the load of the impactor during the test. The load curves of the two heaviest impacts exhibit an abrupt decrease, which reveals the dissipation of energy inside the sample through damage processes.

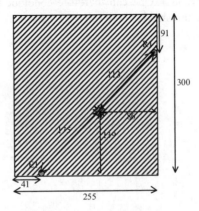

Figure 4.28. *Sensors and impact locations on the CFRP plate*

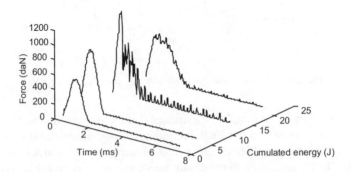

Figure 4.29. *Experimental time history of the impactor load during four successive impacts of energy 2, 4, 8 and 8 J*

The actual presence and the size and shape of the induced damage have been checked using a classical ultrasonic C-scan procedure. This consisted of scanning the impacted area with a high-resolution 50 MHz focused ultrasonic transducer. Windowing the time records makes it possible to visualize echoes from various depths. The overall thickness of the 22-ply plate being equal to 6 mm, the thickness η of the unitary ply is about 270 μm. Figure 4.30 shows C-scan images of the impacted area at depths $-\eta$–3η–5η–7η–9η and -11η from the surface. According to the stacking sequence mentioned above, these depths correspond to the interfaces between orthogonal plies. In agreement with the literature, the delamination is found to occur at the interface between plies having different orientations rather than between plies with the same orientation. Moreover, for each interface, this delamination exhibits a classical two-lobe pattern, oriented in the fiber direction of the lower ply. In the present case, each pattern is then orthogonal to the previous one and, the deeper the delamination is, the larger it is because of the larger bending of the sample during the impact.

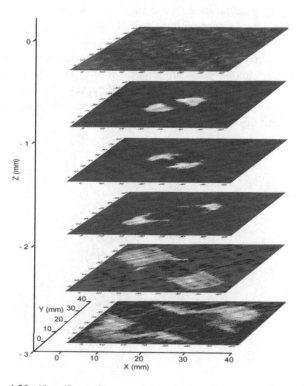

Figure 4.30. *40 × 40 mm C-scans at the interfaces between orthogonal plies (from top to middle of the plate)*

L-scans (for Leaky Lamb-mode scanning image) of the impacted area were also performed. This technique made possible both a second image of the damage and the ability to check the sensitivity of a given Lamb mode to this particular kind of internal flaw. Indeed, traditional echographic imagery such as the C-scan is particularly effective for precisely determining the conformation of the internal defects with respect to the structure (see Figure 4.30), but suffers from the need for a small sampling period, both in time and space. This makes the data processing slow and laborious. Maslov and Kundu proposed the use of leaky Lamb mode imagery, the main advantage of which is that it requires significantly coarser sampling, because of the large wavelength of the plate modes. This results in an important time saving. Moreover, the same authors showed the superiority of the L-scan technique in comparison with C-scan for the detection of particular defects, providing that a distinctive Lamb mode is chosen [MAS 97]. To achieve mode selection, the frequency is selected by filtering and the velocity is selected by tilting the transmitter and receiver at the same angle θ to the normal of the sample (see Figure 4.31). The value of θ is selected on the basis of the well-known Snell–Descartes laws. The axes of the transducers do not coincide on the surface of the plate but are separated by a sufficient distance with the aim of only detecting leaky Lamb modes. This avoids any interference between leaky waves and the direct reflected beam. Consequently, for a given angle, the frequencies for which leaky Lamb modes are generated correspond to the peak amplitudes of the transmitted spectrum. This technique, known as defocusing, has been retained because it shows least sensitivity to local thickness variation, which was caused by the presence of the impact-induced surface indentation of the sample.

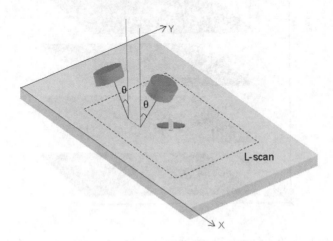

Figure 4.31. *Experimental setup for L-scan*

The L-scan image results from scanning the pitch–catch arrangement shown in 04.31 over the whole sample surface, and assigning indexed colors to peak spectrum magnitudes. The sensors used have a central frequency of 500 kHz, which thus corresponds to a 3 MHz·mm frequency–thickness product in the tested composite CFRP laminate. The modes that can be propagated for this frequency–thickness value are A_0, S_0, A_1, S_1 and S_2. As Lamb waves interrogate the whole thickness of the plate, L-scans are not able to precisely image each particular interface, but this technique nevertheless makes it possible to quickly test the sensitivity of a given Lamb mode to any kind of internal defect. For instance, Figure 4.32 shows the L-scan of the damaged CFRP sample for the first symmetric Lamb mode S_0. This is a fast and reliable damage inspection technique as it is far less time-consuming than the conventional high-frequency and high-resolution C-scans previously shown.

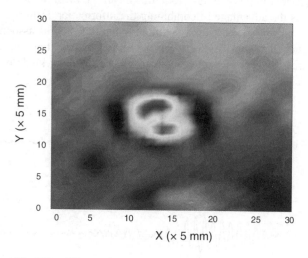

Figure 4.32. *150 × 150 mm around the impacted area leaky Lamb mode S_0 L-scan*

The structure was then interrogated with propagating Lamb modes generated by the surface-bonded piezoelectric patches. This consisted of exciting the sensor with a five-cycle toneburst of adjustable frequency, in order to select a particular Lamb mode, and recording the transmitted waveforms on the receiver, which is identical to the emitter. Figure 4.33 shows the S_0 transmitted waveforms in the undamaged case and after the damaging process. Waveforms corresponding to the first two impacts were not represented, as they were almost the same as the reference waveform. The S_0 mode was retained, as it appeared to give better results than the first anti-symmetric Lamb mode A_0 for the detection of these kinds of "in-plane" flaws. Nevertheless, it is difficult to find a reliable discriminating parameter through the observation of the time signals. Even if it is possible to predict the time of flight of a particular Lamb mode through the study of the general problem of guided waves in this anisotropic medium, leading to the calculation of the dispersion curves, the complicated configuration of the source, sample, edges and sensor position makes the identification of every contribution of the waveform ambiguous. Hence, it was decided to use spectral analysis. The Fast Fourier Transform of each signature was computed and, to take into account the instrumental dispersion of the results, a set of 100 unaveraged reference signatures was recorded in the undamaged plate, and 10 unaveraged signatures were recorded after each impact. In Figure 4.34, the maximum magnitudes of the transmitted waveform spectra are plotted in the complex plane. The grid is made up of curves representing several attenuation levels (the 0 dB reference level corresponds to the average amplitude of the measurements in the undamaged case) and rays that correspond to the phase shift induced by a time delay equal to one sampling point from the mean of the references. This 2-D representation has the advantage of combining the information from two significant parameters: the transmitted spectrum maximum magnitude and its associated phase. This enables better discrimination between the unfaulted condition and the various damage signatures. Indeed, this often leads to all the different kinds of signatures being distinctly clustered.

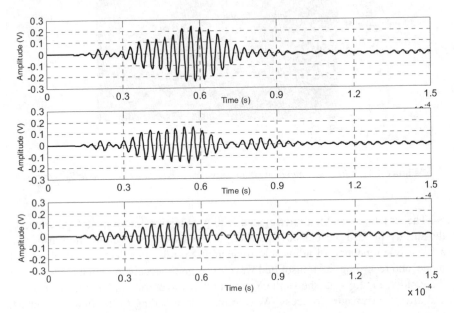

Figure 4.33. *Time signals transmitted between emitter and receiver in the undamaged case (top), after the first 8 J impact (middle) and after the second 8 J impact (bottom)*

Figure 4.34 shows this representation, for the undamaged plate and then for the two light impacts followed by the two heavy ones previously depicted in Figure 4.28. It is not possible to discriminate the sets of points representing the first two impacts from the reference set, but this result is consistent with the fact that the energy of these impacts was below the damage threshold. The high-energy impacts are easily detected as the overall amplitude of the set of stars representing the first 8 J impact is attenuated more than 1 dB and its phase is shifted from the reference set. Moreover, one can easily see that a second heavy impact at the same location and of the same energy damages the composite further, as the overall amplitude of the signature of the second 8 J impact is more attenuated and also more phase shifted.

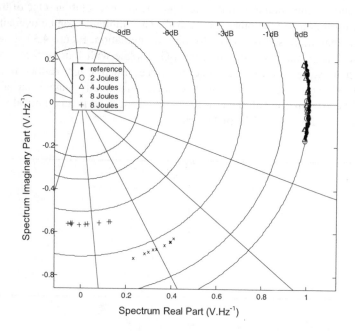

Figure 4.34. *Transmitted spectrum maximum in CFRP plate*

4.3.5.3.2. Damage location through Lamb wave echo measurement

The problem of damage location has also been assessed. In the same types of sample, the introduced damage was a simple hole of diameter 10 mm, with the objective of evaluating the efficiency of the method. Using pulse/echo measurements, a graphical construction was used to identify the points of intersection between the curves that represent the possible damage location.

Through ultrasonic material characterization, the GRP sample was found to be transversely isotropic, that is to say its stiffness tensor is invariant with respect to any rotation around the normal of the plate. In other words, the velocity of the Lamb modes is constant, leading to the possible location curves being circles of radius:

$$c\tau_i = \sqrt{(x_i - x)^2 + (y_i - y)^2}, \; i = 1...4 \qquad [4]$$

where c is the velocity of either A_0 or S_0, τ_i is the TOF of an echo, (x_i, y_i) are the coordinates of the emitter and (x, y) are the coordinates of the unknown damage. Experimentally, the intersection of such circles from two emitters generally gives two intersection points, rather than the ideal unique correct solution. One of them is the expected solution and the other is a "ghost". The ambiguity can be overcome by using at least one more sensor and performing triangulation. Figure 4.35 shows the results of such a graphical construction for a GRP plate with four surface-mounted sensors, which were interrogated three by three. Each cross results from the intersection of two possible S_0 and/or A_0 echoes. This finally enables an efficient location of the 10 mm hole: all the crosses could be enclosed in an elliptically shape surface whose area is 6.9% of the plate surface.

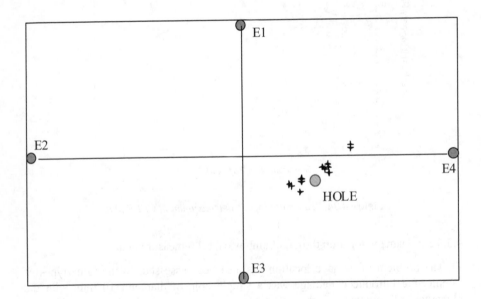

Figure 4.35. *10 mm hole location in a 250 × 450 mm GRP plate using four sensors*

In CFRP, a graphical construction was again used to identify potential damage locations. This was significantly complicated by the anisotropy of the material, which required the velocity curve versus the angle of propagation direction to be computed for each fundamental Lamb mode. As there are no analytical solutions for the angular dependence of Lamb wave velocities, there is no simple relationship between the TOF of an echo and the possible location of its source. That is why no equation similar to equation [4] is derived here, the only worthwhile method for localization through TOF measurements in such an anisotropic material being the graphical construction of the possible source loci. S_0 velocity was measured in various directions, and the velocity curve was fitted to the experimental points under the assumption of a quadratic symmetry (Figure 4.36). One can also see in this figure that A_0 (dashed line) has relatively less angular dependence than S_0 (solid line). 04.37 makes possible better visualization of the angular dependence of the phase velocity of the A_0 mode (Figure 4.37).

Figure 4.38 shows the effective damage location in the CFRP sample with the help of six PZT discs, 10 mm in diameter, bonded onto the plate edges. The area of the dashed box in Figure 4.18 is 6.25% of the probed area. This kind of well-controlled damage in the shape of through-holes is not representative of the expected defects in aircraft panels, but the holes were introduced as a preliminary step towards proving the operability of the technique in such anisotropic cases. Advanced localization techniques presented earlier (see Figure 4.38) should also apply here for purposes of structurally integrated location.

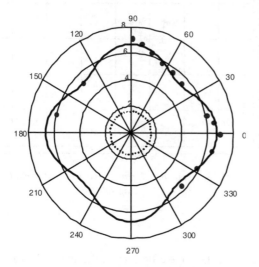

Figure 4.36. *Velocity curves (km/s) of S_0 (solid line) and A_0 (dotted line) in CFRP cross-ply laminate*

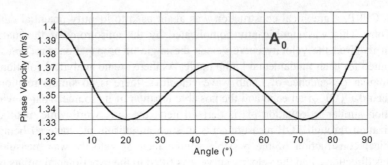

Figure 4.37. *Angular dependence of the A_0 Lamb mode velocity (at frequency 150 kHz)*

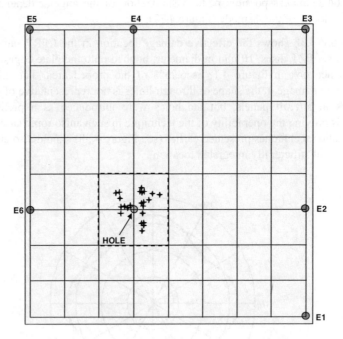

Figure 4.38. *10 mm hole location in a 300 × 300 mm CFRP sample using six sensors*

4.3.5.3.3. Conclusion

The results presented show the efficiency of the attached piezoelectric element in generating and sensing Lamb-like modes for the detection of distant localized flaws. The sensitivity of the fundamental Lamb modes to real damage produced by low-speed impacts can be related to the plate properties and damage mechanisms.

The interest of such a multi-modal technique has been pointed out, since a particular Lamb mode can be arbitrarily selected by tuning the excitation frequency of the bonded sensor. Consequently, it is possible to improve the sensitivity to a given defect in a given structure. The shape of the measured waveform is highly dependent on the interaction of the selected mode with the induced damage, but also on the geometry and the anisotropy of the structure to be tested, because these parameters influence the number of guided waves that can propagate and interfere at the receiver. So, for the specific case of the structures described here, the evolution of the maximum spectral amplitude and its associated phase were observed. Indeed, these parameters of the propagating modes had previously been defined as the most significant. Moreover, the use of the same sensors on the same samples was proved to be efficient for the location of simple damage.

4.3.6. *Available industrial acousto-ultrasonic systems with piezoelectric sensors*

Two SHM technologies based on piezoelectric-sensors are industrially available. The first is the SMART Layer® developed by Stanford University and put on the market by Acellent Technologies, Inc. This consists of a conformable network of piezoelectric patches emitting and/or receiving plate waves. The second one is the SWISS system, developed by EADS Germany/Siemens NDT, which uses a phased array of bulk wave transducers working in pulse–echo mode. Aside from the fact that they use different types of waves, the two systems achieve SHM from two different viewpoints. In the first case, the sensitivity of the method is distributed throughout the structure, and having a number of sensors means that enhanced beam forming does not take place in specific directions. In this context also, the geometry of plates makes the intensity of the wave inversely proportional to the distance traveled between emitter and receiver. The second system tends to increase the sensitivity and the operating range of one single highly specialized sensor through beam steering and focusing (the basis of which has been discussed above; see section 4.3.2.1).

4.3.6.1. *SMART Layer*

Professor Fuo-Kuo Chang, at Stanford University, conceived the Stanford Multi-Actuator–Receiver Transduction (SMART) Layer. The network of distributed thin piezoelectric patches (emitters and receivers) of this technology is embedded into a thin dielectric film, which can be surface mounted onto metallic structures, or embedded in composite structures before curing (see Figure 4.39). In parallel with the hardware design, software dedicated to data acquisition, data processing and diagnosis was developed. The complete system is commercially available from Acellent Technologies and can be adapted to various types of applications on demand. The hardware and software products can be purchased separately or together [LIN 02a]. The usual configurations are:

– The SMART Layer®: the film is available in various configurations: as thin strips with the sensors in one row; as a flat 2-D layer of typical dimensions up to 0.6 m × 0.9 m; and as 3-D customized layers to fit specific structures [LIN 01c]. The standard thicknesses of the film are 60 μm and 125 μm and of the piezoelectric patches 250 mm and 750 mm. The distance between the piezoelectric transducers is adjusted according to the attenuation of the material of the structure on which the layer is fixed. Generally, the acceptable spacing falls between 100 μm and 250μm.

– The SMART Suitcase®: This is the test instrument specifically designed to drive the piezoelectric emitters of the SMART Layer (with voltages up to 200 V at frequencies up to 1 MHz) and to record high-resolution data from the receivers, with a 50 MHz sampling rate. The standard launched waveform is a five-cycle sine burst, but arbitrary waveforms can be programmed. The standard setting allows simultaneous operation of 30 input/output channels. Finally, the SMART Suitcase provides a PC platform for data storage, analysis and display. Data collection uses either a stand-alone Windows executable file or routines that can be called within other programs, MATLAB in particular.

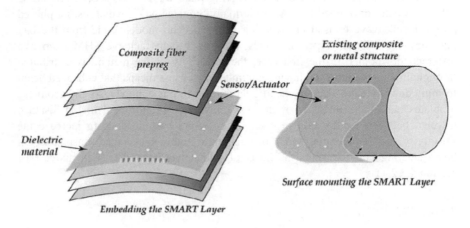

Figure 4.39. *Installation of SMART Layer in uncured composite laminate or onto an existing structure*

Although the SMART Layer is mainly used for composite structure (the thin layer is particularly well suited to this kind of application), it can also be used for crack monitoring in metallic structures. References [BOL 02, LIN 02a, IHN 02] present results obtained with the SMART Layer on aluminum structures. For instance, the detection of fatigue cracks around rivet holes is covered in reference

[LIN 02a]. For that particular application, a couple of narrow strips are mounted close to the rivet line, on both sides of the panel. The strips, which each comprise one single row of piezoelectric transducers, are positioned so that each patch lies between two adjacent rivets. With this configuration, the diagnostic wave passes across the row of rivets. The degradation of the signals indicates the presence of a growing crack and is related to the crack length. Reference [IHN 02] presents fatigue tests of an aluminum plate with a notch (the dimensions of the plate are approximately 0.3 m × 0.4 m × 3 mm). Three pairs of piezoelectric transducers, driven at frequencies between 60 and 600 kHz, are used to follow the progression of the crack. The distance between emitter and receiver is 0.18 m. The monitoring of the crack growth is due to the definition of a damage index based on the attenuation of the transmitted wave. The choice to use a transmission configuration, instead of a reflection configuration, means that the pairs of piezoelectric transducers have to be kept a relatively short distance apart. The monitoring of long cracks or large areas where the starting location of the crack is a priori unknown would thus require a large number of piezoelectric sensors.

The same reference gives a second example of the application of SMART Layer to metallic structures: the monitoring of cracks at bolted–bonded lap joints. Here, two lines of rivets are separated by 22 mm. The distance between rivets in the same line is 17.2 mm. Between two particular successive rivets in one of the lines an artificial edge crack, 1.27 mm long, has been introduced. The strip of piezoelectric emitters is positioned between the two rivet lines. The strip of receivers is moved 120 mm away from the strip of emitters. Signal generation and sensor data storage, using the SMART Suitcase, took 22 s to cover 19 rivets. The damage index resulting from the measurements during a constant amplitude fatigue test under tensile loading is compared to that from eddy-current and ultrasonic NDE. Probabilities of damage detection (POD) were evaluated for various threshold levels of the damage index, to find finally a suitable threshold value that enables the detection of crack lengths down to 5 mm and avoids false alarms when compared to conventional NDE assessments. More detailed description of the methodology and the signal processing can be found in [BOL 01] and [IHN 01].

When applied to composite structures, the SMART Layer makes it possible to identify impact events (passive technique) using the algorithms developed by Tracy and Chang [TRA 98a, TRA 98b] and very recent genetic algorithms developed by Haywood *et al.* [HAY 05] for simple plates; Seydel and Chang [SEY 99, SEY 01a, SEY 01b] have given a solution for stiffened plates. The Smart Layer also makes it possible to identify subsequent delaminations using the algorithms of Wang and Chang [WAN 99, WAN 01], and more recently the algorithm of Dugnani and Malkin [DUG 03] (active acousto-ultrasonic technique). The SMART Layer has been applied to a variety of composite structures, some with very complicated shape (see reference [VON 01], which presents an automotive application), some of a complex nature such as multi-layered thick composites used for armored vehicles

[PAR 03], or of a very large size (see for example reference [DUG 03], which presents an aeronautical structure equipped with 72 piezoelectric transducers included in several SMART Layers).

Acellent Technologies Inc. also develops hybrid SMART Layers equipped with both piezoelectric transducers and fiber-optic sensors [LIN 02a, LIN 02b]. In the intended application, the piezoelectric transducers will be used as actuators to input a controlled excitation to the structure and the fiber-optic sensors will be used to capture the corresponding structural response, the advantage being mode separation between actuators and sensors. A preliminary test, made with a SMART Layer equipped with just one fiber optic-sensor measuring strain, shows that the fiber-optic sensors can be successfully integrated into a SMART Layer without loosing sensitivity.

Finally, the recent development of the synthetic time-reversal technique [WAN 03a], which applies to distributed sensing technologies, will generate more interest in the SMART Layer concept.

4.3.6.2. *SWISS (Smart Wide-area Imaging Sensor System)*

Using an approach similar to medical ultrasonic imaging, the SWISS system is composed of small permanently mounted ultrasonic phased array sensors with equipment from Siemens NDT. In this way, the sensor installation area is very small compared to the area monitored. Measurement of the scattered waves allows the reconstruction of the monitored structure. At present, the technique can only be operated on the ground. On-ground interrogation of sensors may be started on demand via suitable links to on-board sensors. Significant parts of the equipment of the Siemens Alok PHased array Integrated and Reliable UT equipment (SAPHIR[plus]) may be fully ground-based, with a suitable link to an on-board interface. (SAPHIR[plus] is a third generation phased transducer array system developed by former Siemens NDT, and now commercialized by intelligeNDT, a subsidiary of Framatome ANP, an Areva and Siemens Company.) SAPHIR[plus] analysis software allows the scans to be presented as angle-corrected C-scans, tomographic displays or reconstructed images that take account of beam divergence. Printouts are self-explanatory as they contain all the relevant parameters (see Figure 4.40).

Detecting small areas of damage (for example, 1–2 mm cracks, corrosion) has been the major objective for the design, development and demonstration of SWISS. The other objective of these developments was to prove the ability to detect in parts with complex geometry. The functional and structural robustness of sensor instrumentation has been demonstrated in a fatigue ground test covering more than 60,000 simulated flight hours under a realistic fighter aircraft spectrum. This approach is evaluated and the results are reported in several references [NEU 01, KRE 01, KRE 03a, KRE 03b].

The system can image fatigue cracks or corrosion, which can be at a relatively long distance from the sensors (up to 100 mm, the distance being limited by adjacent boundaries of the part, such as an adjacent rivet row, and the required resolution). In references [NEU 01] and [KRE 01] the images of cracks originating from rivet holes in lines of rivets were obtained. The presence and the orientation of these cracks are clearly visible on the images. From these experiments, it can be inferred that 3–5 mm long cracks can be reliably detected in this configuration, and that initiation and propagation of such damage can thus be monitored. Reference [KRE 01] presents interesting results of corrosion imaging.

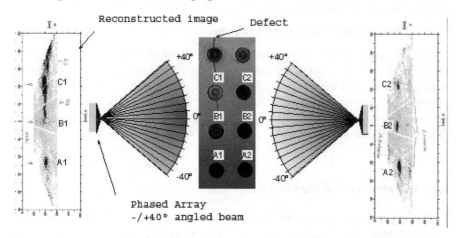

Figure 4.40. *Imaging of a crack originating from rivet holes in line of rivets, (by courtesy of intelligeNDT Systems & Services GmbH & Co. www.IntelligeNDT.com)*

Reference [KRE 03b] presents results for a 3-D part (also briefly reported in reference [NEU 01] and [KRE 03a]), in which it seems possible to have information up to 200 mm, and presents the detection of an unforeseen crack at 80 mm from the sensors.

These experiments have demonstrated that SWISS system:

– can be easily adapted for component applications, since phase array equipment and sensors have, in the meantime, been well tested in many NDT applications;

– needs little installation weight and surface (surface in comparison with that required for a classical strain-gauge).

Until now, the SWISS system has been oriented towards on-demand on-ground inspection. To become a fully, permanently installed on-board system, the hardware will have to be miniaturized and the higher layers of the SAPHIR[plus] system software upgraded to operate completely automatically, to avoid the need for human image interpretation. This task is difficult but of prime importance since the amount of information is tremendous. Furthermore, it would be interesting to assess the ability of the system to generate Lamb waves.

Finally, as stated in [KRE 03b], the system could be extended by:

– the simultaneous acquisition of acoustic emission information. In reference [KRE 03b], comparative measurements of in-flight noise and on-ground AE confirm that the impact detection algorithm developed before the development of the SWISS System can be implemented as a simple software upgrade;

– the development of a digital interface open to other available systems.

Finally, other SHM systems are available, although not especially based on piezoelectric devices. These include the SHM®-Industrial system from Oceana Sensor. SHM®-Industrial is an interface to Oceana Sensor's ICHM® suite of products, which provides device management and setup, data fusion and local decision support, as well as display and storage. ICHM® is a data acquisition and processing module built on Bluetooth™ wireless technology.

4.4. Electromechanical impedance

In the previous sections, the information about the condition of materials was extracted from the analysis of the varying voltage recorded between the piezoelectric element electrodes. These waveforms are representative of the interaction of elastic Lamb waves with the tested structure.

Another possible implementation of piezoelectric discs consists of analyzing their electrical impedance. This approach is commonly referred as the electromechanical (E/M) impedance technique.

4.4.1. E/M impedance for defect detection in metallic and composite parts

The electromechanical impedance technique has been tested for crack (or artificial local defect) detection in metallic structures by very few authors, even though they accept that in far-field, piezoelectric patches will not be able to capture local defects, unlike the wave propagation method, which can successfully assess the damage [ZAG 02]. Even at medium ranges, the electromechanical impedance technique has difficulties in detecting the simple presence of cracks. This has been clearly demonstrated by Giurgiutiu in some logical experiments and using various data processing techniques (overall-statistics damage metrics approach, features-based probabilistic neural network) reported in several references [ZAG 01, GIU

02b, GIU 02c, ZAG 02, GIU 03b]. Reference [TSE 02], from another group, presents some academic experiments with 5 mm diameter machined holes in aluminum plate. One of the experiments detects the presence of such an artificial defect at far-field, but with a geometry which gives more weight to one dimension. Furthermore, a hole is not very representative of a crack.

In conclusion, it can be stated that, despite the existence of several references relating to the monitoring of cracks in metallic structures, the electromechanical impedance method is definitely not suitable.

A few papers have been found describing applications of the electromechanical impedance method to the SHM of composite structures: [GIU 98, BOI 02, BOI 04].

The E/M technique has been used by Pardo de Vera and Güemes [PAR 97] with a simplified impedance technique (RC bridge) on GFRP, but only for the detection of an artificial defect (hole) in a quasi 1-D sample.

The technique has been used to detect bond breakage in a composite sandwich structure (rotor blade) at the University of South Carolina [GIU 98].

More recently, Bois and Hochard [BOI 02, BOI 04] presented modeling and measurement for the E/M technique applied to the characterization of delamination in C/epoxy composite. However, the application is limited to quasi 1-D samples.

Finally, it can be stated that applications on truly 2-D samples are needed to provide better evaluation of the possibilities of the technique and to show that the E/M technique is only really sensitive to delaminations in near-field configurations, which limits its interest.

4.4.2. *The piezoelectric implant method applied to the evaluation and monitoring of viscoelastic properties*

When driven with a constant intensity current of variable frequency, a piezoelectric element vibrates according to its eigenmodes. The modulus of the electrical impedance of the piezoelectric element exhibits several narrow peaks in the frequency domain. These peaks denote the radial or in-plane vibration modes of the ceramic patch in the low frequency range and the thickness or out-of-plane vibration modes in the higher frequency range [MON 00]. The thickness vibration mode can be used to monitor the homogeneous aging of the structure through electrical impedance measurement [PER 94, SAI 98] while the radial vibration modes are used to generate and detect Lamb-like waves [MON 00].

The condition of the structure to be evaluated depends on the homogeneously distributed degradation of its mechanical properties during its life cycle (aging) as well as on the initiation and growth of localized damage such as cracks or delaminations. Techniques are required that are able to recover quantitative information about both phenomena. The SHM technique of the piezoelectric insert can be a good option since it enables recovery of both kinds of information. The embedding of a piezoelectric element in the composite structure avoids the limitations of conventional techniques, such as coupling reproducibility [PER 94, SAI 98, MON 99, MON 00]. The shape of the piezoelectric patches can be designed in order to decouple the frequency ranges of their thickness vibrations from their in-plane vibration (radial vibration in the case of piezoelectric disc-shaped patches).

As stated above, when the frequency of the applied voltage is within the frequency range of the thickness vibration mode of the sensor, a bulk wave propagates forwards and backwards through the thickness of the host structure and interacts with the piezoelectric element. The variation of the sensor's electrical impedance with frequency can be measured and related to the mechanical and physical properties of both the surrounding medium and the sensor itself [PER 94, SAI 98]. The properties measured by this method are concerned with a limited volume in the close vicinity of the sensor. Nevertheless, if the aging of the composite structure is assumed to be homogeneous, the in-service monitoring of the electrical impedance is representative of the degradation of the whole material.

Analytic modeling [PER 94, SAI 98] shows that the electrical impedance of an element depends on the physical properties of the element itself as well as on the acoustic properties of the surrounding medium. The acoustic parameters, namely the velocity and the attenuation, are directly linked to the elastic properties and microstructure of the material. Then for NDE purposes, the resolution of the so-called inverse problem makes it possible to recover the ultrasonic properties of the surrounding plates from the measurement of the electric impedance of the piezoelectric sensor. This can be achieved through a non-linear optimization algorithm, as shown below.

4.4.2.1. Theoretical model

The piezoelectric ceramic considered here is a disk of negligible thickness relative to its diameter. As a consequence, the radial and thickness vibration modes will occur in two very different frequency ranges and they will be considered as independent of one another.

In the following, only the thickness mode of vibration will be taken into account and consequently a one-dimensional approach is convenient to model the frequency variation of the electrical impedance in relation to the axial vibration modes. A schematic of the problem is presented in Figure 4.41. The active element of thickness e is considered to be of infinite lateral dimensions. It is coupled to two different viscoelastic media (denoted by the subscripts 2 and 3 respectively) of thickness e_2 and e_3 respectively. Those two media are themselves coupled to two semi-infinite media (subscripts 1 and 4). The physical properties of each medium will be denoted with the same subscript as those used for the medium concerned. Complex quantities will be denoted by an asterisk.

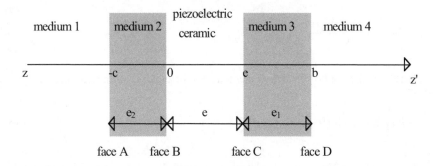

Figure 4.41. *Schematic of the one-dimensional model*

According to previous hypotheses, any vector quantity has to be symmetrical with respect to every *zz'* axis normal to the ceramic faces. This implies that only the components parallel to *zz'* are different from zero and that they are only dependent on *z* and time.

The aim of the theoretical study is to establish the dependence between the electrical impedance of the inserted piezoelectric ceramic, when it is harmonically excited, and the acoustic parameters of the surrounding medium. These parameters, namely the velocity and the attenuation, are directly linked to the elastic properties and microstructure of the host material. The whole study is conducted in the Laplace domain, where *p* will denote the Laplace operator.

The electrical potential *V(p)* created between the two faces of the active element is expressed as follows:

$$V(p) = \int_{0}^{e} E(z,p)\,dz \qquad\qquad [5]$$

where $E(z, p)$ is the *z* component of the electric field.

From the fundamental laws of piezoelectricity [SAI 98], the electric field $E(z, p)$ is related to the particle displacement $U(z, p)$ and to the electrical induction $D(z, p)$ by:

$$E(z,p) = -h_{33}\frac{\partial U(z,p)}{\partial z} + \beta_{33}D(z,p) \qquad\qquad [6]$$

where h_{33} and β_{33} are the z components of the piezoelectric tensor and the dielectric impermeability of the active element.

The electric induction $D(z, p)$ is linked to the excitation current $I(p)$ by:

$$D(z,p) = \frac{1}{S_e}\frac{I(p)}{p}$$ [7]

where S_e is the area of the active element. So the electric voltage becomes:

$$V(p) = h_{33}\left(U(0,p) - U(a,p)\right) + \frac{\beta_{33}I(p)}{S_e p}$$ [8]

$U(0, p)$ and $U(e, p)$ are the displacement amplitudes of the B and C faces of the active element respectively (Figure 4.41). These quantities $U(z, p)$ are derived from the acoustic wave propagation equation, which is expressed in the Laplace domain as:

$$U(z,p) = \frac{C_{33}^*}{p^2\rho}\frac{\partial^2 U(z,p)}{\partial z^2} = \frac{1}{\lambda^{*2}}\frac{\partial^2 U(z,p)}{\partial z^2}$$ [9]

where ρ is the density of the considered medium and C_{33}^* the axial component of the complex viscoelastic coefficient whose real part is associated with the elasticity properties and whose imaginary part is related to the viscosity of the medium.

The complex wavenumber λ^* given by expression:

$$\lambda^* = \sqrt{\frac{\rho p^2}{C_{33}^*}}$$ [10]

can be related to the propagation conditions by:

$$\lambda^* = \frac{p}{C(p)} - \alpha(p)$$ [11]

where $C(p)$ and $\alpha(p)$ denote the wave velocity and attenuation respectively.

The complex acoustic impedance of the considered propagation medium is defined by $Z^* = \rho p / \lambda^*$.

A general solution of the differential equation (equation [9]) is given by:

$$U(z,p) = k_1 e^{-\lambda^* z} + k_2 e^{\lambda^* z}$$
[12]

The constants k_1 and k_2 are determined from the continuity relations of the normal components of both displacement and stress at each interface and from the initial conditions upon electrical excitation.

The electrical impedance of the active element is finally expressed analytically as follows:

$$Z_{\acute{e}l}^*(p) = \frac{V(p)}{I(p)} = \beta_{33} \frac{(e)}{S_e p} - \frac{h_{33}^2}{S_e p} \left\{ \frac{\gamma_6}{\gamma_5} \left(e^{\lambda^*(e+c)} - e^{\lambda^* c} \right) \right.$$

$$+ \left[\frac{\frac{\gamma_6}{\gamma_5} \left[pZ^* \left(e^{\lambda_2^* c} - R_1^* e^{-\lambda_2^* c} \right) - pZ_2^* \left(e^{\lambda_2^* c} + R_1^* e^{-\lambda_2^* c} \right) \right] e^{\lambda^* c} - \left(e^{\lambda_2^* c} - R_1^* e^{-\lambda_2^* c} \right)}{e^{-\lambda^* c} \left[pZ^* \left(e^{\lambda_2^* c} - R_1^* e^{-\lambda_2^* c} \right) + pZ_2^* \left(e^{\lambda_2^* c} + R_1^* e^{-\lambda_2^* c} \right) \right]} \right]$$
[13]

$$\left. \times \left(e^{-\lambda^*(e+c)} - e^{-\lambda^* c} \right) \right\}$$

with:

$$\gamma_5 = e^{-\lambda^* c} \left[pZ^* \left(e^{\lambda_2^* c} - R_1^* e^{-\lambda_2^* c} \right) - pZ_2^* \left(e^{\lambda_2^* c} + R_1^* e^{-\lambda_2^* c} \right) \right]$$

$$\times \left[pZ^* \left(R_4^* e^{-\lambda_3^*(2b+e+c)} + e^{-\lambda_3^*(e+c)} \right) + pZ_3^* \left(R_4^* e^{-\lambda_3^*(2b+e+c)} - e^{-\lambda_3^*(e+c)} \right) \right]$$

$$- e^{-\lambda^* c} \left[pZ^* \left(e^{\lambda_2^* c} - R_1^* e^{-\lambda_2^* c} \right) + pZ_2^* \left(e^{\lambda_2^* c} + R_1^* e^{-\lambda_2^* c} \right) \right]$$

$$\times \left[pZ^* \left(R_4^* e^{-\lambda_3^*(2b+e+c)} + e^{-\lambda_3^*(e+c)} \right) - pZ_3^* \left(R_4^* e^{-\lambda_3^*(2b+e+c)} - e^{-\lambda_3^*(e+c)} \right) \right]$$

and:

$$\gamma_6 = \left(e^{\lambda_2^* c} - R_1^* e^{-\lambda_2^* c} \right) e^{-\lambda^*(e+c)} \left[pZ^* \left(R_4^* e^{-\lambda_3^*(2b+e+c)} + e^{-\lambda_3^*(e+c)} \right) \right]$$

$$\left. \left|+pZ_3^*\left(R_4^*e^{-\lambda_3^*(2b+e+c)}-e^{-\lambda_3^*(e+c)}\right)\right]-\left(R_4^*e^{-\lambda_3^*(2b+e+c)}+e^{-\lambda_3^*(e+c)}\right)e^{-\lambda^*c}\left[pZ^*\left(e^{\lambda_2^*c}-R_1^*e^{-\lambda_2^*c}\right)\right.\right.$$

$$\left.\left.\left|+pZ_2^*\left(e^{\lambda_2^*c}+R_1^*e^{-\lambda_2^*c}\right)\right]\right.$$

In these expressions R_1 and R_4 are the complex reflection coefficients for the displacements at the A and D interfaces respectively and are given as functions of the acoustic impedances by:

$$R_1^*=\frac{Z_1^*-Z_2^*}{Z_1^*+Z_2^*}\qquad\text{and}\qquad R_4^*=\frac{Z_3^*-Z_4^*}{Z_3^*+Z_4^*}$$

As the experimental work is carried out in the frequency domain, the quantity $Z_{el}^*(p)$ has to be expressed in the Fourier domain. This function exhibits a pole for $p=0$, for which the remainder $R(0)$ of the function is different from zero. The translation from the Laplace to the Fourier domain yields two terms. The first one results in the substitution of p by $j\omega$, while the second one is the Dirac function centered on the pole and multiplied by the half remainder of the pole:

$$Z_{el}^*(j\omega)=\left[Z_{el}^*(p)\right]_{p=j\omega}+\delta(0)\frac{R(0)}{2}\qquad\qquad[14]$$

where ω is the angular frequency $\square(0)$ the Dirac function at $\omega=0$. So when $\omega\neq0$, it becomes:

$$Z_{el}^*(j\omega)=\left[Z_{el}^*(p)\right]_{p=j\omega}\qquad\qquad[15]$$

So, whatever the frequency, from equation [14] it is possible to obtain an analytical expression for the electrical impedance of the embedded piezoelectric element vibrating in a pure thickness mode. Equation [13] shows that the electric impedance of the piezoelectric element depends not only on the electrical, mechanical and geometrical parameters of the ceramic but also on the viscoelastic and geometrical properties of the surrounding medium.

4.4.2.2. *Validation of the model*

This section is intended to validate this model by comparison of the calculated electric impedance with the experimental data obtained from several materials with different acoustic properties coupled to each face of the piezoelectric element. Two materials are considered in view of their differing acoustic properties:

– PMMA (poly-methyl-methacrylate), an amorphous polymer that can be considered as homogeneous, so the ultrasonic attenuation due to viscosity is weak and the wave velocity is practically non frequency dependent;

– PA (polyamide) and PP (polypropylene), which are semi-crystalline polymers in which the high-level attenuation is principally due to scattering; such materials will then be very dispersive for acoustic properties.

The studied structures consist of a piezoelectric element bonded onto each of the faces of plates of different materials. As shown by equation [13], the acoustic parameters of both the ceramic plate and the surrounding medium have to be known to calculate the electrical impedance $Z_{el}^*(j\omega)$. In our study, the analytical frequency dependence laws of the acoustic properties are measured by classical pulsed ultrasonic spectroscopy. Then the validity of the model is improved by comparing the experimental results with the calculated data.

4.4.2.3. *Ceramic characterization*

The axial components of the piezoelectric and dielectric parameters are determined by using an impulse method developed in the laboratory [PER 93, SAI 96A]. The acoustic parameters such as velocity and attenuation are determined by using a classical transmission technique implemented on PZT samples of appropriate dimensions.

The PZT can be considered as a non-dispersive material and hence the measured velocity is assumed to be constant in the frequency range of interest (1.5 to 3 MHz), while the attenuation law is assumed to be given by $\alpha(\omega) = A\omega^2$. The measured parameters for the PZT element are reported in Table 4.1.

	Symbol	Value	Unit
Thickness	e	0.99	mm
Radius	r	10	mm
Density	ρ	7800	kg m^{-3}
Piezoelectric coefficient	h_{33}	2.0×10^9	V m^{-1}
Dielectric impermeability	β_{33}	1.3×10^8	m F^{-1}
Ultrasonic velocity	C	4500	m s^{-1}
Ultrasonic attenuation	A	5.5×10^{-14}	Np.m^{-1} Hz^{-2}

Table 4.1. *Ceramic characteristics*

4.4.2.4. *Ultrasonic characterization of materials*

The two materials which are coupled to each face of the ceramic used for this validation are PMMA and PA. The acoustic propagation velocity and attenuation of these materials have been characterized over a broad frequency range using ultrasonic spectroscopy [JAY 96], which allows very accurate determination of the evolution of velocity and attenuation as a function of frequency.

The calculated results from experimental data are presented in Figure 4.42 for a frequency range from 1.5 to 3 MHz.

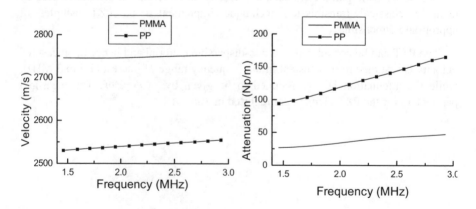

Figure 4.42. *Velocity and attenuation versus frequency for PMMA and PP measured by spectroscopy*

The ultrasonic data obtained by classical spectroscopy show a power law variation of the attenuation in the frequency range of interest, while the velocity is quasi linearly dependent on the frequency.

These experimental data can therefore be fitted according to the following expressions:

$$C(\omega) = C_0 (1 + a\omega)$$

where C_0 is the extrapolated value of the velocity at zero frequency and:

$$\alpha(\omega) = A\omega^b$$

This simple expression for the attenuation law has been chosen in order to limit the number of parameters to be determined.

These laws will then be introduced into equations [10] to [14]. These relations are close approximations of the physical phenomenon in the frequency range of interest but do not apply for other frequencies.

The fitted values obtained from experimental data for these different coefficients representing the velocity and attenuation dispersion laws are presented in Table 4.2.

	Symbol	PMMA	PP
Ultrasonic velocity	C_0	2736 m s^{-1}	2506 m s^{-1}
Ultrasonic attenuation	a	3.15×10^{-10} s	1.05×10^{-9} s
	A	9.57×10^{-6}	1.16×10^{-4}
	b	0.922	0.847

Table 4.2. *Values of parameters describing velocity and attenuation in PMMA and PP*

4.4.2.5. *Simulated and experimental results*

If these properties of the different materials are known, the model can be experimentally validated by comparison of the simulated results with experimental measurements of the electric impedance of the active element in several configurations [SAI 98]. A single example will be shown here.

The studied structure is made of a PZT5 ceramic bonded between two PMMA layers of different thickness (5.95 mm and 18.64 mm respectively). The whole structure is placed in air, so that the acoustical properties of the media 1 and 4 are those of air.

The electric impedance measurement is realized with the help of an HP 4194A impedance meter. Figure 4.43 presents a comparison of the measured and simulated moduli of the electric impedance as a function of frequency.

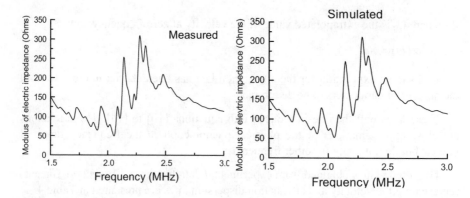

Figure 4.43. *Comparison between measured and simulated spectra for a ceramic between two PMMA plates of 5.95 mm and 18.64 mm in thickness*

Each surrounding layer of finite thickness supports acoustic resonances that affect the damping of the free vibration of the active element. This phenomenon is responsible for the modulations observed in electrical impedance spectra. The frequency of the associated peaks is closely related to the thickness of each layer and to the acoustic wave velocities in it. The amplitude depends directly on the ultrasonic attenuation and the thickness of the material.

Excellent agreement has been found between simulated and measured impedances. The model can thus be considered as valid for that kind of experiment and it may therefore be used to solve the inverse problem, which consists of determining the acoustic properties of the materials surrounding the active element from an electrical impedance measurement. It should be noted that the developed method is suitable for many kinds of layered structures made up of absorbing media and appears to be quite general. In particular, this approach has already been used for the *in situ* monitoring of polymer-based materials.

In the following example, the sample considered is composed of a piezoelectric disk with a resonant frequency of 2.25 MHz perfectly bonded between two parallel-faced plates made of PMMA and PA respectively.

In Figure 4.44, the dispersion laws for the acoustic velocity and attenuation obtained from the measured electric impedance in the 1.5 MHz to 3 MHz frequency range are compared with those measured by ultrasonic spectroscopy.

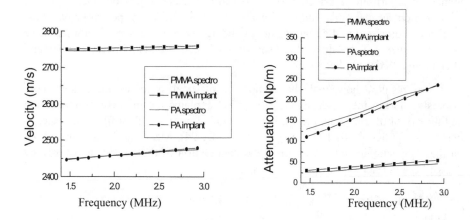

Figure 4.44. *Comparison between spectroscopy and implant results: velocity and attenuation for PMMA and PA*

The good agreement that can be observed validates this technique for determining the acoustic properties of materials.

Although velocity and attenuation are extracted separately from the model, they are actually linked by the Kramers–Kronig laws, which govern the causality of the acoustic waveforms. This enables the user to check the reliability of the recovered properties. The model can also take into account geometrical defects such as a non-parallelism of the plates [SAI 98].

4.4.2.6. *Application in SHM: aging monitoring*

The first practical application of this electrical impedance measurement concerns the *in situ* monitoring of the properties of a composite structure during its polymerization or curing process.

After the composite has been created, the piezoelectric element remains inserted in the composite structure and it is easy to monitor the variations in its resonance properties throughout the lifetime of the composite structure. The resonance properties will be an indicator that can show the changes of the mechanical properties of the composite with time or as a result of the action of external factors. In particular, in the case of polymer-based composites, water is an important factor in the degradation of their mechanical properties.

For instance, a polymer-based composite (epoxy resin + 57% w glass material) is immersed in water at high temperature (70°C). The samples are tested at regular time intervals. The electrical impedance of the inserted elements is measured and the acoustic parameters of the polymer-based composite are determined by using the numerical algorithm described above. To validate these results obtained from the inserted element, the same samples are also tested by ultrasonic spectroscopy. These results are compared in 4.45 and again a good agreement is observed between these two techniques. The results that are shown have been obtained after curing (0), and after 17 and 28 days in hot water.

These results show that the composite material becomes more and more dispersive during the water aging. This effect explains the increase in attenuation and the origin of this phenomenon is the degradation of the glass fiber interface, as observed by microscopy [VER 84, API 82, CHA 94]. Mechanical tests carried out on the same samples have confirmed the degradation of the composite, mostly by glass-fiber interface ruptures.

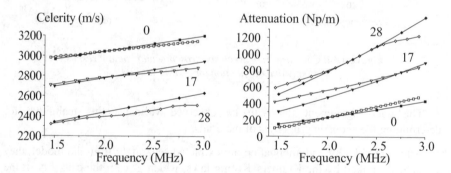

Figure 4.45. *Acoustic parameters of the composite during water aging, after 0, 17 and 28 days in hot water (filled symbols: piezoelectric implant; plain symbols: ultrasonic spectroscopy)*

4.4.3. *Conclusion*

The analysis of the electrical impedance of a piezoelectric element inserted in a structure or bonded onto its surface leads in general to local measurements. The electromechanical method used in the very low frequency range is suited to detection of flaws such as delaminations or debondings in the immediate vicinity of the active element.

In the frequency range of the thickness resonance modes of the piezoelectric element, the viscoelastic properties of the surrounding materials can be evaluated and monitored in real time throughout the lifetime of the structure (from its curing to its destruction.). This second method has been studied more extensively than the E/M method.

4.5. Summary and guidelines for future work

Piezoelectric sensors have been shown to be very well-suited for SHM.

The same set of piezoelectric elements can be utilized for aging monitoring, for damage detection, location and identification, and finally to record the AE activity of the structure under test. In the latter case, methods based on neural networks allow the identification of the damage mechanism.

The multifunctional character of piezoelectric sensors is quite relevant in its application to the on-line health monitoring of aeronautical structures. In practice, the use of the same set of sensors to perform at least three kinds of measurement reduces the weight of the monitoring system and hence simplifies the maintenance process.

In terms of the goal of lifetime improvement of a structure, the piezoelectric sensor could also be used to perform an active or semi-passive vibration control.

In practice, when damage is initiated in a composite structure, it will propagate faster if the vibration level increases. This aspect is currently under investigation at the GEMPPM laboratory in cooperation with the LGEF of INSA de Lyon.

Furthermore, it will be of a great interest to develop wireless frameworks of sensors allowing both weight savings and simplification of the insertion process, especially in the case of the articulated parts of a complex structure.

In conclusion, piezoelectric inserts still have great potential for development in the field of SHM. In the near future we will see more and more multifunctional SHM systems able to collect different kinds of signals and physical parameters with the same patches.

4.6. References

[AGO 00] AGOSTINI V., DELSANTO P.P., ZOCCOLAN D., "Flaw Detection in Composite Plates by Means of Lamb Waves", Proc. of the 15th World. Congr. on Non Destructive Testing, Roma, Italy, 2000.

[ALL 97] ALLEYNE D.N., CAWLEY P., "Long range propagation of Lamb waves in chemical plant pipework", *Materials Evaluation*, 1997, pp. 504-508.

[API 82] APICELLA A., MIGLIARESI C., NICODEMO L., NICOLAIS L., IACCARINO L., ROCCOTELLIAL. S., "Water sorption and mechanical properties of a glass-reinforced polyester resin", *Composites*, vol. 13, pp. 406-410, 1982.

[BAR 99] BAR-COHEN Y., MAL A., LIH S.S., CHANG Z., "Composite materials stiffness determination and defects characterisation using enhanced leaky Lamb wave dispersion data acquisition method", *Proc. of SPIE*, vol. 3586, pp. 250-255, 1999.

[BAR 94] BARRÉ S., BENZEGGAGH, M.-L., "On the use of acoustic emission to investigate damage mechanisms in glass-fiber-reinforced polypropylene", *Comp. Science Tech.*, 52, pp. 369-376, 1994.

[BAS 03] BASHEER M.R., RAOV, DERRISO M., "Self-organizing wireless sensor networks for Structural Monitoring", 4th IWSHM, 2003, pp. 1193-1207.

[BEA 83] BEATTIE A.G., "Acoustic emission, principles and instrumentation", *Journal of Acoustic Emission*, vol. 2, no. 1/2, pp. 95-128, 1983.

[BEA 97] BEARD S., CHANG F.-K., "Active damage detection in filament wound composite tubes using built-in sensors and actuators", *J. Intelligent Material Systems & Structures*, 8, 10, pp. 891-897, 1997.

[BLA 97] BLANAS P., WENGER M.-P., SHUFORD A.-J., DAS-GUPTA D.-K., "Active Composite Materials and Damage Monitoring", 1st EWSHM, 1997, pp. 199-207.

[BOI 02] BOIS C., HOCHARD C., "Measurements and modeling for the monitoring of damaged laminate composite structures", 1st EWSHM, 2002, pp. 425-432.

[BOL 01] BOLLER C., IHN J.-B., STASZEWSKI W.J., SPECKMANN H., "Design principles and inspection techniques for long-life endurance of aircraft structures", 3rd IWSHM, 2001, pp. 275-283.

[BOL 02] BOLLER C., TRUTZEL M., BETZ D., "Emerging technologies for monitoring military aircraft structures", 1st EWSHM, 2002, pp. 809-816.

[CHA 94] CHATEAUMINOIS A., VINCENT L., CHABERT B., SOULIER J.P., "Study of the interfacial degradation of a glass-epoxy composite during hygrothermal ageing using water diffusion measurements and dynamic mechanical thermal analysis", *Polymer*, vol. 35, no. 22, pp. 4766-4774, 1999.

[CHI 91] CHIMENTI D.E., MARTIN R.W., "Nondestructive evaluation of composite laminates by leaky Lamb waves", *Ultrasonics*, 29(1), pp. 13-21, 1991.

[COV 03] COVERLEY P.T., STASZEWSKI W.J., "Impact damage location in composite structures using optimized sensor triangulation procedure", *Smart Mater. Struct.*, 12, 5, pp. 795-803, 2003.

[DAL 00] DALTON R.P., LOWE M.J.S., CAWLEY P., "Propagation of guided waves in aircraft structures", *Review of Progress in QNDE*, vol. 19, pp. 225-232, 2000.

[DAL 01] DALTON R.P., CAWLEY P., LOVE M.J.S., "The potential of guided waves for monitoring large areas of metallic aircraft fuselage structure", *L. Non Destructive Evaluation*, vol. 20, 1, pp. 29-46, 2001.

[DAT 03] DATTA S., KIRIKERA G., SCHULTZ M., SUNDARESAN M., "Continuous sensors for structural health monitoring", *Proc. SPIE*, vol. 5046, pp. 164-175, 2003.

[DIA 04] DIAMANTI K., HODGKINSON J.M. and SOUTIS C., "Detection of low-velocity impact damage in composite plates using Lamb waves", *Structural Health Monitoring*, vol. 3, no. 1, pp. 33-41, 2004.

[DIE 75] DIEULESAINT E., ROYER D., *Ondes élastiques dans les solides*, Masson, Paris, 1975.

[DIT 03] DITTRICH K.W., MUELLER G., "In-flight measurement of acoustic background noise for the development of impact detection algorithms", *Proc. SPIE*, vol. 5046, 2003, pp. 263-271.

[DUG 03] DUGNANI R., MALKIN M., Damage Detection on a Large Composite Structure, 4th IWSHM, 2003, pp. 301-309.

[DUP 99] DUPONT M., OSMONT D., GOUYON R., BALAGEAS D.L., "Permanent Monitoring of Damaging Impacts by a Piezoelectric Sensor Based Integrated System", 2nd IWSHM, 1999, pp. 561-570.

[EGU 98] EGUSAY S., IWASAWAZ N., "Piezoelectric paints as one approach to smart structural materials with health-monitoring capabilities", *Smart Mater. Struct.*, 7, pp. 438-445.

[EIT 84] EITZEN D.G., WADLEY H.N., "Acoustic emission: establishing the fundamentals", *Journal of Research of the National Bureau of Standards*, 1984, vol. 89, no. 1, pp. 75-100.

[ELY 95] ELY T.M., HILL E. K., "Longitudinal splitting and fiber breakage characterization in graphite epoxy using acoustic emission data", *Materials Evaluation*, pp. 288-294, 1995.

[FIN 92] FINK M., "Time Reversal of ultrasonic fields - Basic principles", *IEEE Transactions Ultrasonics, Ferroelectric and Frequency Control*, (1992) 39, p. 555.

[FIN 97] FINK M., "Time reversed Acoustics", *Physics Today*, (1997) 20, 34.

[FIN 00] FINLAYSON R.D., FRIESEL M., CARLOS M., COLE P., LENAIN. J.C., "Health Monitoring of Aerospace Structures with Acoustic Emission and Acousto-Ultrasonics", Proc. of the 15th World Conference on Non-Destructive Testing, 15-21 Oct. 2000, Roma, Italy, ISSN: 1435-4934.

[FIS 98] FISHER, M.E., "Renormalization group theory: its basis and formulation in statistical physics", *Review of Modern Physics* 70, PP. 653-681, 1998.

[FRO 04] FROMME P., WILCOX P., LOWE M., CAWLEY P., "Development of a permanently attached guided ultrasonic waves array for structural integrity monitoring", Proc. of the 16th WCNDT 2004, Montreal, Canada.

[GAR 02] GARG D.P., ZIKRY M.A., ANDERSON G.L., STEPP D., "Health monitoring and reliability of adaptative heterogeneous structures", *Structural Health Monitoring*, 1, 1, pp. 23-40, 2002.

[GAY 91] GAY D., *Matériaux composites* (3e ed.), Hermes-Science Publications, Paris, 1991.

[GEN 04] GENG R.S., "Application of acoustic emission for aviation industry – problems and approaches", Proc of the 16th WCNDT 2004 - World Conference on NDT, Aug 30 - Sep 3, 2004, Montreal, Canada.

[GHO 01] GHOSHAL A., MARTIN W.N., SUNDARESAN M.J., SCHULZ M.J., FERGUSON F., "Active Wave Propagation and Sensing in Plates", 3rd IWSHM, 2001, pp. 1291-1300.

[GHO 02] GHOSHAL A., MARTIN W.N., SCHULZ M.J., CHATTOPADHYAY A., PROSSER W.H., "Wave propagation sensing for damage detection in plates", *Proc. of SPIE* Vol. 4693, 2002, pp. 300-311.

[GHO 03] GHOSHAL A., PROSSER W.H., KIRIKERA G.,. SCHULZ M. J, HUGHES D.J., ORISAMOLU W., "Concepts and Development of Bio-Inspired Distributed Embedded Wired/Wireless Sensor Array Architectures for Acoustic Wave Sensing in Integrated Aerospace Vehicles", 4th IWSHM, 2003, pp. 1161-1168.

[GIU 01] GIURGIUTIU V., BAO J., SHAO W., "Active sensor wave propagation health monitoring of beam and plate structures", *Proc. of SPIE*, vol. 4327, 2001, pp. 234-245.

[GIU 02a] GIURGIUTIU V., ZAGRAI A., BAO J.J., "Piezoelectric wafer embedded active sensor for aging aircraft structural health monitoring", *Structural Health Monitoring*, 1(2002),1, pp. 41-62.

[GIU 02b] GIURGIUTIU V., "Current issues in vibration-based fault diagnostics and prognostics", *Proc. of SPIE*, vol. 4702, 2002, pp. 101-112.

[GIU 02c] GIURGIUTIU V., "Damage metric algorithms for active-sensor structural health monitoring", 1st EWSHM, 2002, pp. 433-441.

[GIU 03a] GIURGIUTIU V., "Embedded ultrasonic NDE with piezoelectric wafer sensors", RS-I2M - 3/2003 Ultrasonic Methods, pp. 149-180.

[GIU 03b] GIURGIUTIU V., YU L., "Comparison of short-time Fourier transform and wavelet transform of transient and tone burst wave propagation signals for Structural Health Monitoring", 4th IWSHM, 2003, pp. 1267-1274.

[GOD 98] GODIN N., R'MILI M., MERLE P., BABOUX J.C., Comptes Rendus des 11e Journées Nationales sur les Composites (JNC11), Arcachon, 18-20 Novembre 1998.

[GOD 04] GODIN N., HUGUET S., GAERTNER R., SALMON L., "Clustering of acoustic emission signals collected during tensile tests on unidirectional glass/polyester composite using supervised and unsupervised classifiers", *NDT&E Internat.*, 37, (2004), pp. 253-264.

[GOD 05] GODIN N., HUGUET S., GAERTNER R., "Integration of the Kohonen's self-organising map and k-means algorithm for the segmentation of the AE data collected during tensile tests on cross-ply composites", *NDT&E International*, 38, 2005, pp. 299-309.

[GRI 04] GRISSO B.L., "Considerations of the Impedance Method, Wave Propagation and Wireless Systems for Structural Health Monitoring", Master's thesis, Faculty of the Virginia Polytechnic Institute and State University, 2004, 120 pp.

[GUO 93] GUO N., CAWLEY P., "The interaction of Lamb waves with delaminations in composite laminates", *J. Acoust. Soc. Am.*, vol. 94, no. 4, pp. 2240-2246, 1993.

[GUY 92] GUY P., "Contribution à l'étude de la propagation des ondes de Lamb dans les plaques et les milieux plans stratifies", PhD thesis, Denis Diderot University, Paris, 1992. 192 pp.

[HAY 01] HAYWARD G., HAILU B., FARLOW R., GACHAGAN A., McNAB A.," The design of embedded transducers for structural health monitoring applications", *Proc. of SPIE* vol. 4327, 2001, pp. 312-323.

[HAY 05] HAWOOD J., COVERLEY P.T., STASZEWSKI W.J., WORDEN K., "An automatic impact monitor for a composite panel employing smart sensor technology", *Smart Mater. Struct.*, 14, 2005, pp. 265-271.

[HUA 97] HUANG Q., NISSEN G.L., "Structural Health Monitoring of DC-XA LH2 Tank Using Acoustic Emission", 1st EWSHM, 1997, pp. 301-309.

[HUG 00] HUGUET S., GODIN N., GAERTNER R., SALMON L., VILLARD D., ECCM-9, Brighton UK, 4-7 June 2000.

[HUG 02] HUGUET S. "Application de classificateurs aux données d'émission acoustique: identification de la signature acoustique des mécanismes d'endommagement dans les composites à matrice polymère", PhD thesis, INSA Lyon, France, 2002, 153 p.

[IHL 00] IHLER E., ZAGLAUER W., HEROLD-SCHMIDT U., DITTRICH K.W., WIESBECK W., "integrated wireless piezoelectric Sensors", *Proc. of SPIE*, vol. 3, no. 991, 2000, Newport Beach, California, pp. 44–51.

[IHN 01] IHN J.B., CHANG F.K., "Built-in diagnostics for monitoring crack growth in aircraft structures", 3rd IWSHM, 2001, pp. 284-295.

[IHN 02] IHN J.B., CHANG F.-K., "Multicrack growth monitoring at riveted lap joints using piezoelectric patches", *Proc. of SPIE*, vol. 4702, 2002, pp. 29-40.

[IKE 99] IKEGAMI R., "Structural Health Monitoring: assessment of aircraft customer needs", 2nd EWSHM, 1999, pp. 12-23.

[JAC 99] JACQUESSON M., PAULMIER P., GODIN N., FOUGERES R., ICCM-12, Paris 5-9 July 1999 p. 452.

[JAN 94] JANSEN D.P., HUTCHINS D.A., MOTTRAM J.T., "Lamb wave tomography of advanced composite laminates containing damage", *Ultrasonics*, vol. 32 (2), 1994, pp. 83-89.

[KEH 02] KEHLENBACH M., DAS S., "Identifying damage in plates by analyszing Lamb wave propagation characteristics", *Proc. of SPIE*, vol. 4702, 2002, pp. 364-375.

[KES 02] KESSLER S.S., SPEARING S.M., ATALLA M.J., "*In-situ* damage detection of composite structures using Lamb wave methods", 1st EWSHM, 2002, pp. 374-381.

[KES 03a] KESSLER S.S., JOHNSON C.E. and DUNN C.T., "Experimental application of optimized Lamb wave actuating/sensing patches for Health Monitoring of Structures", 4th EWSHM, 2003, pp. 429-436.

[KES 03b] KESSLER S.S., DUNN C.T., "Optimization of Lamb actuating and sensing materials for health monitoring of composite structures", *Proc. of SPIE*, vol. 5056, 2003, pp. 123-133.

[KIR 03a] KIRIKERA G.R., DATTA S., SCHULZ M.J., WESTHEIDER B., GHOSHAL A., SUNDARESAN M.J., "An artificial central nervous system for Structural Health Monitoring of orthotropic materials", 4th EWSHM, 2003, pp. 1292-1299.

[KIR 03b] KIRIKERA G., DATTA S., SCHULTZ M., GHOSHAL A., SUNDARESAN M., FEASTER J., HUGHES D., "Recent advances in an artificial neural system for structural health monitoring", *Proc. of SPIE*, vol. 5046, 2003, pp.152-163.

[KOS 05] KOSEL T., GRABEC I., KOSEL F., "Intelligent location of two simultaneously active acoustic emission sources", *Aerospace Science and Technology*, 9, 2005, pp. 45-53.

[KRE 01] KRESS K.-P., DITTRICH K., GUSE G., "Smart wide-area imaging sensor system (SWISS)", *Proc. of SPIE*, vol. 4332, 2001, pp. 490-496.

[KRE 03a] KRESS K.-P, "Integrated Imaging Ultrasound SWISS", 4th EWSHM, 2003, pp. 852-859.

[KRE 03b] KRESS K., BADERSCHNEIDER H., GUSE G., "Imaging Ultrasonic Sensor System SWISS completed 60000 simulated flight hours to check structural integrity of aircraft subcomponent", *Proc. of SPIE*, vol. 5046, 2003, pp. 284-290.

[KWU 02a] KWUN H., LIGHT G.M., KIM S., PETERSON R.H., SPINKS R.L., "Permanently installable, active guided-wave sensor for structural health monitoring", 1st EWSHM, 2002, pp. 390-397.

[KWU 02b] KWUN H., LIGHT G., KIM S., SPINKS R., "Magnetostrictive sensor for active health monitoring in structures", *Proc. of SPIE*, vol. 4702, 2002, pp. 282-288.

[LAM 17] LAMB H., "On waves in an elastic plate", Proceedings of the Royal Society of London Containing Papers of a Mathematical and Physical Character, 93(651), pp. 293-312.

[LEE 02] LEE B.C., STASZEWSKI W.J., "Local interaction modeling for acousto-ultrasonic wave propagation", *Proc. of SPIE*, vol. 4693, 2002, pp. 279-288.

[LEE 03a] LEE B.C., STASZEWSKI W.J., "Modeling of Lamb waves for damage detection in metallic structures: Part I. Wave propagation2, *Smart Mater. Struct.*, 12(2003),5, pp. 804-814.

[LEE 03b] LEE B.C., STASZEWSKI W.J., "Modeling of Lamb waves for damage detection in metallic structures: Part II. Wave interaction with damage", *Smart Mater. Struct.*, 12(2003), 5, pp. 815-824.

[LEM 99] LEMISTRE M., GOUYON R., KACZMAREK H., BALAGEAS D., "Damage localization in composite plates using wavelet transform processing on Lamb waves signals", 2nd EWSHM, 1999, pp. 861-870.

[LEM 00] LEMISTRE M.B., OSMONT D.L., BALAGEAS D.L., "Active health system based on wavelet transform analysis of diffracted Lamb waves", *Proc. of SPIE*, vol. 4073, Aug. 2000, pp. 194-202.

[LEM 01] LEMISTRE M., BALAGEAS D., "Structural health monitoring system based on diffracted Lamb wave analysis by multiresolution processing", *Smart Mater. Struct.*, 10(2001), 3, pp. 504-508.

[LEM 03] LEMISTRE M.B., BALAGEAS D.L., "A hybrid electromagnetic acousto-ultrasonic method for SHM of carbon/epoxy structures", *Structural Health Monitoring*, 2(2003), 2, pp. 153-160.

[LIK 03] LIKAS A., VLASSIS N., VERBEEK J., "The global k-means clustering algorithm", *Pattern Recogn 2003*; 366(2), pp. 451–61.

[LIN 99] LIN X., PAN E., YUAN F.G., "Imaging the damage in the plate with migration technique", 2nd EWSHM, 1999, pp. 731-742.

[LIN 01a] LIN X., YUAN F.G., "Experimental study of applying migration technique in Structural Health Monitoring", 3rd EWSHM, 2001, pp. 1311-1320.

[LIN 01b] LIN X., YUAN F.G., "Damage detection of a plate using migration technique", *J. Intelligent Material Systems & Structures*, 12(2001),7, pp. 469-482.

[LIN 01c] LIN M., QING X., KUMAR A., BEARD S.J., "Smart layer and SMART suitcase for structural health monitoring applications", *Proc. of SPIE*, vol. 4332, June 2001, pp. 98-106.

[LIN 02a] LIN M., KUMAR A., QING X.-L., BEARD S.J., "Advances in utilization of structurally integrated sensor networks for health monitoring in commercial applications", *Proc. of SPIE*, vol. 4701, 2002, pp. 167-176.

[LIN 02b] LIN M., POWERS W.T., QING X.L., KUMAR A., BEARD S.J., "Hybrid piezoelectric/fiber optic SMART layers for Structural Health Monitoring", 1st EWSHM, 2002, pp. 641-648.

[LIN 03] LIN M., KUMAR A., QING X., BEARD S.J., RUSSELL S.S., WALKER J.L., "Monitoring the integrity of filament wound structures using built-in sensornetworks", Proceedings of SPIE on Smart Structures and Material Systems, March 2003.

[LYN 04a] LYNCH J.P., SUNDARARAJAN A., LAW K.H., SOHN H., FARRAR C.R., "Design of a wireless active sensing unit for Structural Health Monitoring", Proceedings of SPIE's 11th Annual International Symposium on Smart Structures and Materials, March 14-18, 2004, San Diego, CA, vol. 5394, pp. 157-168.

[LYN 04b] LYNCH J.P., "Linear classification of system poles for structural damage detection using piezoelectric active sensors", SPIE 11th Annual International Symposium on Smart Structures and Materials, San Diego, CA, USA, March 14-18, 2004.

[MAR 01] MARTIN Jr W.N., GHOSHAL A., LEBBY G., SUNDARESAN M.J., SHULZ M.J., PRATAP P., "Artificial nerves for structural condition monitoring", 3rd EWSHM, 2001, pp. 1486-1495.

[MAR 02] MARTIN W., GHOSHAL A., SUNDARESAN M., LEBBY G., SCHULZ M., PRATAP P., "Artificial nerve system for structural monitoring", *Proc. of SPIE*, vol. 4702, 2002, pp. 49-62.

[MAR 03] MARTIN T., READ I., FOOTE P., "Automated notification of structural damage for Structural Health Management using acoustic emission detection", 4th EWSHM, 2003, pp. 828-835.

[MAS 97] MASLOV K., KUNDU T., "Selection of Lamb modes for detecting internal defects in composite laminates", *Ultrasonics*, vol. 35, pp. 141-150, 1997.

[MAS 98] MASERAS-GUTIERREZ M.A., STASZEWSKI W.J., FOUND M.S., WORDEN K., "Detection of impacts in composite materials using piezoceramic sensors and neural network", *Proc. of SPIE*, vol. 3329, pp. 491-497.

[MIT 99] MITCHELL K., SANA S., BALAKRISHNAN V., RAO V., POTTINGER H.,"Micro sensors for Health Monitoring of Smart Structures, Smart Electronics and MEMS", *Proc. of SPIE*, vol. 3,673, 1999, pp. 351-358.

[MIT 00] MITCHELL K., SANA S., LIU P., CINGIRIKONDA K., RAO V.S., POTTINGER H.J., "Distributed computing and sensing for Structural Health Monitoring systems, Smart Structures and Materials 2000: Smart Electronics and MEMS", *Proc. of SPIE*, vol. 3, no. 990, 2000, pp. 156-166.

[MON 99] MONNIER T., JAYET Y., GUY P., BABOUX J.C., GOBIN P.F., "Non-destructive evaluation in composite structures using embedded piezoelectric sensors", Proceedings of the 10th International Conference on Adaptive Structures and Technologies (ICAST '99), pp. 577-584, Paris, 1999.

[MON 00] MONNIER T., JAYET Y., GUY P., BABOUX J.C. "The piezoelectric implant method: implementation and practical applications", *J. of Smart Mater. Struct.*, vol. 9, no. 3, 2000, pp.267-272.

[MON 00] MONNIER T., JAYET Y., GUY P., BABOUX J.C., "Aging and damage assessment of composite structures using embedded piezoelectric sensor", QNDE, Montreal 99, Review of Progress in QNDE, eds. D.O. Thompson and D.E. Chimenti, Plenum, New York, AIP-Conference-Proceedings, vol. 19B, 2000, pp. 1269-1276.

[MOU 99] MOULIN E., "Contribution à l'étude de la génération d'ondes de Lamb par transducteurs piézoélectriques integers. Application à la modélisation de matériaux sensibles", PhD thesis, University of Valenciennes and of Hainaut Cambrésis, 1999, p. 154..

[MOU 03] MOULIN E., BOURASSEAU N., ASSAAD J., DELEBARRE C., "Lamb-wave beam-steering for integrated health monitoring", *Proc. of SPIE*, vol. 5046, 2003, pp.124-131.

[NAY 95] NAYFEH A.H., "Waves propagation in layered anisotropic media, with applications to composites", North Holland series in Applied Mathematics and Mechanics, Elsevier, 1995

[NEC 04] NECHAD H., HELMSTETTER A., EL GUERJOUMA1 R., SORNETTE D., "Andrade and Critical Time-to-Failure Laws in Fiber-Matrix Composites: Experiments and Model", Condensed Matter e-prints, 0404035, 2004.

[NEC 05] NECHAD H., HELMSTETTER A., EL GUERJOUMA1 R., SORNETTE D., "Creep Ruptures in Heterogeneous Materials", Phys. Rev. Lett. 94, 045501, 2005.

[NEU 01] NEUMAIR M., KRESS K.P. and BUDERATH M., "State of the art and experiences with Structural Health Monitoring", 3rd EWSHM, 2001, pp. 206-220.

[OBR 02] O'BRIEN E.W., "Reduction of scatter factors by use of Structural Health Monitoring", 1st EWSHM, 2002, pp. 294-297.

[OSM 00a] OSMONT D., DUPONT M., GOUYON R., LEMISTRE M.L., BALAGEAS D.L., "Piezoelectric transducer network for dual-mode (active/passive) detection, localization, and evaluation of impact damages in carbon/epoxy composite plates", *Proc. of SPIE.*, vol. 4073, Aug. 2000, pp. 130-137.

[OSM 00b] OSMONT D.L., DUPONT M., GOUYON R., LEMISTRE M.B., BALAGEAS D.L., "Damage and damaging impact monitoring by PZT sensor-based HUMS", *Proc. of SPIE.*, vol. 3986, June 2000, pp. 85-92.

[OSM 01] OSMONT D.L., DEVILLERS D., TAILLADE F., "Health monitoring of sandwich plates based on the analysis of the interaction of Lamb waves with damages", *Proc. of SPIE*, vol. 4327, 2001, pp. 290-301.

[OSM 02] OSMONT D., BARNONCEL D., DEVILLERS D., "Health monitoring of sandwich plates based on the analysis of the interaction of Lamb waves with damages", 1st EWSHM, 2002, pp. 336-343.

[PAG 02a] PAGET C., LEVIN K., GRONDEL S., DELEBARRE C., "Damage detection in composites by wavelet-coefficient technique", 1st EWSHM, 2002, pp. 313-320.

[PAG 02b] PAGET C.A., LEVIN K., DELEBARRE C., "Actuation performance of embedded piezoceramic transducer in mechanically loaded composites", *Smart Mater. Struct.*, 11(2002), 6, pp. 886-891.

[PAG 03a] PAGET C.A., ATHERTON K., O'BRIEN E.W., "Triangulation algorithm for damage location in aeronautical composite structures", 4th EWSHM, 2003, pp. 363-370.

[PAG 03b] PAGET C.A., GRONDEL S., LEVIN K., DELEBARRE C., "Damage assessment in composites by Lamb waves and wavelet coefficients", *Smart Mater. Struct.*, 12 (2003), 3, pp. 393-402.

[PAG 04] PAGET C.A., ATHERTON K., O'BRIEN E., "Damage assessment in a full-scale aircraft wing by modified acoustic emission", Proc. of the 2nd European Workshop on Structural Health Monitoring, July 7-9, 2004, Munich, Germany.

[PAR 97] PARDO DE VERA C., GUEMES J.A., "Embedded self-sensing piezoelectric for damage detection", Proceedings of the International Workshop on Structural Health Monitoring, September 18-20, 1997, Stanford, CA 445-455.

[PAR 04] PARK H.W., SOHN H., LAW K.H., "Damage detection in composite plates by using time reversal active sensing", Proc. of the 3rd International Conference on Advances in Structural Engineering and Mechanics, Seoul, Korea, September 2-4, 2004.

[PAR 05] PARK H.W., SOHN H., LAW K.H. and FARRAR C.R., "Time reversal active sensing for Structural Health Monitoring of a composite plate" (submitted for publication) *Journal of Sound and Vibration*, 2005.

[PED 01] PEDEMONTE P., STASZEWSKI W.J., AYMERICH F., FOUND M., PRIOLO P., "Signal processing for passive impact damage detection in composite structures", *Proc. of SPIE*, vol. 4326, August 2001, pp. 169-178.

[PER 93] PERRISSIN-FABERT I., "Suivi de polymérisation par implant piezoelectrique simulation et correlations experimentales", PhD thesis, INSA Lyon, France, 1993.

[PER 94] PERRISSIN-FABERT I., JAYET Y., "Simulated and experimental study of the electric impedance of a piezoelectric element in a viscoelastic medium", *Ultrasonics*, 1994, vol. 32, no. 2, pp. 107-112.

[PIE 00] PIERCE S.G., CULSHAW B., MANSON G., WORDEN K., STSZEWSKI W.J., "The application of ultrasonic Lamb wave techniques to the evaluation of advanced composite structures", *Proc. of SPIE.*, vol. 3986, June. 2000, pp. 93-103

[PRA 04] PRASAD S.M., BALASUBRAMANIAM K., KRISHNAMURTHY C.V., "Structural health monitoring of composite structures using Lamb wave tomography", *Smart Mater. Struct.*, 13 (2004) Technical note N73–N79.

[PRO 91] PROSSER W.H., "The Propagation Characteristics of the Plate Modes of Acoustic Emission Waves in Thin Aluminum Plates and Thin Graphite/Epoxy Composite Plates and Tubes", NASA Technical Memorandum 104187, 1991.

[PRO 99] PROSSER W.H., GORMAN M.R., HUMES D.H., "Acoustic emission signals in thin plates produced by impact damage", *Journal of Acoustic Emission*, vol. 17(1-2), June 1999, pp. 29-36.

[PUR 04] PUREKAR A.S., PINES D.J., SUNDARARAMAN S., ADAMS D.E., "Directional piezoelectric phased array filters for detecting damage in isotropic plates", *Smart Mater. Struct.*, 13, 2004, pp. 838–850.

[ROS 96] ROSE L., JIAO D., SPANNER J., "Ultrasonic Guided Wave NDE for Piping", Materials Evaluation, 1996, pp. 1310-1313.

[SAI 96a] SAINT-PIERRE N., "Mise en oeuvre de la méthode de l'implant piézoélectrique: application au suivi du cycle de vie d'un matériau composite", PhD thesis, INSA Lyon, France, 1996.

[SAI 96b] SAINT-PIERRE N., JAYET Y., PERRISSIN-FABERT I., BABOUX J.C. "Influence of bonding defect on the electric impedance of a piezoelectric embedded element", *J. Phys.D. Appl. Phys.*, 1996, vol. 29, pp. 2976-2982.

[SAI 98] SAINT-PIERRE N., JAYET Y., GUY P., BABOUX J.C., "Ultrasonic evaluation of dispersive polymers by the piezoelectric embedded element method: modeling and experimental validation", *Ultrasonics*, vol. 36, pp. 783-788, 1998.

[SAI 05] SAICHEV A., SORNETTE D., "Andrade, Omori, and time-to-failure laws from thermal noise in material rupture", Phys. Rev. E 71, 016608, 2005.

[SAN 00] SANINGER J., REITHLER L., GUEDRA DEGEORGES D., DUPUIS J.P., TAKEDA N., "Learning by Experience Methodology for Aircraft Structural Health Monitoring Using Acoustic Emission", COST F3 Conference, Madrid, 2000, pp. 781-787.

[SAN 01] SANIGER J., REITHLER L., GUEDRA-DEGEORGES D., TAKEDA N., DUPUIS J.-P., "Structural health monitoring methodology for aircraft condition-based maintenance", *Proc. of SPIE*, vol. 4332, June 2001, pp. 88-97.

[SAN 02] SANIGER J., DUPUIS J.-P., "Composite damage localisation using integrated acoustic emission technology", 1st EWSHM, 2002, pp. 298-303.

[SCH 99] SCHILLER D., SCHERLING D., WEHMANN G., "Experience with conventional and new ndi-inspections on fiber reinforced plastic (FRP) structures", 43rd annual ATA NDT forum, Atlanta, Georgia, USA, 1999, p. 32.

[SCH 02] SCHULZ M., KIRIKERA G., DATTA S., SUNDARESAN M., "Piezocerramic and nanotube materials for health monitoring", *Proc. of SPIE*, vol. 4702, 2002, pp. 17-28.

[SEA 97a] SEARLE I., ZIOLA S., MAY S.," Damage detection experiments and analysis for the F-16", 1st EWSHM, 1997, pp. 310-324.

[SEA 97b] SEARLE I., ZIOLA S., SEIDEL B., "Crack detection on a full-scale aircraft fatigue test2, *Proc. of SPIE*, vol. 3042, Smart Structures and Materials 1997: Smart Sensing, Processing, and Instrumentation, Richard O. Claus, Editor, June 1997, pp. 267-277.

[SEA 98a] SEALE M.D., SMITH B.T., PROSSER W.H., "Lamb wave assessment of fatigue and thermal damage in composites", *J. Acoust. Soc. Am.*, 1998, vol. 103, no. 5, pp. 2416-2424.

[SEA 98b] SEALE M.D., SMITH B.T., PROSSER W.H., ZALAMEDA J.N., "Lamb wave assessment of fiber volume fraction in composites", *J. Acoust. Soc. Am.*, 1998, vol. 104, no. 3, pp. 1399-1403.

[SEA 99] SEALE M.D., MADARAS E.I., "Lamb wave characterisation of the effects of long-term thermal-mechanical ageing on composite stiffness", *J. Acoust. Soc. Am.*, 1999, vol. 106, no. 6, pp. 1346-1352.

[SEY 99] SEYDEL R.E., CHANG F.-K., "Implementation of a real-time impact identification technique for stiffened composite panels", 2nd EWSHM, 1999, pp. 225-236.

[SEY 01a] SEYDEL R., CHANG F.-K., "Impact identification of stiffened composite panels: I. System development", *Smart Mater. Struct.*, 10(2001), no. 2, pp. 354-369.

[SEY 01b] SEYDEL R., CHANG F.-K., "Impact identification of stiffened composite panels: II. Implementation studies", *Smart Mater. Struct.*, 10(2001) no. 2, pp. 370-379.

[SOH 03a] SOHN H., PARK G., WAIT J.R., LIMBACK N.P., "Wavelet based analysis for detecting delamination in composite plates", 4th EWSHM, 2003, pp. 567-574.

[SOH 03b] SOHN H., FARRAR C.R., HAMEZ F.M., SHUNK D.D., STINEMATES D.W., NADLER B.R., 2003, "A Review of Structural Health Monitoring Literature: 1996-2001", Los Alamos National Laboratory Report, LA-13976-MS.

[SOH 04] SOHN H., PARK H.W., LAW K.H., FARRAR C.R., "Damage detection in composite plates by using an enhanced time reversal", *Journal of Aerospace Engineering*, ASCE, 2004

[SPE 04] SPENCER Jr. B.F., RUIZ-SANDOVAL M.E., KURATA N., "Smart sensing technology: opportunities and challenges", *Journal of Structural Control and Health Monitoring*, 2004.

[STA 99] STASZEWSKI W.J., BIEMANS C., BOLLER C., TOMLINSON G.R., "Impact damage detection in composite structures – recent advances", 2nd EWSHM, 1999, pp. 754-763.

[STA 00] STASZEWSKI W.J., WORDEN K., WARDLE R., TOMLINSON G.R., "Fail-safe sensor distributions for impact detection in composite materials", *Smart Mater. Struct.*, 9 (2000), 3, pp. 298-303.

[STA 01] STASZEWSKI W.J., WORDEN K., "Overview of optimal sensor location methods for damage detection", *Proc. of SPIE*, vol. 4326, August 2001, pp. 179-187.

[STR 98] STRASER E.G., KIREMIDJIAN A.S., MENG T.H., REDLEFSEN L., "A modular wireless network platform for monitoring structures", Proceedings of the International Modal Analysis Conference, 1998, pp. 450-456.

[SU 02] SU Z.Q., YE L., BU X.Z., "Evaluation of delamination in laminated composites based on Lamb wave modes: FEM simulation and experimental verification", 1st EWSHM, 2002, pp. 328-335.

[SU 04] SU Z.Q., YE L., "Fundamental Lamb mode-based delamination detection for CF/EP composite laminates using distributed piezoelectrics", *Structural Health Monitoring*, 2004, vol. 3, pp. 43-68.

[SUN 94] SUN K.J., "Disbond detection in bonded aluminium joints using Lamb wave amplitude and time of flight", *Review of Progress in QNDE*, 1994, vol. 13, pp. 1507-1514.

[SUN 00] SUNG D.-U., OH J.-H., KIM C.-G., HONG C.-S., "Impact monitoring of smart composite laminates using neural network and wavelet analysis", *J. Intelligent Material Systems & Structures*, 11(2000), 3, pp. 180-190.

[SUN 02] SUNDARESAN M.J., GHOSHAL A., SCHULZ M.J., "A continuous sensor for damage detection in bars", *Smart Mater. Struct.* 11, pp. 475-488.

[SUN 03] SUNDARARAMAN S., ADAMS D.E., RIGAS E.J., "Structural damage characterization through beamforming with phased arrays", 4th EWSHM, 2003, pp. 634-641.

[TAI 00] TAILLADE F., KRAPEZ J.-C., LEPOUTRE F., BALAGEAS D., "Shearographic visualization of Lamb waves in carbon epoxy plates interation with delaminations", *Eur. Phys. J.*, AP 9 (2000), p. 69-73.

[TAN 95] TAN K.S., GUO N., WONG B.S., TUI C.G., "Comparison of Lamb waves and pulse echo in detection of near-surface defects in laminate plates", *NDT & E International*, 1995, 28(4), pp. 215-223.

[TAN 03] TANNER N.A., WAIT J.R., FARRAR C.R., SOHN H., "Structural Health Monitoring using modular wireless sensors", *Journal of Intelligent Material Systems and Structures*, vol. 14, no. 1, 2003, pp. 43-55.

[TIT 03] TITTMANN B.R., "Recent results and trends in health monitoring with surface acoustic waves (SAWs)", *Proc. of SPIE*, 5045, 2003, pp. 37-46.

[TRA 98a] TRACY M., CHANG F.-K., "Identifying impacts in composite plates with piezoelectric stain sensors. Part I: Theory", *J. Intelligent Material Systems & Structures*, 9(1998),11, pp. 920-928.

[TRA 98b] TRACY M., CHANG F.-K., "Identifying impacts in composite plates with piezoelectric stain sensors. Part II: Experiment", *J. Intelligent Material Systems & Structures*, 9(1998), 11, pp. 920-928.

[TSE 02] TSENG K.K.-H., NAIDU A.S.K., "Non-parametric damage detection and characterization using smart piezoceramic material", *Smart Mater. Struct.*, 11(2002), 3, pp. 317-329.

[VER 84] VERDU J., *Vieillissement des plastiques*, Ed. AFNOR Technique, 1984.

[VIK 67] VIKTROV I.A., *Rayleigh and Lamb Waves, Physical Theory and Applications*, 1967, Plenum Press, New York.

[VON 01] VON PANAJOTT A., "The Structural Health Monitoring process at BMW", 3rd EWSHM, 2001, pp. 703-712.

[WAN 99] WANG C.S., CHANG F.-K., "Built-in diagnostics for impact damage identification of composite structures", 2nd EWSHM, 1999, pp. 612-621.

[WAN 01] WANG C.S., WU F., CHANG F.-K., "Structural health monitoring from fiber-reinforced composites to steel-reinforced concrete", *Smart Mater. Struct.*, 10(2001), 3, pp. 548- 574.

[WAN 03a] WANG C.-H., ROSE J.T., CHANG F.-K., "Computerized time-reversal method for structural health monitoring", *Proc. of SPIE*, vol. 5046, 2003, pp. 48-58.

[WAN 03b] WANG L., YUAN F.G., "Imaging of multiple damages in a composite plate by prestack reverse-time migration technique", 4th EWSHM, 2003, pp. 658-665.

[WIL 99] WILCOX P.D., DALTON R.P., LOWE M.J.S., CAWLEY P., "Mode Selection and transduction for Structural Monitoring using Lamb waves", 2nd EWSHM, 1999, pp. 703-712.

[WIL 01] WILCOX P.D. *et al.*, "Mode and transducer selection for long range Lamb wave inspection", *J. Intelligent Material Systems & Structures*, 2001, vol. 12, no. 8, pp. 553-566.

[WOR 04] WORDEN K., DULIEU-BARTON J.M., "An overview of intelligent fault detection in systems and structures", *Structural Health Monitoring*, vol. 3, no. 1, pp. 85-98, 2004.

[XIA 99] XIAORONG Z., BAOQI T., SHENFANG Y., "Study on delamination detection in cfrp using wavelet signal singularity analyses", 2nd EWSHM, 1999, pp. 830-839.

[XU 95] XU W.J., "Etude par ondes acoustiques de la qualité des interfaces dans une structure multicouche", PhD thesis, University of Valenciennes and Hainaut Cambrésis, 1995, p. 200.

[ZAG 01] ZAGRAI A., GIURGIUTIU V., "Electro-mechanical impedance method for crack detection in thin plates", *J. Intelligent. Mat. Systems & Structures*, 12 (2001), 10, pp. 709-718.

[ZAG 02] ZAGRAI A., GIURGIUTIU V., "Health monitoring of aging aerospace structures using the electromechanical impedance method", *Proc. of SPIE*, vol. 4702, 2002, pp. 289-300.

[ZHA 04] ZHAO X., KWAN C., LUK K.M., "Wireless nondestructive inspection of aircraft wing with ultrasonic guided waves", Proc. of the 16th WCNDT 2004, Montreal, Canada.

Chapter 5

SHM Using Electrical Resistance

5.1. Introduction

High-performance composites have been extensively used in high-tech areas, such as the aerospace and automobile industries. Numerous primary structural parts are made of these materials. The classical methods for periodic maintenance use many NDE techniques (ultrasonics (A-Scan or C-Scan), X-ray radiography, infrared thermography, holographic interferometry, eddy current), which require extensive human involvement and expensive procedures. Moreover, this kind of periodic inspection cannot give any information on accidents and failures which occur between two successive overhauls. There is thus growing interest in the development of "sensitive" materials or structures which integrate sensors providing real-time information about the material itself or its environment. The use of such "sensitive materials" offers good opportunities, for the realization of "on line health monitoring systems" able to operate throughout the life, processing and usage of the materials or structures. Continuous health monitoring of materials would result in improved durability and safety of structures.

Online health monitoring sensors must satisfy three requirements. They must be small so as not to damage the structure, they must be embedded in remote and inaccessible areas of the structure, and they must be able to transmit information to a central processor. This information must be in direct relation to the physical process being monitored and the properties and performance which are to be maintained. Evidently they must compete in sensitivity with conventional NDE techniques and be able to monitor a large enough area of the structure.

Chapter written by Michelle SALVIA and Jean-Christophe ABRY.

One of the most natural ways to obtain a "sensitive" material is to use the material itself or part of this material-system as the sensor. Evidently, that opportunity exists in the case of Carbon Fiber Reinforced Polymers (CFRP) or hybrid glass–carbon FRP [SCH 89, THI 94, MUT 92, MUT 01, SCH 01, MEI 01]. Since carbon fibers are electrical conductors ($\rho \approx 1.5 \ 10^{-5}$ Ω m) embedded in an insulating matrix ($\rho \approx 10^{13}$ to 10^{15} Ω m) the measurement of the global electrical resistance appears to be a valuable technique for monitoring fiber fractures in unidirectional materials, and the delamination process connected with a modification of the resistive tracks in the laminates [CEY 96a, WAN 97a, ABR 99, ABR 01]. These articles aim to show that macroscopic detection of damage is possible using electrical resistance measurements by placing electrodes at the edges of the sample and by monitoring the variations of the longitudinal electrical resistance.

5.2. Composite damage

Failure in metals is known to stem from the initiation and growth of a single dominant crack. The crack initiation progress takes much longer than the crack growth (Figure 5.1, curve a). For composite materials, damage is characterized by the initiation, then by the multiplication of cracks, which differ according to the kind of laminate, such as: fiber breakage, matrix microcracking, fiber debonding from the matrix and ply separation in a laminated composite (delamination). The crack initiation takes place very early and the multiplication and growth of cracks take up most of the life of the composite (Figure 5.1, curve b).

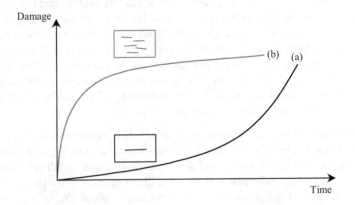

Figure 5.1. *Damage versus time (a) Metals; (b) Composite materials*

In the case of a 0° unidirectional composite under longitudinal tensile load, the damage is initiated by carbon fiber breakage since the fibers are brittle (Figure 5.2).

Cross-ply laminates undergo more complex damage processes involving transverse cracks, rebounding of the fiber matrix, delamination and fiber failure, depending on the stacking sequence (Figure 5.3).

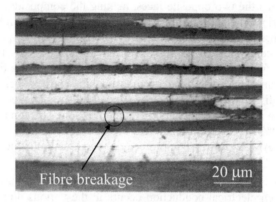

Figure 5.2. *Unidirectional composite: fiber failures*

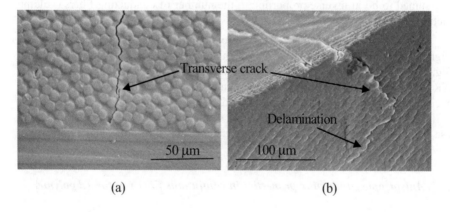

(a)   (b)

Figure 5.3. *Multidirectional laminates: (a) [90°/0°]; (b) [±45°]*

5.3. Electrical resistance of unloaded composite

5.3.1. *Percolation concept*

Two kinds of composite can be considered:

– Randomly distributed fiber (carbon short fiber (SFC) or carbon nanotube (CNT)) reinforced polymers: in this case, conduction properties are explained within the framework of the percolation concept [KIR 73]. The electrical conductivity of these composites changes by several orders of magnitude (insulator–conductor transition) from a critical reinforcement rate or percolation threshold corresponding to the appearance of a continuous path of interconnected conducting particles that spans the two opposite faces, making the composite conducting. This concept of electrical contacts is in fact rather ambiguous: they can be real physical contacts between solid surfaces geometrically connected or inter-particle tunneling junctions which carry most of the current through a thin polymer film. As well as the fiber volume fraction, the percolation threshold depends on the fiber aspect ratio and the fiber orientation distribution: a composite with a larger fiber aspect ratio and more randomly distributed fibers becomes conductive at a smaller volume fraction of fiber, resulting in a smaller threshold volume fraction. The conduction threshold is about 8–10 wt% for SFC reinforced matrix [JAN 92] and lies between 0.5 and 1 wt% [ALL 02, SAN 99] for CNT reinforced polymer due to the high aspect ratio of CNTs.

– Continuous carbon fiber reinforced polymers: these composites are anisotropic materials in which electrical conduction occurs in the various directions, but with different resistivity values, depending on the direction considered:

– for current flow along the carbon fibers: in this direction, the resistance can be assumed to be in inverse proportion to the number of conducting fibers (volume fraction),

– for conduction in the off-axis direction as a result of contacts between neighboring fibers, carbon fibers not being perfectly aligned in the composite (Figure 5.1), percolation effects can be assumed to be the main process [STR 79, STR 82]. The percolation limit for the attainment of high conductivity is found to be much higher for continuous fiber reinforced composite (about 40 wt%), compared to SFC filled systems and depending on the sequence stacking. Nevertheless, in the case of industrial composites, the fiber rate is above this threshold conduction point.

5.3.2. *Anisotropic conduction properties in continuous fiber reinforced polymer*

Unidirectional HR (High Resistance) carbon fiber reinforced epoxy systems, made by prepreg tape lay up and autoclave curing (1 hour 30 at 120°C) and varying in volume fraction, namely: 43%, 49% and 58% [ABR 99] were first considered. The location of the electrodes (copper layer (thickness ≈ 0.1 mm) realized by electrolytic deposition) for the measurement of the longitudinal resistance and the transverse resistances (configuration (a) through the width of the sample; configuration (b) through the thickness of the sample) are given in Figure 5.4.

Longitudinal resistance   Transverse resistance

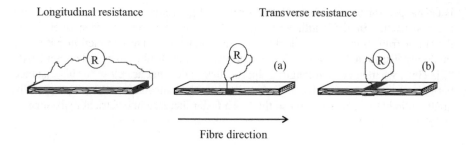

Fibre direction

Figure 5.4. *Location of the electrodes for the measurement of the resistance*

Figure 5.5 shows the longitudinal resistance of the UD (Uni-Directional) composites (2 mm thick and 10 mm wide) as a function of the fiber volume fraction (V_f) and the specimen's length (L which varies from about 10 mm to 100 mm). Whatever the value of V_f, there is a linear relationship between the longitudinal resistance and the length of the specimen, while the intercept with the resistance axis is constant. Moreover, the slope of this relation increases as the fiber fraction decreases. Such behavior can be described by considering the composite acts in the longitudinal direction as an electrical circuit of resistances in parallel [PRA 90]. Resistance R of the composite is thus given by:

$$R = \frac{\rho_f}{bhV_f}L + R_c \qquad\qquad [1]$$

where ρ_f is the fiber resistivity and L is the distance between the two electrodes. R_c is the contact resistance.

Transverse resistance is much higher than the longitudinal resistance (by a factor of 200 or more) and varies linearly with the inverse of the specimen length up to a critical value ($1/L_c$), which is quasi-independent of V_f and after which the resistance increases all the more rapidly as $1/L$ is greater, whether configuration (*a*) or (*b*) is considered (Figure 5.6). This behavior can be explained by the fact that transverse conductivity involves point contacts between adjacent fibers. When L is much greater than b (configuration (*a*)) or h (configuration (*b*)), the surface of the sample where a copper electrode is deposited can be assumed to be at the same electrical potential, because of the large difference between the longitudinal and the transverse resistances (macroscopically homogeneous material). Thus, the effective surface of the electrodes can be considered as the overall surface, i.e. hL (configuration (*a*)) or bL (configuration (*b*)). Therefore, the variation of transverse resistance can be described by equation [1] where ρ_L is replaced by ρ_{tw} or ρ_{tt} and L

is exchanged for *b* (configuration (*a*)) or *h* (configuration (*b*)), where ρ_{tw} and ρ_{tt} are the resistivities in the width and the thickness of the sample respectively. The deviation from linearity is linked to the fact that transverse conduction paths are randomly distributed with a critical length between each fiber to fiber contact [PAR 02]. Thus, when specimen length *L* approaches scales that are close to this distance between contact points, a decrease of *L* involves the cancellation of a proportionally greater number of contacts, and so there is a faster increase in electrical resistance.

Longitudinal resistance

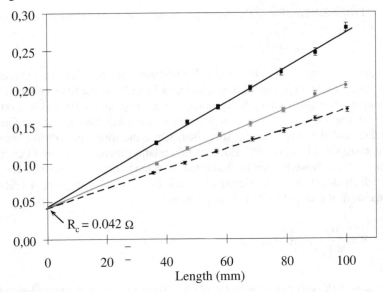

Figure 5.5. *Longitudinal resistance vs sample length.* (——■——) $V_f = 0.43$, (– –•– –) $V_f = 0.49$, (——♦——) $V_f = 0.58$

The resistivities obtained for each electrode configuration and each V_f (for $L > L_c$) are given in Table 5.1.

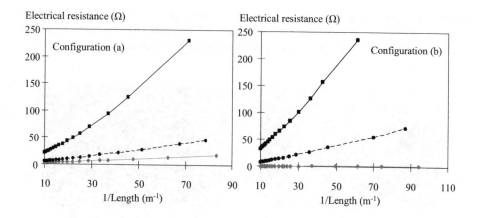

Figure 5.6. *Transverse resistance vs the inverse of the length:* (—■—) $V_f = 0.43$, (–-●-–) $V_f = 0.49$, (—◆—) $V_f = 0.58$

V_f	Longitudinal resistivity, ρ_L (Ωm)	Transverse resistivity (Ωm)	
		Configuration (a),ρ_{tw}	Configuration (b), ρ_{tt}
40%	4.7 10^{-5}	4.7 10^{-1}	16.0
50%	3.7 10^{-5}	1.1 10^{-1}	2.8
60%	2.9 10^{-5}	4.2 10^{-2}	4.8 10^{-2}

Table 5.1. *Resistivities of unloaded UD GFRP composites [ABR 99]*

Table 5.1 shows:

– for a given volume fraction, the high anisotropy of the electrical properties of the unidirectional composite due to the difference in the conduction process in the directions parallel and perpendicular to the fibers, as mentioned previously;

– the great influence of V_f, especially in the case of transverse resistivity in the thickness of the sample, ρ_{tt} (10^3 higher for the 40% composite than for the 60% one);

– for the lower V_f, the unidirectional composites cannot be regarded as transversely isotropic in terms of electrical conduction, because of the significant difference between the two components of the transverse conductivity.

These last two features can be related to the microstructure of the composites. Microscopic observations of cross-sections of composites with the lower V_f reveal large matrix zones resulting from the processing method (one or two layers of resin are placed between every two layers of prepregs in order to obtain the desired fiber content) (Figure 5.7). As a consequence, composite plies are partially isolated by matrix inter-plies in which very few electrical contacts are present (V_f at the interface between plies is probably close to the percolation threshold).

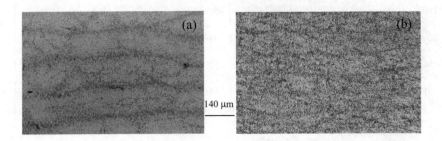

Figure 5.7. *Polished cross-section of the unidirectional composites.*
(a) $V_f = 0.43$; (b) $V_f = 0.58$

The electrical resistivity has been measured on the off-axis and by different authors [KAD 94, CEY 96a, ABR 99]. Experimental results for electrical resistivity obtained on UD 0°, UD 90° and [±45°] laminates made by lay-up of UD fabrics (HR carbon fiber/epoxy) and achieved by hot press molding are given in Table 5.2. The fiber volume fraction is 60% and the copper electrodes are glued to the surface at the ends of the specimens. The conductivity is found to be higher for angle-ply laminates than for UD 90°. In fact, owing to the stacking sequence, each fiber of a ply at the interface between plies has less chance of contacting the fibers of the other lamina in the unidirectional case than in the angle-ply configuration. Therefore, the number of contact points between the two plies is lower for the unidirectional sample than for the angle-ply samples.

Material	UD 0°	UD 90°	[±45°]
Resistivity (Ω m)	$5\ 10^{-5}$	10^{-2}	$5\ 10^{-4}$

Table 5.2. *Resistivities of different unloaded GFRP laminates ($v_f \approx 60\%$)*
[CEY 96b]

5.3.3. Influence of temperature

Several studies have been conducted in order to characterize the influence of the temperature on the electrical properties of carbon fiber reinforced composite materials. The electrical resistance of a composite material reinforced with conductive fibers depends on temperature in a complex manner:

– the carbon fiber shows semiconducting behavior, i.e. the resistivity decreases with increasing temperature [DON 84]. For continuous CFRP, it was noted that the longitudinal resistivity decreases when the temperature increases [SCH 89, ABR 98b]. The decrease is about 2% between 25°C and 70°C (Figure 5.8),

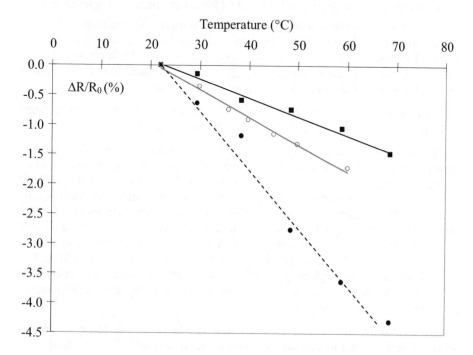

Figure 5.8. *Longitudinal electrical resistance vs temperature:*
(—■—) unidirectional composites ($V_f = 0.43$); (—○—) [0°/90°] laminates (from [SCH 89]);
(--●--) carbon fiber bundle

– when the composite is heated, the percolation network structure is modified due to the differences between the thermal expansion of the polymer matrix and the conductive reinforcement. Some percolation paths are broken, and new percolation can be formed if the fibers are very close to each other (i.e. high volume fraction) [CHE 94]. This may explain the results obtained on both black and SCF filled systems and on continuous CFRPs in the off-axis direction. The resistivity of the first materials increases with the temperature (Positive Temperature Coefficient effect) whereas the second ones display a semi-conducting behavior [CHU 99]. For example, a [±45°] laminate shows a relative decrease of electrical resistance ($\Delta R/R_0$)

of about 5% when the temperature increases from room temperature to 150°C [CEY 96b].

5.4. Composite strain and damage monitoring by electrical resistance

Concerning the conduction behavior of CFRPs, it can be then concluded that:

– in the case of continuous fiber reinforced composites, electrical conduction occurs not only in the direction of the fibers, but also in the off-axis directions as a result of contacts between neighboring fibers and plies in the laminate when the volume fraction of fibers is sufficiently high to allow the formation of a continuous conduction path, which is the case for industrial composites,

– for randomly distributed SFC or NCT composites, a low enough concentration of filler in the composite is needed to make a conductive network.

When composites are damaged, the temperature increases locally. The increase in temperature is dependent on strain or frequency rate, stress amplitude, structure geometry, loading mode, characteristics of the material (intrinsic damping, thermal conductivity) and the damage state. In particular, during fatigue tests, local heating is revealed to be related to the friction of broken strands in their housing [SAL 97]. Nevertheless, Schulte [SCH 89] has shown on CFRP cross-ply laminates subjected to fatigue tensile loading (frequency = 10 Hz) that the maximum temperature rise of the material is less than 25°C just before failure of the sample (end of life). This increase of temperature induces a variation of electrical resistance which falls below 0.5%. The influence of temperature can be neglected if the electrical resistance variations related to damage are higher.

By monitoring the variations of electrical resistance, the detection of various kinds of damage in carbon/epoxy laminates can be detected if the electrodes are judiciously positioned.

5.4.1. *0° unidirectional laminates*

5.4.1.1. *Piezoresistivity and strain sensing*

If the electrical resistance of a material depends on external straining, and if this phenomenon is reversible, the material exhibits "piezoresistivity". Several publications [SHU 96] have shown that UD carbon fiber composite materials strained in the fiber direction in the structural integrity state exhibit piezoresistive behavior. In fact, geometrical changes in PAN fibers imply a linear increase in resistance [DON 84] up to failure following the following relationship:

$$\frac{\Delta R}{R_0} = K\varepsilon \qquad\qquad [2]$$

where ΔR is the resistance variation, R_0 the unloaded fiber resistance, ε the applied strain and K a constant varying between 1 and 2. Moreover, and as mentioned previously, the longitudinal electrical resistance of 0° unidirectional CFRP due to current flow along the fibers can be modeled by relationship [1] and is four orders of magnitude greater than the electrical resistance through the thickness (Table 5.1 or 5.2). One application of such behavior is to use the reinforcement as an intrinsic strain gauge [RAS 88]. Rask *et al.* carried out a three-point bending test on a hybrid carbon glass fiber reinforced polymer (15 mm long, 2.5 mm wide and 0.20 mm thick) consisting of UD CFRP layers (0.055 mm each) separated by a bi-directional GFRP core. Electric resistance is measured on the tensile side and on the compression side as a function of deflection using two electrodes bonded onto the carbon layers with a conducting adhesive. The maximum deflection is chosen in order to avoid fiber failures. The material exhibits reversible linear relative resistance change ($\Delta R/R$) with strain (Figure 5.9) in both ways (tension and compression) and the authors conclude that CFRP is able to sense its own strain.

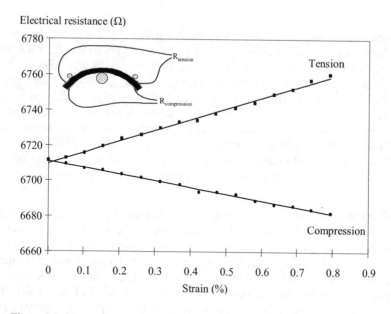

Figure 5.9. *Electrical resistance vs strain of surface in tension and compression (from [RAS 88])*

5.4.1.2. *Damage sensing*

Damage sensing must be distinguished from strain sensing: damage is irreversible but strain can be reversible. During mechanical loading, fiber fractures occurring in the composite will cause apparent V_f to decrease, hence increasing R, following equation [1] (Figure 5.10).

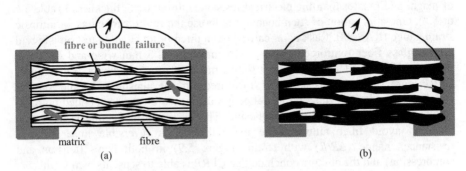

Figure 5.10. *0° unidirectional CFRP: (a) damage mechanisms and (b) its electrical analogue*

Since the pioneering work of Schulte and Baron [SCH 89], this idea has been validated by many investigations on UD CFRP subjected to tensile [PRA 90, THI 94, WAN 96] or flexural loading [CEY 96a, CEY 96b, ABR 98a, ABR 98b]. These studies demonstrate that macroscopic damage during quasi-static loading can be detected by electrical resistance measurements. Examples of results can be seen in Figure 5.11a showing both stress and electrical resistance (measured using electrodes located on the tensile side of the sample) versus strain in the case of a 0°UD specimen subjected to flexural loading up to failure [ABR 99]. A post-buckling bending test is used in order to overcome fast and randomly distributed failure occurring on the compressive side in the case of CFRPs tested in three-point bending [LAR 96].

The composite shows a quasi-linear mechanical behavior until macroscopic damage appears at 2.05% strain. At the same time, there is a slight linear increase in resistance when the strain is less than 0.6%, at which point a deviation from linearity is observed and then the slope rises progressively at higher strain. After that, the sample is gradually damaged by failure of the fiber bundles followed by delamination on the tensile side of the sample (Figure 5.11b). As the damage propagates macroscopically through the depth of the sample, the resistance rises in bursts.

In order to use such resistance measurements for sensing purposes, i.e. long before any macroscopic damage occurs, the microscopic phenomena occurring in the linear part of the stress–strain curve must be closely studied. For this purpose, the authors [ABR 99] performed successive loading–unloading cycles at increasing strain up to the ultimate stress. During these cycles, the variations of both electrical resistance and stress were monitored as a function of strain. Moreover, acoustic emission analysis experiments were carried out in parallel (Figure 5.12). The acoustic emission technique was used to detect and possibly to identify damage mechanisms in CFRP. This was achieved by analyzing AE parameters such as the amplitude of the event, and, to a lesser degree, the energy and the duration of the event. However, comparing amplitude values found during different tests and with different samples is sometimes difficult. In spite of this, several authors ([BER 88, BER 90, BAR 94]) agree that low amplitudes are correlated with matrix cracking, medium amplitudes with delamination and high amplitudes with fiber breakage (Figure 5.12).

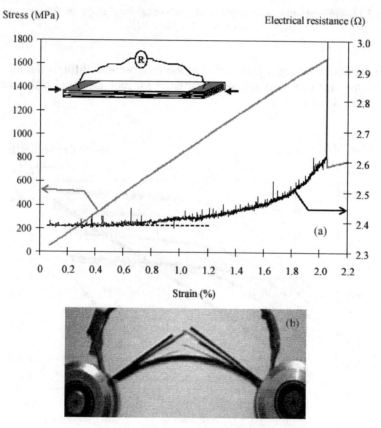

Figure 5.11. *0° UC CFRP (V$_f$=0.43): (a) changes in stress and electrical resistance as a function of the applied strain during flexural monotonic loading, (b) failure feature view*

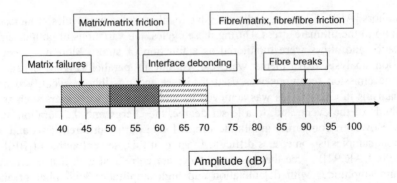

Figure 5.12. *Relation between amplitude zone and composite damage mechanisms during AE analysis, from [BAR 94]*

No significant changes were noted in the macroscopic stiffness during cycles, but both the maximum and minimum resistance for each increased with successive strain cycles (Figure 5.13).

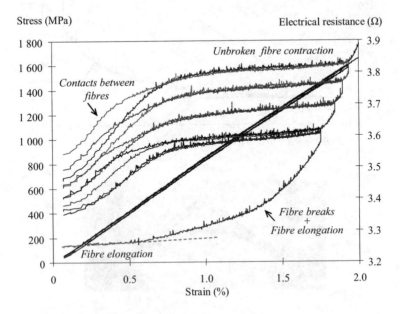

Figure 5.13. *Electrical resistance measurements in the linear part of the stress–strain curve (UD 0° – V_f = 43%)*

The loading phase of the first cycle is stopped at 80% of the stress to failure, that is, well above the onset of the non-linearity of the resistance–strain curve. The linear rise at the lower strain level is linked to the elongation of the fibers on the tensile side of the sample, as mentioned previously. Indeed, no acoustic emission event is detected until the strain reaches 0.6%. Above this threshold, numerous events are detected which are characterized by high amplitude (Figure 5.14).

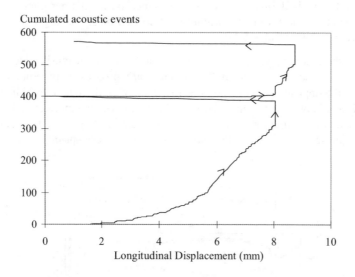

Figure 5.14. *Acoustic emission monitoring during two subsequent loading/unloading cycles. (V$_f$ = 0.43; Maximum applied strain: 1.77% (first cycle) and 1.83% (second cycle))*

According to previous investigations mentioned above, these events can be attributed to microscopic damage associated with isolated fiber failures occurring for very small strains even if the modulus does not seem weakened. During unloading a slight decrease occurs, followed by a quite rapid decrease of the resistance, and no acoustic emission was detected. At each ensuing cycle, the sample is strained up to a slightly greater maximal strain than the preceding one. The new loading curve appears superimposed on the unloading one of the previous cycle until the previous maximal strain is reached. Beyond, a sharp increase in the electrical resistance may be observed. One explanation for greater decrease in resistance at the end of the unloading phase for each cycle could be that the ends of previously broken fibers come into contact, resulting from fibers slipping in their socks; then contacts are broken again during the first part of the loading phase of the subsequent cycle. In fact, when a continuous stretched fiber breaks, it retracts into the matrix and swells (elastic return). Thus, this hypothesis seems rather unlikely.

This fast decrease of electrical resistance during unloading phases is probably related more to the establishment of contacts between fibers. Recently, Park *et al.* [PAR 02, PAR 03] introduced the concept of electrical ineffective length by analogy with the ineffective length in the case of a transfer load at the ends of a broken fiber [PIG 80]. As the fibers are not perfectly aligned in a composite, there is a chance of contact between the failed fiber and surrounding unbroken fibers. In addition to these contacts, conduction is then possible through the broken fiber. The electrical ineffective length is defined as the mean distance between fiber contacts (Figure 5.15). Obviously, the lower the volume fraction of fibers, the larger the electrical ineffective length (see Table 5.1).

During the loading phase, the waviness of the fibers on the tensile side decreases and above a certain strain level the contacts are progressively broken. During unloading, the opposite phenomenon occurs ($\varepsilon < 0.8\%$): there is re-establishment of some contacts between previously broken fibers and both the ineffective length and the apparent volume fraction of failed fibers decrease, leading in the last stage of the unloading phase, to an accelerated decrease in electrical resistance.

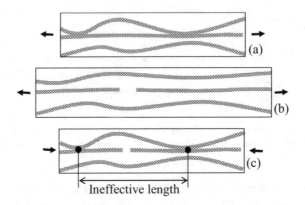

Figure 5.15. *(a) Elongation of fiber; (b) fiber breaks and elongation of fiber;*
(c) re-establishment of some contacts and the notion of ineffective length

It is well known that the multiplication of fiber breaks during flexural loading results in a superficial crack in the central zone of the sample and then to delamination of successive layers and to catastrophic damage. So an important point to know is the proportion of broken fibers before macroscopic damage. For the considered composite ($V_f = 0.43$), the first ply on the tensile side is isolated by an insulating matrix layer and thus conduction occurs in the external ply. Considering also that the percolation effect is negligible (which assumption is probably valid above 0.8% (see above)) and using relationship [1], the corresponding volume fraction of unbroken fibers can therefore be expressed as follows, where h can be approximated to the thickness of a single ply:

$$V_f = \frac{\rho_f L}{bh(R - R_c)} \qquad [3]$$

and the volume fraction of broken fibers (V_{bf}) is given by:

$$V_{bf}(\varepsilon) = \frac{V_f - V_f(\varepsilon)}{V_f} = \frac{R_0 - R(\varepsilon)}{R(\varepsilon) - R_c} \qquad [4]$$

This relationship was applied to the previous example and showed a broken fiber rate of about 10% before failure of the first ply.

Thus, even though the variation of resistance with the strain is non-linear, such measurements allow the strain to be monitored. It is interesting to note that the sensitivity is very precise for the low strains, that is, in the use range of the composite parts. Moreover, the evolution of the global resistance level appears to be a valuable technique that can give information about the early stages of damage. This technique can be defined as a warning for the health monitoring approach during loading. This technique is, in fact, more sensitive if the volume fraction is low because there are fewer fiber to fiber contacts. Indeed, Muto *et al.* [MUT 92, MUT 01] have shown the same results in the case of UD carbon fiber–glass fiber hybrid reinforced polymers with 0.3 to 0.4% of carbon fiber volume content, and this can be used as a warning in security walls [MUT 01].

If a structure is subjected to dynamic loading during use, for instance rotor blades in wind turbines or helicopter rotor blades, fatigue is a frequent cause of damage. Fatigue tests reveal the same trends as monotonic loading [WAN 97b, ABR 98a, IRV 98]. In Figure 5.16a, the relative changes in the stiffness and in the electrical resistance are reported as a function of the number of cycles for a specimen with a volume fraction of fibers of 43% subjected to compression bending.

The graph in Figure 5.16b shows that there is a linear correlation between two parameters: $D_L = 1 - F/F_0$, the load damage criterion and $D_R = 1 - R_0/R$, the electrical resistance damage criterion. The monitoring of the electrical resistance is thus a very good *in situ* indicator of the damage state in composites under fatigue loading. The resistance increase becomes less important than the load loss when the volume fraction increases. This is because, as the volume fraction increases, as in the previous case, the fiber to fiber contact rate is higher.

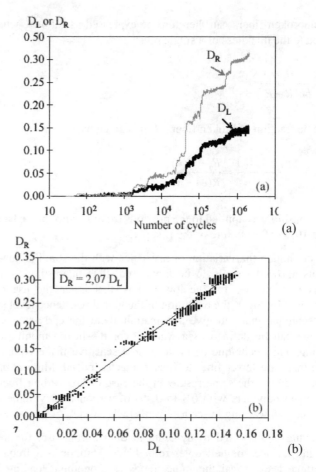

Figure 5.16. *0° unidirectional sample (V_f = 43%) — Fatigue compression bending test:
(a) D_L and D_R vs number of cycles; (b) D_R vs D_L*

5.4.2. *Multidirectional laminates*

The damage mechanisms of multidirectional laminates depend on the stacking
sequence. Let us first consider multidirectional laminates consisting of a few plies
with fibers parallel to the load ([0°/90°], for example): for such composites with
intra-ply matrix cracking in the transverse ply (Figure 5.3a) and inter-laminar failure
(Figure 5.3b), fiber breaks can occur. As shown previously, the measurement of the
global electrical resistance appears to be a valuable technique that can be used to
monitor the fiber fractures. Transverse conduction also exists by contacts between
neighboring fibers and plies. Consequently, conductivity variation measurement
must also enable the detection of transverse and longitudinal intra-ply matrix

cracking and delamination by modifications to the fiber–fiber conduction paths and the gradual cutting of the resistive tracks.

In fact, for such composites [ABR 01], it was shown that only the fiber failure occurring in the longitudinal lamina can be clearly detected using electrical resistance measurements. Nevertheless, because, for these composites, the longitudinal plies are responsible for carrying a large part of the load, it could be possible to postulate a critical threshold giving a warning in health monitoring approaches.

Figure 5.17(a) illustrates how matrix cracks can cause changes in conduction in the case of an angle-ply laminates [±45°] fiber carbon reinforced PEEK [RIS 98] subjected to a three-point bending test.

Regarding the stress-strain curve, two zones can be identified:

– Zone I: (0% < ε < 0.4%): the stress versus strain curve is linear. There is no AE activity. The very slow increase in electrical resistance can be explained by the fact that the fibers tend to move away from each other when the strain increases. This phenomenon is reversible, with a delay due to the visco-elastic behavior of the matrix.

– Zone II: ($\varepsilon > 0.4\%$): the mechanical behavior becomes non-linear and the activity of acoustic emission appears and increases regularly. At the same time, the electrical resistance exhibits a three-stage behavior: rapid increase, a very small decrease and finally an increase region. The amplitude of the acoustic events is centered around 45–50 dB (Figure 5.17b), which is characteristic of matrix failures. At this stage, optical observations on edges polished by metallographic methods shows intra-ply transverse matrix cracks near the central zone of the sample (Figure 5.17c). In fact, for [± 45°] laminates, high intra-laminar shear occurs close to or at the fiber matrix interface. This is linked to the rotation of fibers that tend to lie closer to the load axis, leading to transverse failures and then possibly delamination.

The onset of the non-linearity of the stress–strain curve indicates the threshold nucleation of the cracking processes and probably visco-plastic deformation of the matrix, although it is difficult to separate the two processes. The global increase in the resistive characteristic of the material is due to the breakage of conducting paths between contacting fibers linked to the presence of matrix cracks. The slowing down of the resistance at about $\varepsilon = 3\%$ is probably due to shear plastic deformation.

In the case of PEEK/carbon laminates there is no occurrence of delamination due to the marked visco-plastic behavior of the matrix, unlike in carbon reinforced epoxy laminates [CEY 96a]. In this work, it was shown that surface resistance measurement is not always suitable for the detection of delamination.

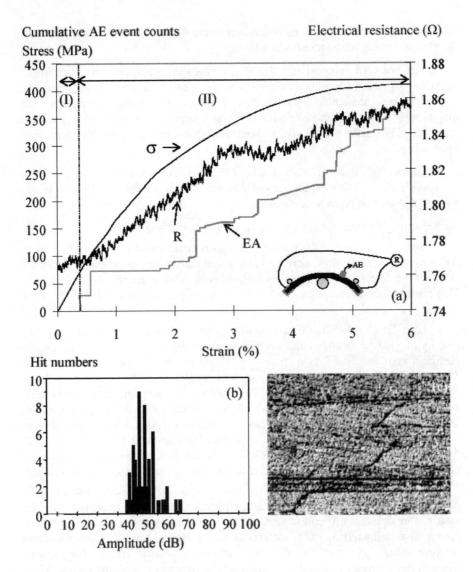

Figure 5.17. *[±45°] PEEK/carbon laminate subjected to three-point bending monotonic test: (a) stress, electrical resistance and cumulative AE event counts vs strain; (b) AE amplitude distribution; (c) post-mortem free edge view*

Some studies using carbon reinforced CFR epoxy laminates [KRA 96] indicate that the through-thickness measurement of the dielectric response of materials using AC impedance spectroscopy (Z', Z") can provide an alternative way of detecting such damage. As mentioned previously, CFRPs can be considered as an assembly of electrically conducting materials (fibers) and an insulator (matrix). According to this assumption, CFRP laminates can be said to behave like an RC circuit [SCH 95] such that:

$$R = \frac{Z'^2 + Z''^2}{Z'} \qquad [5]$$

$$C = \frac{-Z''}{(Z'^2 + Z''^2)\omega} \qquad [6]$$

where ω is the angular frequency. An example of AC results (100 kHz and using a voltage amplitude equal to ±0.5 mV) carried out in the direction perpendicular to the fiber layers (through thickness) on [±45°] laminate subjected to a monotonic bending test is given in Figure 5.18.

The stress–strain relationship exhibits a quasi-linear behavior during the initial stages of the loading ($\varepsilon < 0.7\%$). In this domain, no significant changes in C and R are detected. When the strain is increased above 0.7%, the development of a non-linear mechanical response is associated with an increase in R and a decrease in C. It can also be noted that the rate of change of the dielectrical parameters is enhanced as the strain is increased to $\varepsilon = 2.6\%$ (dotted line).

The occurrence of changes in R and C thus appear to be strongly correlated with the non-linear behavior observed on the stress–strain curve and can then be attributed to both visco-plastic deformation of the matrix and transverse cracking:

– The resistive component of the material is connected to the number of conducting paths between contacting fibers and is then highly sensitive to visco-plastic flow and intra-ply and inter-ply (delamination) cracks. The marked increase in the resistance growth rate is then probably due to the occurrence of delamination.

– The macroscopic damage can also induce a change in the capacitive response of the composite. Crack opening leads to the formation of cavities filled with air or vacuum. If it is kept in mind that capacitance is proportional to the material's permittivity and that the permittivity of an epoxy matrix is about seven times that of air, it becomes clear that cracks will enhance capacitive effects in the composite. The capacity remains unaffected by visco-plastic flow. This is probably the reason why the resistive part increases at strain levels just below plastic flow, rather than the capacitance.

As cracking proceeds, the upper ply becomes progressively uncoupled from the adjacent ply as a result of inter-laminar failure, and slip between plies occurs. At this damage level, conduction paths and void rates vary little or not at all. As a consequence, resistance and capacitance growth rates flatten out and decrease.

SEM observation (Figure 5.3b) of the specimens after testing supports these conclusions. Both inter-ply delamination and transverse cracking were observed, which is consistent with the occurrence of changes in C and R during loading.

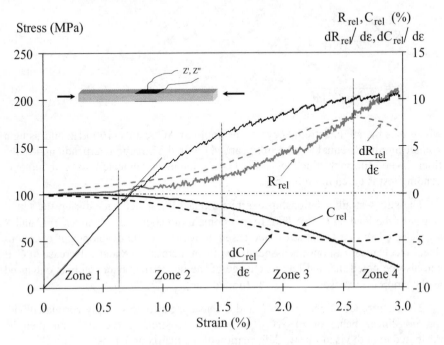

Figure 5.18. *Monotonic test of [±45°] epoxy/carbon laminate subjected to compression bending: stress, electrical resistance and capacitance versus strain*

Measurements of the electrical resistance (DC experiments) and AC properties (resistance and capacitance) provide an accurate means of monitoring the *in situ* evolution of damage nucleation and growth in cross-ply CFRP laminates, particularly for internal damage. A critical change of resistance or capacitance variation can be defined as a warning for the health monitoring approach during monotonic or cyclic loading. It is important to note the complementary aspects of surface and through-thickness measurements: the surface measurements technique is mainly sensitive to fiber failures if the laminate consists of a few plies with fibers parallel to the load, while through-thickness measurements provide information essentially on the development of matrix cracks (intra-ply matrix cracks and inter-ply delamination). The combined use of these two methods will enable the detection, as well as the qualitative identification, of various damage mechanisms.

5.4.3. *Randomly distributed fiber reinforced polymers*

Most of these materials are conductive SFC or NCT reinforced polymer composites, just above the percolation threshold. When such a composite is subjected to an applied tensile load, it shows variable conductivity (decrease) and it becomes non-conductive around a critical strain, exhibiting switching behavior. The decrease in conductivity is linked to random local breaks of the initially percolating network due to matrix deformation. The recovery of the conductivity versus time is quasi-complete with a delay due to the visco-elastic behavior of the polymer host (time dependent) [FLA 00]. These materials are widely used for their piezoresistive properties as touch control switches and strain sensors for various electronic applications. The polymer is either an elastomer or an epoxy resin. Because of their conductivity, these materials can be used for anti-static applications (electrostatic charge dissipation).

The self-monitoring of strain using resistance measurements has also been reported for concrete containing short carbon fibers [CHE 93] and recently for silicon carbide whiskers reinforced silicon nitride [WAN 97c].

Moreover, the use of SFC reinforced polymer as a coating on various substrates in order to play the role of probe to detect local cracks is quite promising [WAN 98].

5.5. Damage localization

For SHM of real structural parts, a great deal of development needs to be carried out for the identification of the damage location and size and for the signature of the damage mechanisms. Several projects have been carried out on this topics since the pioneering work of Kemp [KEM 94] using the electrical potential technique. This method consists of making measurements of potential distribution along current lines going through a composite structure before and after damage using a network of contacts attached to one face of the composite. In Kemp's study, a cross-ply laminate $[(0/90)_4]_s$ (300 × 300 mm; thickness 2 mm), lightly grit-blasted in order to allow electrical contact using silver cement, was instrumented with a (6 × 6) array of sensing wires. Current input was centrally located at opposite edge of the laminate. Current flow is parallel to the surface fiber direction. The plate was subjected to a low-energy impact test. Such loading can cause sub-surface delamination cracks (Barely Visible Impact Damage, [KEM 94]) in structural parts, which is difficult to detect by visual examination but must be located during maintenance inspection. Kemp used a drop-weight impact device to conduct impact tests from 2 J to 8 J. Contours consisting of equipotential changes on the surface of the laminate between initial and impacted panels (called a "difference plot") were obtained through regression analysis of experimental data to generate a finer grid. Significant changes were observed in the potential fields in the vicinity of the impact from 6–8 J energy levels, allowing the localization of the damage site. At such a level the damage comprises various mechanisms, all leading to reduce conduction: delamination between plies with different orientations (0° and 90°) (reduction of inter-fiber contacts due to ply separation), matrix cracks and fiber breaks (decrease of apparent fiber volume fraction). Thus it is not clear which component is responsible for the change in potential.

More recently, Angelidis *et al.* [ANG 05] used the same technique for detecting the location and the size of low-velocity impact damage in CFRP plate structures. In this study, which was more detailed than Kemp's study, two stacking sequence were used: cross-ply ([(0/90)4]s) and QI ([02/452/902/452]s). On each plate (300 × 300 mm; thickness 2 mm) the potential measurements were taken using an 11 × 11 set of probes implemented on one face. In this study, different locations of input/output current were investigated (Figure 5.19a), in particular at the mid-points of the horizontal edges (current flow in the direction parallel to the surface fiber direction) (α), or at mid-points of the vertical edges (current flow in the direction perpendicular to surface fiber direction) (β) on the instrumented side or on opposite side. The authors showed that potential measurements were not affected whether the current locations were on the instrumented surface or on the opposite side and the tests were performed using the latter configuration.

The plates were impacted with energies varying from 2 J to 8 J. On cross-ply composites C-Scan analysis showed that the damage threshold is at 4 J and made it possible to estimate the shape and the size of the damage. Since at 6 J the delaminated area exhibited a classical oblong pattern for (0/90)$_s$, the stacking sequence with the principal axis was oriented in the direction of the fibers [ANG 04]. In fact, delamination occurred in the inter-lamina between plies having different orientations and the delaminated area showed a peanut-like shape with its major axis having delamination in the direction of the fibers of the lower ply. Moreover, the larger delamination is created in the inter-lamina between the 90-degree ply and the 0-degree ply, which is located near the laminate surface opposite the impacted side.

As the impact affects the current paths (fiber to fiber contacts, fiber failures) the potential field must be modified. By superposing a C-san image on the equipotential contour measured using current introduction at mid-points parallel to the surface fiber direction of the damaged cross-ply plate after successive impacts from 2 J up to 8 J, only a small local disturbance due to impact damage was noted (Figure 5.19b). The authors showed that a better appreciation of the damage can be obtained by the analysis of equipotential changes between damaged and undamaged cross-ply panels, as previously used by Kemp. This point is illustrated in Figure 19c, which shows the relative potential resulting from the impact along the middle of the plate in the fiber surface direction.

Even though there was no change after the 4 J energy level impact, changes in potential increased with impact energy level (for the impact energy level range tested for the considered stacking sequence) and there was a correlation between potential change and damage size (Figure 5.19c). The authors also noted that the extent of detectable potential changes in the surface fiber direction was about to double the damage size in the same direction if the decrease in the number of electrodes is allowed for.

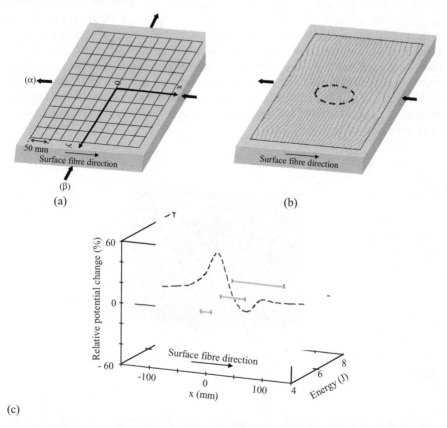

(a)   (b)

(c)

Figure 5.19. *Cross-ply epoxy/carbon laminate subjected to impact loading (from [ANG 04] and [ANG 05]): (a) potential measurement probes and current input/output locations; (b) potential field after successive 2, 4, 6 and 8 J impacts in the centre of the laminate (damaged area within dotted line); (c) relative potential change due to impact damage (—), and damage size (⊱⊶⊰) along the middle of the plate on a section parallel to the surface fiber*

Nevertheless, Angelidis *et al.* pointed out a difficulty: when the current flows perpendicular to the surface fiber, no modification of potential change was observed and thus the damage could not be detected. In the same way, tests performed on a laminate with a hole drilled in the middle of the plate showed that hole did not modify the potential contour when current input was perpendicular to the surface fiber direction.

The same results were obtained on QI laminates.

Scheuler *et al.* [SCH 01] proposed a conductivity mapping technique in order to detect damage in a UD composite plate with edge contacts instead of a surface grid. No implemented sensors are required, the authors using razor blades pressed to the specimen edge as electrodes. They used electrical impedance tomography (EIT), a method that has a long history in geology and biomedical applications, to reconstruct the damage state from a series of measurements between adjacent edge contacts (Figure 5.20). A large number of measurements and complex and time-consuming calculations (a complete analysis is required for each new measurement) are necessary to reconstruct the damage state of the composite, but the location and the size of a hole drilled in the specimen calculated from EIT measurement are in good agreement with actual values.

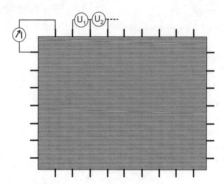

Figure 5.20. *Impedance tomography (EIT) principle*

Hou *et al.* [HOU 02] have recently used a new technique, embedding in the laminates [0°/90°] thin copper wire electrodes (120 μm) in the fiber direction, at 10 mm intervals in both 0° and 90° plies. Each sensing ply is isolated from the rest of the composite using a glass/epoxy layer or epoxy film, thus allowing damage localization. The technique was validated on a composite plate subjected to low-velocity impact by comparison with conventional methods (X-rays). By subtracting the initial resistance of each sensor from the resistance of the damaged structure, it enables precise impact damage location to be determined. The major drawback of this technique is the embedding of heterogeneity in the composite and the major interest of the carbon fiber sensor is that this kind of sensor is compatible with the fabrication of the material from which the structure is made and does not adversely affect the behavior of the structure being monitored.

5.6. Conclusion

Carbon sensors could be employed to monitor *in-situ* the structural integrity of composite industrial components (primary structures) such as aircraft wings, helicopter blades, in real time, and possibly costing little compared with current composite structure inspection techniques. Nevertheless, a great deal of development still needs to be carried out in this area in the case of real structural parts.

In particular, a network of electrodes must be judiciously implemented throughout the structure in order to extract the appropriate local information and the location of the damage. These electrodes need to be structurally robust according to in-service life conditions. This is particularly true for impact damage during overhauls, for example.

The estimation of the condition of the structure is based on data comparison between unloaded and healthy structure reference data and continuous measurements in real time during service. A pre-stored resistance (or/and capacitance) acquisition on the real structure before loading is then required, because composite materials are essentially inhomogeneous.

Moreover, there is a need for simultaneous resistance/capacitance and temperature measurements coupled to a reliable thermo-electrical modeling of the monitored material. Without any load or damage, the resistivity of such a material varies with temperature. This variation is due, first, to the difference between the expansion coefficients of the material components which modify fiber to fiber contacts and, second, to the semiconductor nature of carbon fiber, even if this effect is small in the temperature range in which composite structures are used.

Furthermore, the sensing system must also be compatible with the operating environment (electrical, magnetic environment) that the structure will experience.

5.7. References

[ABR 98a] ABRY J.C., BOCHARD S., CHATEAUMINOIS A., SALVIA M., GIRAUD G., "*In situ* monitoring of flexural fatigue damage in CFRP laminates by electrical resistance measurements", Proceedings of the 4th ESSM and 2nd MIMR conference, 6-8 July 1998, Harrogate, G.R. Tomlinson and W.A. Bullough Eds., pp. 389-396.

[ABR 98b] ABRY J.C. "Suivi *in situ* d'endommagement dans les materiaux composites carbone/epoxy par mesure des variations de propriétés électriques", PhD thesis, Marseille University, France, 1998.

[ABR 99] ABRY J.C., BOCHARD S., CHATEAUMINOIS A., SALVIA M., GIRAUD G., "*In situ* detection of damage in CFRP laminates by electrical resistance measurements", *Composites Science and Technology*, vol. 59, no. 6, 1999, pp. 925-935.

[ABR 01] ABRY J.C., CHOI Y. K., CHATEAUMINOIS A., GIRAUD G., SALVIA M., "In situ monitoring of damage in CFRP laminates by means of AC and DC measurements", *Composites Science and Technology*, vol. 61, no. 6, 2001, pp. 855-864.

[ALL 02] ALLAOUI A., BAI S., CHENG H.M., BAI J.B., "Mechanical and electrical properties of a MWNT/epoxy composite", *Composites Science and Technology*, vol. 62, 2002, pp. 1993-1998.

[ANG 04] ANGELIDIS N., "Damage sensing in CFRP composites using electrical potential techniques", *PhD thesis*, Cranfield University, UK, 2004.

[ANG 05] ANGELIDIS N., KHEMIRI N., IRVING P.E. "Experimental and finite element study of the electrical potential technique for damage detection in CFRP laminates", *Smart Materials and Structures*, vol. 14, no. 1, 2005, pp.147-154.

[BAR 94] BARRE S., BENZEGGAGH M.L., "On the use of acoustic emission to investigate damage mechanisms in glass-fiber-reinforced polypropylene", *Composites Science and Technology*, vol. 52, no. 3, 1994, pp. 369-376.

[BER 88] BERTHELOT J.M., "Relation between amplitudes and rupture mechanisms in composite materials", *J. of Reinforced Plastics and Composites*, vol. 7, 1988, pp. 284-299.

[BER 90] BERTHELOT J.M., RHAZI J., "Acoustic emission in carbon fiber composites", *Composites Science and Technology*, vol. 37, 1990, pp. 411-428.

[CEY 96a] CEYSSON O., SALVIA M., VINCENT L., "Damage Mechanisms Characterisation of Carbon Fiber/Epoxy Laminates by both Electrical Resistance Measurements and Acoustic Emission Analysis", *Scripta Materialia*, vol. 34, no. 8, 1996, pp. 1273-1280.

[CEY 96b] CEYSSON O., "Caractérisation du comportement en fluage de matériaux composites carbone époxyde: étude de l'endommagement", PhD thesis, Ecole Centrale Lyon, France, 1996.

[CHE 93] CHEN PU-WOEI, CHUNG D.D.L., "Carbon fiber reinforced concrete for smart structures capable of non-destructive flaw detection", *Smart Mat. Struct.*, vol. 2, 1993, pp. 22-30.

[CHE 94]. CHEN X.B., ISSI J.P., CASSART M., DEVAUX J., BILLAUD D., "Temperature dependence of the conductivity in conducting polymer composites", *Polymer*, vol. 35, no. 24, 1994, pp. 5256-5258.

[CHU 99] CHUNG D.D.L., WANG S., "Carbon fiber polymer matrix composite as a semiconductor and concepts of optoelectronic and electronic devices made from it", *Smart Mater. Struct.*, vol. 8, 1999, pp. 161-166.

[DON 84] DONNET J.B., BANSAL R.C., *Carbon Fibers*, New-York and Basel, Marcel Dekker Inc., 1984.

[FLA 00] FLANDIN L., BIDAN G., BRECHET Y., CAVAILLE J.Y., "New nanocomposite materials made of an insulating matrix and conducting fillers: processing and properties", *Polymer Composites*, vol. 21, 2000, pp. 165-174.

[HOU 02] HOU L., HAYES S.A., "A resistance-based damage location sensor for carbon-fiber composites", *Smart Mater. Struct.*, vol. 11, 2002, pp. 966-969.

[IRV 98] IRVING P.E., THIAGARAJAN C., "Fatigue damage characterization in carbon fiber composite materials using an electrical potential technique", *Smart Mater. Struct.*, vol. 7, 1998, pp. 456-466.

[JAN 92] JANA P.B., CHAUDHURI S., PAL A.K., DE S.K., "Electrical conductivity of short carbon fiber-reinforced polychloroprene rubber and mechanism of conduction", *Polymer Engineering and Science*, vol. 32, no. 6, 1992, pp. 448-456.

[KAD 94] KADDOUR A.S., AL-SALEHI F.A.R., AL-HASSANI S.T.S., HINTON M.J., "Electrical resistance measurement technique for detecting failures in CFRP materials at high strain rates", *Composites Science and Technology*, vol. 51, no. 3, 1994, pp. 377-385.

[KEM 94] KEMP M., "Self-sensing composites for smart damage detection using electrical properties", Proceedings of 2nd European Conference on Smart Structures and Materials (ECSSM 2), 12-14 October 1994, Glasgow, McDonach A., Gardiner P.T., McEwen R.S., Culshaw B. Eds, pp. 136-139.

[KIR 73] KIRKPATRICK S., "Percolation and conduction", *Reviews of Modern Physics*, vol. 45, no. 4, 1973, pp. 574-588.

[KRA 96] KRANBUEHL D., AANDAHL H., HARALAMPUS N., NEWBY W., HOOD D., BOITEUX G., SEYTRE G., PASCAULT J.P., MAAZOUZ A., GERARD J.F. SAUTEREAU H., CHAILAN J.F., LOOS A.C., MACRAE J.D., "Use of in situ dielectric sensing for intelligent processing and health monitoring", Proceedings of the 3rd International Conference on Intelligent Materials, 3-5 June 1996, Lyon, Vol SPIE 2779, Gobin P.F., Tatibouet J.Eds, pp. 112-117.

[LAR 96] LARGE-TOUMI B., SALVIA M., VINCENT L., "Fiber/Matrix Interface Effect on Monotonic and Fatigue Behavior of Unidirectional Carbon / Epoxy Composites, Fiber, Matrix, and Interface Properties", ASTM STP 1290, C.J.Spragg and L.T.Drzal, Eds., American Society for Testing and Materials, 1996, pp. 182-200.

[MEI 01] MEI Z., CHUNG D.D.L., "Thermal history of carbon-fiber polymer-matrix composite, evaluated by electrical resistance measurement", *Thermochimica Acta*, Vol. 369, no. 2, 2001, pp. 135-147.

[MUT 92] MUTO N., YANAGIDA H., NAKATSUJI T., SUGITA M., OHTSUKA Y., ARAI Y., "Design of intelligent materials with self diagnosing function for preventing fatal fracture", *Smart Mat. Struct.*, vol. 1, 1992, pp. 84-90.

[MUT 01] MUTO N., ARAI Y., SHIN S.G., MATSUBARA H., YANAGIDA H., SUGITA M., NAKATSUJI T., "Hybrid composites with self-diagnosing function for preventing fatal fracture", *Composites Science and Technology*, vol. 61, no. 6, 2001, pp. 875-883.

[PAR 02] PARK J.B., OKABE T., TAKEDA N., CURTIN W.A., "Electromechanical modeling of unidirectional CFRP composites under tensile loading condition", *Composites Part A: Applied Science and Manufacturing*, vol. 33, no. 2, 2002, pp. 267-275.

[PAR 03] PARK J.B., OKABE T., TAKEDA N., "New concept for modeling the electromechanical behavior of unidirectional carbon-fiber-reinforced plastic under tensile loading", *Smart Mater. Struct.*, vol. 12, 2003, pp. 105–114.

[PIG 80] PIGGOTT M.R., *Load Bearing Fiber Composites*, Oxford, Pergamon Press, 1980.

[PRA 90] PRABHAKARAN R., "Damage assessment through electrical resistance measurement in graphite fiber-reinforced composites", *Experimental Techniques*, vol. 14, no. 1, 1990, pp. 16-20.

[RAS 88] RASK O.N., ROBINSON D.A., "Graphite as an imbedded strain gauge material", *SAMPE Journal*, January/February, 1988, pp. 52-55.

[RIS 98] RISSON T., "Comportement au fluage de matériaux composites à renfort carbone et matrice époxyde et PEEK", PhD thesis, Ecole Centrale Lyon, France, 1998.

[SAL 97] SALVIA M., FOURNIER P., FIORE L., VINCENT L., "Flexural fatigue behaviour of UDGFRP – experimental approach", *International Journal of Fatigue*, vol. 3, 1997, pp. 253-262.

[SAN 99] SANDLER J., SHAFFER M.S.P., PRASSE T., BAUHOFER W., SCHULTE K., WINDLE A.H., "Development of a dispersion process for carbon nanotubes in an epoxy matrix and the resulting electrical properties", *Polymer*, vol. 40, 1999, pp. 5967-5971.

[SCH 89] SCHULTE K., BARON C., "Load and failure analysis of cfrp laminates by means of electrical resistivity measurements", *Composites Science and Technology*, vol. 36, 1989, pp. 63-76.

[SCH 95] SCHULTE K., WITTICH H., "The Electrical Response of Strained and/or Damaged Polymer Matrix Composites", Proceedings of the 10th International Conference on Composite Materials (ICCM-10), 14-18 August 1995, Whistler – Canada, Vol. 5, Street K. and Poursartip A. Eds, pp. 315-325.

[SCH 01] SCHUELER R., JOSHI S.P., SCHULTE K., "Damage detection in CFRP by electrical conductivity mapping", *Composites Science and Technology*, vol. 61, no. 6, 2001, pp. 921-930.

[SHU 96] SHUI X., CHUNG D.D.L., "A piezoresistive carbon filament polymer-matrix composite strain sensor", *Smart Mater. Struct.*, Vol. 5, 1996, pp. 243-246.

[STR 79] STRIEDER W., JO T., "Percolation in a thin ply of unidirectional composite", *Journal of Composite Materials*, Vol. 13, 1979, pp. 72-83.

[STR 82] STRIEDER W., LI P., JOY T., "Random lattice electrical conductivity calculations for a graphite/epoxy ply of finite thickness", *Journal of Composite Materials*, vol.16, 1982, pp. 53-64.

[THI 94] THIAGARAJAN C., SUTHERLAND I., TUNNICLIFFE D., IRVING P.E., "Electrical potential techniques for damage sensing in composite structure", Proceedings of the 2nd European Conference on Smart Structures and Materials, 12-14 October 1994, Glasgow, Scotland, pp. 128-131.

[WAN 96] WANG X., CHUNG, D.D.L., "Continuous carbon fiber epoxy-matrix composite as a sensor of its own strain", *Smart Materials and Structures*, vol. 5, 1996, pp. 796-800.

[WAN 97a] WANG X., CHUNG D.D.L., "Sensing delamination in a carbon fiber polymer-matrix composite during fatigue by electrical resistance measurement", *Polymer Composites*, vol. 18, no. 6, 1997, pp. 692-700.

[WAN 97b] WANG X., CHUNG D.D.L., "Real time monitoring of fatigue damage and dynamic strain in carbon fiber polymer-matrix composite by electrical resistance measurement", *Smart Materials and Structures*, vol. 6, 1997, pp. 504-508.

[WAN 97c] WANG S., CHUNG D.D.L., "Self-monitoring of in silicon carbide whiskers reinforced silicon nitride", *Smart Mat. Struct.*, vol. 6, 1997, pp. 199-203.

[WAN 98] WANG X., CHUNG D.D.L., "Short carbon fiber reinforced epoxy coating as a piezoresistive strain sensor for cement mortar", *Sensors and actuators*, A71, 1998, pp. 208-212.

Chapter 6

Low Frequency Electromagnetic Techniques

6.1. Introduction

In the domains of Aeronautics and particularly of Composite Structures, electromagnetic techniques are little used for SHM. This could be explained by the fact that the main researchers working in the field of SHM have mechanical engineering trainings. One possible exception is eddy current techniques, but these are mainly used for metallic structures [GOL 01]. In fact, in the case of composite structures, eddy currents techniques are very difficult to use and generally give poor results, this kind of structure being either purely dielectric or a poor conductor (e.g. carbon epoxy). However, other techniques that also use eddy currents, but which are based on low-frequency holography [MAD 99, GRI 00, GRI 01], give good results on carbon epoxy structures. Nevertheless, this method is not easily transposable into the SHM domain, the external equipment required being too complicated.

A new family of electromagnetic techniques, which makes it possible to obtain good information on the health of structures made of composites (CFRP and GFRP), has been developed recently. These techniques consist of determining the state of health of a structure by measurement of its two main electrical parameters, the electrical conductivity and/or the dielectric permittivity, since damage induces locally significant variations in these two parameters. The goal of this chapter is to explain these methods and to give their field of application.

First, we shall provide some reminders about electromagnetic theory necessary for gaining a full understanding of the electromagnetic techniques developed recently. After that, the various techniques will be explained and their possible applications in

Chapter written by Michel LEMISTRE.

the domain of SHM given. In each case, several examples will be presented and discussed.

All information obtained must be fully exploitable and easily interpretable. A data reduction process is therefore very often necessary, and this is particularly appropriate in the case of electromagnetic techniques, where the noise level is relatively important. This is the reason why a special section is devoted to signal processing and particularly to wavelet transforms, a recent development in signal processing, which make it possible to decrease the noise level considerably and to extract relevant information.

6.2. Theoretical considerations on electromagnetic theory

6.2.1. *Maxwell's equations*

All electromagnetic phenomena can be explained by Maxwell's equations; there are four and one can write them as follows:

$$\nabla \cdot \vec{E} = \frac{\rho}{\varepsilon} \tag{1}$$

$$\nabla \cdot \vec{B} = 0 \tag{2}$$

$$\nabla \times \vec{E} = -\frac{\partial \vec{B}}{\partial t} \tag{3}$$

$$v^2 \nabla \times \vec{B} = \frac{\vec{J}}{\varepsilon} + \frac{\partial \vec{E}}{\partial t} \tag{4}$$

In these equations, $\vec{E}$, $\vec{B}$ and $\vec{J}$ represent the electric field, the magnetic induction and the current density respectively, ρ a charge density and t a time variable; v is the propagation velocity ($v^2 = 1/\varepsilon \mu$), where ε and μ are respectively the absolute dielectric permittivity and the absolute magnetic permeability of the considered medium, defined as follows:

$$\varepsilon = \varepsilon_0 \varepsilon_r \text{ and } \mu = \mu_0 \mu_r \tag{5}$$

In this relation, ε_0 and μ_0 are proportionality coefficients depending to the system of units used. In the international system (SI), $\varepsilon_0 = 8.84*10^{-12}$ F m^{-1} and $\mu_0 = 4\pi*10^{-7}$ H m^{-1}, ε_r (relative dielectric permittivity) characterizes the medium with regard to the electric field, μ_r (relative magnetic permeability) characterizes the medium with regard to the magnetic field. In free space $\varepsilon = \varepsilon_0$, $\mu = \mu_0$ and $v = c = 3*10^8$ m s^{-1}.

Note – one can use $\vec{H}$ to indicate the magnetic field. Its relation to magnetic induction $\vec{B}$ is $\vec{B} = \mu \vec{H}$; likewise $\vec{D}$ can be used to indicate the electric induction with $\vec{D} = \varepsilon \vec{E}$.

The first two equations [1] and [2] are static equations; they define the electric field $\vec{E}$ and the magnetic induction $\vec{B}$ in terms of characteristics of the medium. The next two equations [3] and [4] are "dynamic" equations, which show the interdependence between $\vec{E}$ and $\vec{B}$. In fact, the magnetic field $\vec{H}$ (or its induction $\vec{B}$) do not have a real identity, but are uniquely the "relativistic" consequence of a displacement of electric charges.

6.2.2. Dipole radiation

6.2.2.1. Electromagnetic field radiated by a current distribution

According to the definition of the magnetic potential vector $\vec{A}$, the magnetic field $\vec{H}$ and the electric field $\vec{E}$ outside a conductor having a current density $\vec{J}$ are given by:

$$\vec{H} = \frac{1}{\mu} \nabla \times \vec{A} \qquad [6]$$

$$\vec{E} = \frac{1}{j\omega\varepsilon} \nabla \times \vec{H} = \frac{1}{j\omega\mu\varepsilon} \Delta\left(\vec{A}\right) \qquad [7]$$

The magnetic potential vector is thus given by:

$$\vec{A} = \int_{vol.} \frac{\mu\left(\vec{J} e^{-jkr}\right)}{4\pi r} dv \qquad [8]$$

where r represents the distance between the observation point and the current element $\vec{J} dv$ (volume current density), k being the wave number given by $k = \omega\sqrt{\mu\varepsilon}$.

After transformation in the time domain t, the relation [8] yields:

$$\vec{A} = \int_{vol.} \frac{\mu \vec{J} \cos\left(\omega\left(t - \frac{r}{c}\right)\right)}{4\pi} dv \qquad [9]$$

where c is the propagation velocity in the considered medium.

6.2.2.2. Electric dipole

Let us consider an infinitesimal current element $\bar{I}dl$ (see Figure 6.1). According to equation [8], the magnetic potential vector at point p is given by:

$$\vec{A}(p) = \frac{\mu e^{-jkr}}{4\pi r}\left(\bar{I}dl\right) a \qquad [10]$$

where a is the length of the dipole.

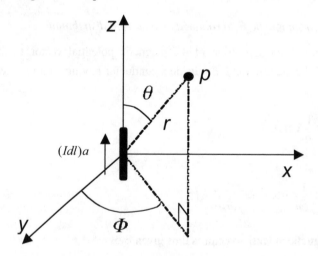

Figure 6.1. *Electric dipole*

In spherical co-ordinates, equation [6] becomes:

$$\vec{H}_\Phi = \frac{Idla}{4\pi}k^2 \sin\theta e^{-jkr}\left[\frac{j}{kr} + \frac{1}{k^2 r^2}\right] \qquad [11]$$

Similarly, for the electric field:

$$\vec{E}_r = Z_0 \frac{\vec{I}dla}{4\pi} k^2 \cos\theta e^{-jkr} \left[\frac{1}{k^2 r^2} - j\frac{1}{k^3 r^3} \right]$$

[12]

$$\vec{E}_\theta = Z_0 \frac{\vec{I}dla}{4\pi} k^2 \sin\theta e^{-jkr} \left[j\frac{1}{kr} + \frac{1}{k^2 r^2} - j\frac{1}{k^3 r^3} \right]$$

Z_0 represents the wave impedance in free space, defined as follows:

$$Z_0 = \frac{|E|}{|H|} = \sqrt{\frac{\mu_0}{\varepsilon_0}} = 120\pi \text{ (Ohm)}$$

[13]

Relations [11] and [12] apply in near-field conditions (i.e. the distance $r <$ the wavelength λ). For far-field conditions ($r > \lambda$), the terms in $1/r^2$ and $1/r^3$ can be neglected and these equations become:

$$\vec{H}_\Phi = j\frac{\vec{I}dl\,ak}{4\pi r} \sin\theta \times e^{-jkr}$$

$$\vec{E}_\theta = Z_0\, j\frac{\vec{I}dl\,ak}{4\pi r} \sin\theta \times e^{-jkr}$$

[14]

$$\vec{E}_r \cong 0$$

6.2.2.3. Magnetic dipole

A circular current element (i.e. an elementary loop) can be considered as a magnetic dipole (see Figure 6.2). This current element produces identical radiation to the electric dipole. However, vectors $\vec{E}$ and $\vec{H}$ are interchanged, so that the equations for the magnetic and electric components for far field and near field are given as follows.

Near field:

$$\vec{E}_\Phi = Z_0 \frac{k\pi a^2 \vec{I}}{4\pi} k^2 \sin\theta \times e^{-jkr} \left[\frac{1}{kr} - j\frac{1}{k^2 r^2} \right]$$

[15]

$$\vec{H}_r = Z_0 \frac{2k\pi a^2 \vec{I}}{4\pi} k^2 \cos\theta \times e^{-jkr} \left[j\frac{1}{k^2 r^2} - \frac{1}{k^3 r^3} \right]$$

[16]

$$\vec{H}_\theta = Z_0 \frac{k\pi a^2 \vec{I}}{4\pi} k^2 \sin\theta \times e^{-jkr} \left[-\frac{1}{kr} + j\frac{1}{k^2 r^2} - \frac{1}{k^3 r^3} \right]$$

Far field:

$$\vec{E}_\Phi = -Z_0 \vec{H}_\theta$$

$$\vec{H}_r = 0$$

[17]

$$\vec{H}_\theta = -\frac{\pi a^2 \vec{I}}{4\pi r} k^2 \sin\theta \times e^{-jkr}$$

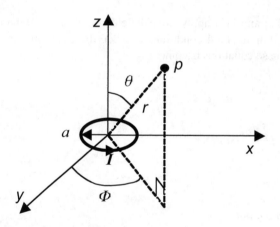

Figure 6.2. *Magnetic dipole*

6.2.3. *Surface impedance*

Let us consider a plane structure made up of material having the following electrical characteristics:

– magnetic permeability $\mu = \mu_0 \mu_r$,

– dielectric permittivity $\varepsilon = \varepsilon_0 \varepsilon_r$,

– electric conductivity σ,

– thickness d.

The structure is illuminated by a plane wave at any incidence (see Figure 6.3).

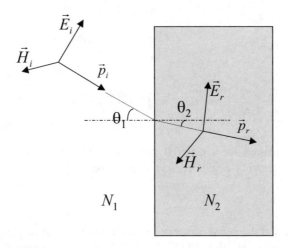

Figure 6.3. *Illumination of a conductive material*

The electric and magnetic components of the incident wave $\vec{p}_i$ are $\vec{E}_i$ and $\vec{H}_i$ respectively, $\vec{E}_r$ and $\vec{H}_r$ being the components of the refracted wave $\vec{p}_r$; θ_1 and θ_2 are the angles of incidence and refraction respectively, N_1 being the refractive index of the external medium, N_2 the refractive index of the structure. θ_1 and θ_2 are linked by the following relation:

$$N_1 \sin \theta_1 = N_2 \sin \theta_2 \qquad [18]$$

N_1 is the index of the external medium, and in free space $N_1 = 1$. Taking into account the complex relative permittivity of the material $\varepsilon_r^* = \varepsilon_r' - j\varepsilon_r''$, N_2 the index of the material can be written:

$$N_2 = \sqrt{\varepsilon_r^*} = \sqrt{\varepsilon_r' - j\frac{\sigma}{\omega\varepsilon_0}} \qquad [19]$$

If the material is considered to be a good conductor (i.e. $\sigma \gg j\omega\varepsilon$), N_2 can be reduced to:

$$N_2 = \sqrt{-j\frac{\sigma}{\omega\varepsilon_0}} \qquad [20]$$

Equation [18] then becomes:

$$\sin \theta_2 = \frac{1}{N_2}\sin \theta_1 \qquad\qquad [21]$$

or:

$$\cos \theta_2 = \left(1 - \left(\frac{1}{N_2}\right)^2 \sin^2 \theta_1\right)^{1/2} = \left(1 + j\frac{\omega \varepsilon_0}{\sigma}\sin^2 \theta_1\right)^{1/2} \qquad\qquad [22]$$

The approximation for a material that is a good conductor allows the second term in parentheses to be neglected, so that equation [22] becomes:

$$\cos \theta_2 \cong 1 \qquad\qquad [23]$$

This shows that the wave penetrates through the material perpendicularly to the surface of the structure ($\theta_2 = 0$). Figure 6.4 shows the configuration of the field inside the structure.

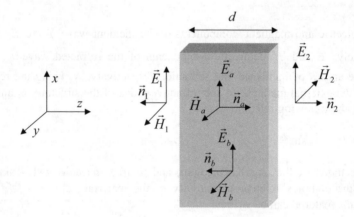

Figure 6.4. *The configuration of the field inside a structure of thickness d*

The two fields $\vec{E}(z)$ and $\vec{H}(z)$ can be written as follows:

$$\vec{E}(z) = \vec{E}_a \exp(-\gamma z) + \vec{E}_b \exp(+\gamma z) \qquad\qquad [24a]$$

$$\vec{H}(z) = \left(\vec{n} \times \vec{E}_a\right)\frac{\exp(-\gamma z)}{Z} + \left(\vec{n} \times \vec{E}_b\right)\frac{\exp(+\gamma z)}{Z} \qquad [24b]$$

with Z the wave impedance inside the material, defined by the relation:

$$Z = \sqrt{\frac{\mu_0}{\varepsilon^*}} = (j+1)\sqrt{\frac{\mu f \pi}{\sigma}} = \frac{j+1}{\sigma \delta} \qquad [25]$$

and γ the propagation constant given by:

$$\gamma = j\omega\sqrt{\varepsilon^* \mu_0} = (j+1)\sqrt{\mu f \pi \sigma} = \frac{j+1}{\delta} \qquad [26]$$

where $\vec{n}$ in equation [24b] is the unit vector in the z direction (i.e. the thickness of the material), f is the frequency of the incident wave and δ is the skin depth defined by the following relation:

$$\delta = \frac{1}{\sqrt{\mu f \pi \sigma}} \qquad [27]$$

Electric and magnetic fields tangential to the structure on each of its faces, $\vec{E}_1$, $\vec{H}_1$ and $\vec{E}_2$, $\vec{H}_2$ (see Figure 6.2) are related by the following equations:

$$\vec{E}_1 = \frac{Z}{th(\gamma d)}\vec{n}_1 \times \vec{H}_1 + \frac{Z}{sh(\gamma d)}\vec{n}_2 \times \vec{H}_2 \qquad [28a]$$

$$\vec{E}_2 = \frac{Z}{sh(\gamma d)}\vec{n}_1 \times \vec{H}_1 + \frac{Z}{th(\gamma d)}\vec{n}_2 \times \vec{H}_2 \qquad [28b]$$

One can define a surface current density $\vec{J}_s$, which is the integral of the volume current density $\vec{J}$ in the thickness of the structure. One can write the boundary equations as:

$$\vec{J}_s = \vec{n}_1 \times \left(\vec{H}_1 - \vec{H}_2\right) = \vec{n}_2 \times \left(\vec{H}_2 - \vec{H}_1\right) \qquad [29]$$

For low frequencies such as $d \ll \delta$ given $|\gamma d| \ll 1$, one can write the following approximations:

$$sh(\gamma d) \cong \gamma d \ \text{ and } \ th(\gamma d) \cong \gamma d \qquad\qquad [30]$$

Equations [28a] and [28b] then become:

$$\vec{E}_1 = \vec{E}_2 = \vec{E}_{tg} = \frac{1}{\sigma d}\vec{J}_s \qquad\qquad [31]$$

where E_{tg} is the electric field tangential to the surface of the structure. It is now possible to define the surface impedance Z_s such as:

$$\vec{E}_{tg} = Z_s \vec{J}_s \qquad\qquad [32]$$

with:

$$Z_s = \frac{1}{\sigma d} \qquad\qquad [33]$$

One can see that, for low frequencies, the current density is distributed uniformly in the material. Thus, one can neglect the skin effect and any possible resonances inside the material. The condition $|\gamma\,d| << 1$ (or $\delta > d$) defines the domain of application of the concept of surface impedance; the material is then called "electrically thin". In the case of $\delta > d$, the material is called "electrically thick" and losses by attenuation inside the material and by reflection are then the main phenomena.

The surface impedance is given in "square ohms" (symbol Ω_c), which is the impedance of a square electrically thin sample, having a conductivity σ. One can represent the equivalent electric diagram for the surface impedance by Figure 6.5.

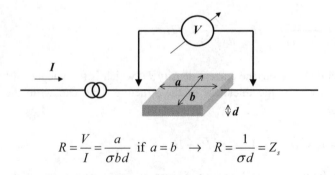

$$R = \frac{V}{I} = \frac{a}{\sigma b d} \ \text{ if } a = b \ \rightarrow \ R = \frac{1}{\sigma d} = Z_s$$

Figure 6.5. *Equivalent electric diagram of the surface impedance*

6.2.4. *Diffraction by a circular aperture*

In the case of an incident plane wave, H.A. Bethe [BET 44] has given an approximate analytical representation of the fields diffracted by a small circular aperture, in a structure considered as infinitely conductive (a metal with good conductivity, such as aluminum, can be used to verify this hypothesis), having an infinite surface (i.e. large relative to the diameter of the aperture), with respect to the following hypothesis:

– the size of the aperture is small relative to the wavelength of the incident field,

– the fields are calculated at a large distance relative to the size of the aperture.

Bethe derived the concept of "short-circuit fields" $\vec{E}_{cc}$ and $\vec{H}_{cc}$, which represent the fields at the aperture loaded with a perfectly conductive material. These fields are defined as follows:

$$\vec{E}_{cc} = 2\vec{E}_0 \text{ and } \vec{H}_{cc} = 2\vec{H}_0 \qquad [34]$$

where $\vec{E}_0$ and $\vec{H}_0$ are respectively the orthogonal electric component and the tangential magnetic component of the incident field. The fields diffracted by the aperture are thus the sum of the fields radiated by an electric dipole having a moment $\vec{P}_e$ and a magnetic dipole having a moment $\vec{P}_m$, which are respectively orthogonal and tangential (see Figure 6.6).

If we introduce the concept of "polarizability", the dipole moments are related to short circuit fields by the following relations:

$$\vec{P}_e = \varepsilon\alpha_e\vec{E}_{cc} \text{ and } \vec{P}_m = -\overline{\overline{\alpha}}_m\vec{H}_{cc} \qquad [35]$$

where α_e and $\overline{\overline{\alpha}}_m$ are respectively the electric polarizability and the magnetic polarizability for an aperture of radius a.

The polarizability depends on the geometry of the structure and the aperture. If we consider a plane aperture of any geometry, the electric polarizability is a scalar, while the magnetic polarizability is a tensor of rank 1. For a plane circular aperture of radius a, the magnetic polarizability is a diagonal tensor having two equal diagonal terms:

$$\alpha_e = \frac{2}{3}a^3 \qquad [36]$$

$$\alpha_{mxy} = \alpha_{myx} = 0 \tag{37}$$

$$\alpha_{mxx} = \alpha_{myy} = \alpha_m = \frac{4}{3}a^3 \tag{38}$$

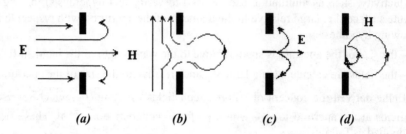

$$(a) \qquad (b) \qquad (c) \qquad (d)$$

Figure 6.6. *Incident fields and equivalent dipoles, (a) orthogonal electric field, (b) tangential magnetic field, (c) equivalent electric dipole, (d) equivalent magnetic dipole*

On the basis of Bethe's hypothesis, K.F. Casey [CAS 81] has calculated the moment of the magnetic dipole for a plane circular aperture having a radius a, loaded with an imperfectly conductive material such as carbon epoxy. The material loading the aperture is inserted without covering the conductive structure. Casey defines the resistance of an "electrical gasket" as R_g, which is the contact resistance between the conducting structure and the material loading the aperture. R_g is has units ohm $\times$ meter (the length of the contact between the material and the structure). Casey's method involves the resolution of an integral equation, and two solutions have been proposed. The first one is exact and leads to a semi-analytic expression for the magnetic dipole including a coefficient that must be calculated numerically. The other one, obtained by an approximation method, leads to the dipolar moment $\vec{P}_m$ being written in the form of a transfer function for a first-order low-pass filter:

$$\frac{\vec{P}_m}{\vec{P}_{m0}} = \frac{1}{1 + j\dfrac{f}{f_c}} \tag{39}$$

In this expression, $\vec{P}_{m0}$ is the magnetic dipolar moment given by a free aperture, f represents the frequency of the incident wave and f_c is the cut-off frequency given by the following:

$$f_c = \frac{3}{8\mu_0} \frac{Z_s}{a} \left(1 + \frac{R_g}{aZ_s}\right) \tag{40}$$

Note that the cut-off frequency is a function of the surface impedance Z_s and thus of the conductivity σ of the material (see equation [33]). This cut-off frequency is called "Casey's frequency".

6.2.5. Eddy currents

If a conductive material is submitted to a variable magnetic induction $\vec{B}$, Maxwell's equation [3] shows that an electric field $\vec{E}$ arises inside the material inducing a current density $\vec{J}$ such that $\vec{J} = \sigma \vec{E}$. This current density generates a magnetic induction $\vec{B}_i$, which is in opposition to $\vec{B}$. The magnetic induction $\vec{B}_i$ can be written using the simplified Maxwell's equation [4]:

$$v^2 \nabla \times \vec{B}_i = \frac{\vec{J}}{\varepsilon}$$

[41]

The magnitude of the electric field $\vec{E}$ inside a conductive material, and thus the current density $\vec{J}$, decreases exponentially:

$$\vec{J} = \vec{J}_0 \, e^{-\sqrt{\frac{\pi \mu \sigma f}{d}}}$$

[42]

where $\vec{J}_0$ is the surface current density, f is the frequency of excitation and d is the penetration depth . This is known as the skin effect phenomenon.

Let us write $\vec{J} = \vec{J}_0 \, e^{-d/\delta}$, in which δ represents the distance at which the current density is attenuated by a factor of e^{-1} (i.e. $1/2.718...$, $\approx 1/3$). This distance δ is called the skin depth, defined by the following:

$$\delta = \frac{1}{\sqrt{\pi \mu \sigma f}}$$

[43]

One can define a characteristic frequency f_s such that $\vec{J} = \vec{J}_0 / e$ (i.e. $d = \delta$); this frequency f_s is called the "skin frequency".

6.2.6. Polarization of dielectrics

Dielectrics are defined as materials that at room temperature (≈ 300 K) have a conductivity $\sigma \leq 10^{-20}$ S m^{-1}. On a macroscopic scale, dielectrics appear to be electrically neutral. However, at microscopic scale, dielectrics show an "assembly" of elementary electric dipoles having a random orientation in space. For a large number of dipoles, one can consider dielectrics to be statistically neutral. If a dielectric material is subject to an electric field, elementary dipoles tend to line up in the direction of the incident electric field. Then, one can define a polarization vector per unit volume $\vec{P}$ as:

$$\vec{P} = Nq\vec{\delta} \qquad [44]$$

with q elementary charges (per atom or molecule), separated by a distance $\vec{\delta}$ and N atoms (or molecules) per unit volume. The product $q\vec{\delta}$ represents the elementary dipolar moment $\vec{p}$ for each atom (or molecule) and this dipolar moment is related to the local electric field $\vec{E}_0$ such as $\vec{p} = \alpha \vec{E}_0$. In this relation, the proportionality factor α is called the "polarizability coefficient"; it is a function of the electrical characteristics of the medium (i.e. the dielectric) and more exactly of its dielectric relative permittivity ε_r. It is defined by the Clausius–Mossoti equation:

$$\alpha = \frac{3}{4\pi N} \left(\frac{\varepsilon_r - 1}{\varepsilon_r + 2} \right) \qquad [45]$$

The polarization phenomenon in a dielectric medium results from three different sources: electronic polarization, ionic polarization and orientation polarization. Each one of these sources involves a polarizability coefficient: α_e, α_i and α_o respectively, the real part of the coefficient α being the sum of the three real parts of the elementary polarizabilities.

Electronic polarization arises because the center of the local electronic charge cloud around the nucleus is displaced under the action of the electric field $\vec{P}_e = N\alpha_e \vec{E}_0$.

Ionic polarization occurs in ionic materials because the electric field displaces positive ions and negative ions in opposite directions $\vec{P}_i = N\alpha_i \vec{E}_0$.

Orientation polarization can occur in materials composed of molecules that have permanent electric dipoles. The alignment of these dipoles depends on temperature and leads to an "orientational polarizability" per molecule $\alpha_o = p^2 / 3KT$, where p is the permanent dipolar moment per molecule, K is the Boltzmann constant and T is the absolute temperature.

Because of the different nature of these three polarization processes, the response of a dielectric solid to an applied electric field will strongly depend on the frequency of the field. The resonance of the electronic excitation occurs in the ultraviolet part of the electromagnetic spectrum; the characteristic frequency of the vibration of the ions is located in the infrared, while the orientation of dipoles requires fields of much lower frequencies (below 10^9 Hz). This response to electric fields of different frequencies is shown in Figure 6.7.

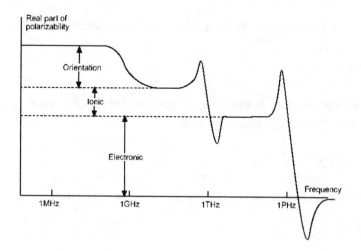

Figure 6.7. *Frequency dependence of the different contributions to polarizability (from Handbook of Chemistry and Physics)*

For a low-frequency excitation (i.e. 1 kHz to 10 MHz), one can consider the response of the dielectric medium to be quasi-static. So, it is possible to describe the electric field inside the dielectric by the following equation:

$$\nabla \cdot \vec{E} = \frac{\rho_l + \rho_p}{\varepsilon_0}$$
[46]

where ρ_l is the density of the free charges and ρ_p is an apparent density of charges due to the polarization phenomenon. The field $\vec{E}$ can be considered as the resultant of two components, the field resulting from the free charges $\vec{E}_f$ and the field resulting from the polarization phenomenon $\vec{E}_p$ such that $\vec{E} = \vec{E}_f + \vec{E}_p$. One can therefore write:

$$\nabla \cdot \vec{E} = \nabla \cdot \vec{E}_f + \frac{\rho_p}{\varepsilon_0}$$
[47]

One can define a polarization vector $\vec{P}$ such that $\rho_p = -\nabla \cdot \vec{P}$ [FEY 86], so that equation [47] can be rewritten as:

$$\vec{E} = E_f + \frac{\vec{P}}{\varepsilon_0} \qquad [48]$$

The vector $\vec{P}$ is related to the incident field $\vec{E}_i$ by the following relation:

$$\vec{P} = \chi_e \varepsilon_0 \vec{E}_i \qquad [49]$$

where χ_e is the electric susceptibility, itself related to the relative electric permittivity ε_r by the following expression:

$$\varepsilon_r = 1 + \chi_e \qquad [50]$$

Taking into account equations [48], [49] and [50], one can write:

$$\vec{E} = \vec{E}_f + (\varepsilon_r - 1) \vec{E}_i \qquad [51]$$

However, in most dielectrics, the term due to the free charges can be neglected and the value of the electric field $\vec{E}$ is reduced to the polarization term $(\varepsilon_r - 1) \vec{E}_i$.

6.3. Applications to the NDE/NDT domain

6.3.1. *Dielectric materials*

Let us consider a dielectric material having an electric relative permittivity ε_r in the presence of an electric field $\vec{E}_i$, the material being considered in a macroscopic manner (i.e. as quasi-isotropic). The electric field $\vec{E}_i$ induces the phenomenon of polarization inside the material, characterized by the vector $\vec{P}$. The total electric field $\vec{E}_T$, measured at point A (see Figure 6.8) can be written as:

$$\vec{E}_T = \vec{E}_i + \frac{\vec{P}}{\varepsilon_0} = \vec{E}_i + (\varepsilon_r - 1) \vec{E}_i \qquad [52]$$

After subtraction of the incident electric field $\vec{E}_i$, one can obtain directly the value of ε_r. Note that the frequency of excitation must be lower than the cut-off frequency of the orientation polarization phenomenon (i.e. 10 MHz).

This method can be applied to all dielectric composite materials such as GFRP, sandwich, etc. An electric probe has been designed [LEM 01] that makes it possible to detect some defects in dielectric composites. This probe performs a differential measurement between two adjacent zones (see Figure 6.9). Figure 6.10 shows an example of detection performed on a sample of a glass epoxy sandwich having a lack of foam in the middle of the lower part. The right-hand view shows the electric image obtained by scanning the material.

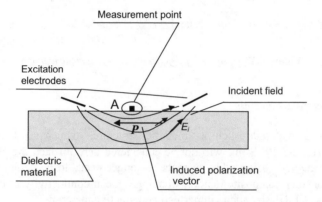

Figure 6.8. *Excitation and measurement configuration applied to a dielectric material*

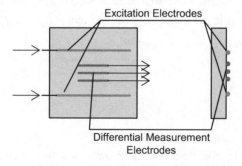

Figure 6.9. *Differential electric probe for detection of damage in dielectric materials*

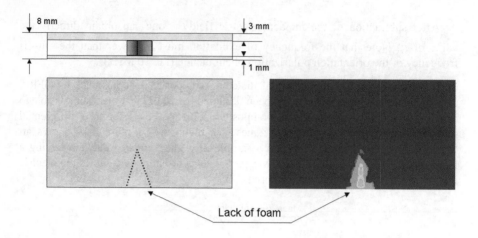

Figure 6.10. *Example of defect detection in dielectric materials*

6.3.2. *Conductive materials*

Conductive materials such as metallic structures that have a conductivity σ between 10^7 and 10^8 S m^{-1} will not be considered here. In fact, the effectiveness of classical methods using eddy currents no longer needs to be demonstrated. We will only consider composite materials having a mean conductivity σ of about 10^4 S m^{-1} such as CFRP, so called "imperfectly conductive materials".

We have seen, in section 6.2.4 (equation [39]) that the magnetic dipolar moment $\vec{P}_m$ of an aperture loaded with an imperfectly conductive material can be written in the form of a transfer function of a first-order low-pass filter. As the dipolar moment is directly proportional to the magnetic field, the terms $\vec{P}_m$ and $\vec{P}_{m0}$ can be replaced respectively by $\vec{H}$ and $\vec{H}_0$ in equation [39]:

$$\frac{\vec{H}}{\vec{H}_0} = \frac{1}{1 + j\dfrac{f}{f_c}} \qquad [53]$$

where $\vec{H}_0$ is the magnetic field measured through a free aperture and $\vec{H}$ is the magnetic field measured through an aperture loaded with a conductive material. Since the cut-off frequency f_c is a function of the surface impedance Z_s, we can directly obtain the value of the conductivity σ of the considered material by using equation [40].

Let us now consider a local excitation by a nearby magnetic field (e.g. with a Hertz loop), and make a local measurement of the resulting magnetic field, as shown in Figure 6.11; one can omit the infinite conductive plane and the contact resistance R_g.

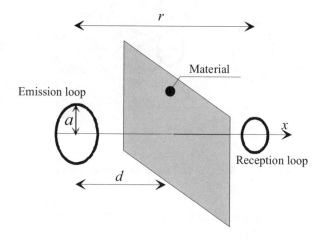

Figure 6.11. *Excitation with a nearby magnetic field*

An analytical calculation gives a new transfer function between $\vec{H}$ and $\vec{H}_0$:

$$\frac{\vec{H}(f,r)}{\vec{H}_0(f,r)} = \left(1+\left(\frac{r}{a}\right)^2\right)^{\frac{3}{2}} \int_0^\infty \frac{u^2}{u+j\dfrac{f}{f_c}} J_1(u) e^{-u\frac{r}{a}} du \qquad [54]$$

where r is the distance between the two loops, a the radius of the emission loop, J_1 the first-order Bessel function and the cut-off frequency f_c is given by the following expression:

$$f_c = \frac{1.4}{\pi \mu_0 \sigma a e} \qquad [55]$$

where a represents the radius of the emission loop and e the thickness of the material. Because the distance between the emission loop and the material d does not influence the transfer function; it is possible to put the material immediately "after" the reception loop, as shown in Figure 6.12. This allows the material to be tested on only a single face.

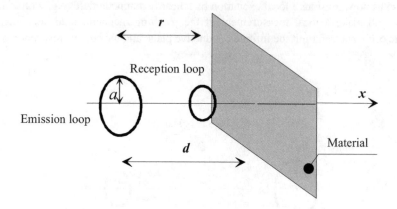

Figure 6.12. *Excitation and measurement on a single face*

However, in this case, the transfer function $\vec{H}/\vec{H}_0$ is given by the sum of the two terms T_1 and T_2, as follows:

$$T_1 = 1 - \frac{\left(1+\left(\dfrac{r}{a}\right)^2\right)^{\frac{3}{2}}}{\left(1+\left(\dfrac{2d-r}{a}\right)^2\right)^{\frac{3}{2}}} \qquad T_2 = \left(1+\left(\dfrac{r}{a}\right)^2\right)^{\frac{3}{2}} \int_0^{\infty} \frac{u^2}{u+j\dfrac{f}{f_c}} J_1(u) e^{-u\frac{2d-r}{a}}\, du \qquad [56]$$

Note that when f tends towards infinity, the function $T_1 + T_2$ tends towards the constant T_1. So, if the distance d increases, the value of the constant T_1 becomes near the value of the transfer function before attenuation (i.e. $f < f_c$). This is the problem of "lift-off", which is well known in eddy current techniques.

The goal of this kind of analysis is not necessarily to measure the exact value of the conductivity of a material under test, but more specifically to detect damage that induces a local variation in this conductivity. A differential magnetic probe [LEM 98, LEM 01], measuring the "contrast" of the conductivity between two adjacent zones, has been designed (see Figure 6.13). This probe compares the magnitude of the magnetic field between the two zones 1 and 2; in the case where the two magnetic fields are different, the voltage V_{out} is non-zero.

The excitation frequency f is set between two characteristic frequencies. On the one hand, it must be lower than the skin frequency f_s (see section 6.2.5), while, on the other, the excitation frequency f_e must be greater than the Casey frequency f_c (equation [55]), in order to produce a significant variation in the measured magnetic field as a result of variation in the local conductivity σ (equation [56]). Figure 6.14 shows the evolution of these two frequencies as a function of the thickness of a quasi-isotropic carbon epoxy multilayer structure.

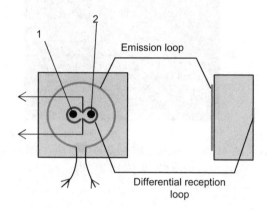

Figure 6.13. *Differential magnetic probe*

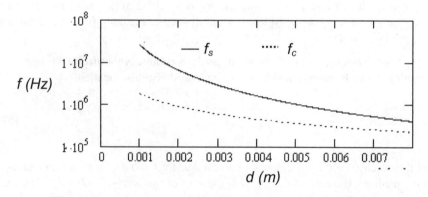

Figure 6.14. *Variation of f_s and f_c as a function of the thickness d*

By scanning a structure, one can build up an image of where damaged areas are clearly shown; an example is given in Figure 6.15. This figure shows the magnetic image (i.e. σ contrast) from a quasi-isotropic carbon epoxy multilayer sample of $60 \times 60 \times 2$ mm including a delamination with fiber breakage. The delamination has been created by a calibrated impact with an energy of 3 J.

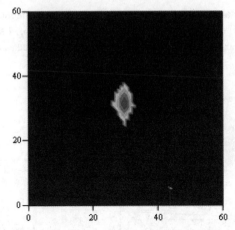

Figure 6.15. *Magnetic image obtained on a carbon epoxy sample
including a delamination*

6.3.3. *Hybrid method*

One can consider a carbon epoxy structure to be made up of two different media. On the one hand, it can be considered as a conductive medium because of the presence of the carbon fibers (conductivity $\sigma \approx 10^4$ S.m^{-1}), while, on the other hand, the resin can be considered as a dielectric medium (relative dielectric permittivity $\varepsilon_r \approx 4$).

The local electric field $\vec{E}_l$ induced inside a conductive structure by magnetic induction $\vec{B}$ can be represented by the following Maxwell's equation:

$$\nabla \times \vec{E}_l = -\frac{\partial \vec{B}}{\partial t} \qquad [57]$$

and this electric field itself induces a current density $\vec{J} = \sigma \vec{E}_l$. However, in a carbon epoxy medium, the current density $\vec{J}$ is the sum of two terms $\vec{J} = \vec{J}_c + \vec{J}_d$. The first term $\vec{J}_c$ (conductive current density) is due to the conductivity of the carbon fibers; the second term $\vec{J}_d$ (displacement current density) is a transient term due to the polarization phenomenon in the epoxy resin (see section 6.3.1). Let us consider the quasi-static hypothesis, with frequency below 10^9 Hz, (i.e. below the cut-off frequency of the orientational polarization phenomenon); with equation [48] taken into account, the measured local electric field $\vec{E}_m$ can be written as follows:

$$\vec{E}_m = \frac{\vec{J}_c}{\sigma} + \frac{\vec{P}}{\varepsilon_0} \qquad [58]$$

where $\vec{P}$ is the polarization vector due to the epoxy resin. From equation [51] we can write:

$$\vec{E}_m = \frac{\vec{J}_c}{\sigma} + \vec{E}_l \left(\varepsilon_r - 1\right)$$

[59]

So the measured electric field $\vec{E}_m$ is a function of the conductivity σ of the medium and also of its relative dielectric permittivity ε_r. This technique, based on the magnetic induction (i.e. eddy currents) and on the analysis of the resulting electric field, is called the "hybrid method".

The possibility of having simultaneous access to the main electric properties of the medium is of great interest with regard to carbon epoxy structures used in the aeronautical domain, because it will enable detection of the main kinds of damage that one can find in these structures. In fact, the damage can be classified into three categories:

– Damage that has a mechanical origin, being produced by impacts, inducing delaminations and, in general, fiber breakage. Such critical damage only occurs in carbon (fibers), so, in terms of electrical properties, they induce a unique local variation in the conductivity.

– Thermal damage, which arises either from proximity to a hot body or from an electrical impact (e.g. sparks, lightning). In the first case one can detect the damage uniquely from variations in conductivity. In the second case, if the burn is small, there is no variation in σ but only in εr due to pyrolysis of the resin. Nevertheless, this kind of damage generally affects both electrical parameters.

– Damage resulting from liquid ingress (water, oil, fuel, etc.), which can be critical because the liquid can start a chemical reaction and weaken the structure. This kind of damage only induces a variation in εr due to the presence of a new medium having a different dielectric permittivity (i.e. the liquid).

On the basis of this hybrid concept, a new hybrid electromagnetic probe has been designed [LEM 04]. This probe can be considered as a combination of the electric probe (section 6.3.1) and the magnetic probe (section 6.3.2), including some improvements. Figure 6.16(a) presents a photo of the probe and Figure 6.16(b) a schematic view of the probe, designed with a dielectric parallelepiped (30 × 30 × 10 mm) with on one face the induction coil while the differential dipole, for the measurement of the electric field, is on the opposite face.

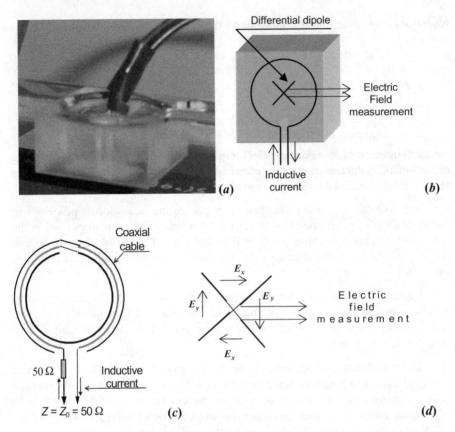

Figure 6.16. *Electromagnetic probe for the hybrid technique:*
(a) Photo of the probe, (b) Schematic view, (c) Induction coil, (d) Electric field measurement

The induction coil appears as a Möbius loop made with a hard coaxial cable (see Figure 6.16(c)). This geometry allows the formation of a double induction loop while keeping the real impedance equal to the characteristic impedance of the coaxial cable (i.e. 50 Ω), throughout the frequency domain used, from 100 kHz to 10 MHz. The measurement unit appears as a double crossed dipole (see Figure 6.16(d)), which makes it possible to perform differential measurements that is sensitive to two orthogonal components of the tangential electric field, $\vec{E}_x$ and $\vec{E}_y$.

The operating frequency f is not as critical as for the magnetic measurement. It is simply necessary to operate with a frequency below the orientational polarization phenomenon cut-off (i.e. $f < 10^9$ Hz; see section 6.2.4). However, since the magnitude of the electric field induced inside material is directly proportional to the frequency of the induced magnetic field (equation [57]), it is preferable to use as high a frequency as possible. Nevertheless, too high a frequency will prevent penetration into the bulk of a material such as a carbon epoxy multilayer, because of the skin effect (see section 6.3.2, Figure 6.14). For this kind of material, a good frequency domain is between 100 kHz and 10 MHz.

A scanning system driven by a computer (see Figure 6.17) enables the building of images of structures under test in which the damage is clearly visible. An example is given in Figure 6.18(a), which shows the electromagnetic image of a quasi-isotropic sample of carbon epoxy (200 × 150 × 4 mm) including various burns. These results are compared with a C-scan ultrasonic investigation in Figure 6.18(b). Only a hybrid electromagnetic investigation is capable of detecting light burns.

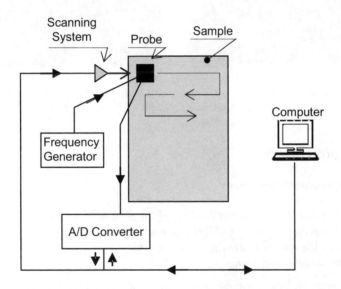

Figure 6.17. *Experimental set-up for electromagnetic imaging using a hybrid technique*

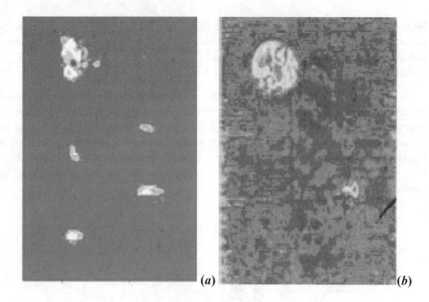

(a)   *(b)*

Figure 6.18. *Investigation of a composite sample including various burns*
(a) electromagnetic image, (b) ultrasonic C-scan image

6.4. Signal processing

6.4.1. *Time-frequency transforms*

In order to facilitate the understanding of the next section, which is devoted to applications developed in the SHM domain, it seems useful to give some explanation of the data reduction process used, particularly concerning the method involving the wavelet transform.

Fourier analysis is one of the major assets of physics and mathematics. It is unavoidable in the field of signal processing for many reasons. The first one is the universality of the concept of "frequency"; a description in term of frequencies often gives a better understanding of a physical phenomenon. A second reason is the mathematical structure of Fourier analysis, which is naturally applicable to transformations such as linear filtering. However, the Fourier transform allows a description of a physical phenomenon in either the frequency domain or the time domain, never in both. On the other hand, knowing the frequency value of a phenomenon requires the knowledge of all of its time history. Conversely, the value of a signal in a time t can be seen as an infinite superposition of everlasting waves. These disadvantages affect the representation of the physical reality in the case of non-stationary phenomena. Today, however, there are several processes that make it possible to have a description of a phenomenon simultaneously in both domains (i.e.

time and frequency). One of the most frequently used is called the wavelet transform, which enables the representation of a phenomenon simultaneously in the time domain and in the frequency domain, which makes a good representation of non-stationary phenomena possible.

6.4.2. *The continuous wavelet transform*

The continuous wavelet transform (CWT), like the Fourier transform, is an integral transform. The analysis or direct transform of a function $f(x)$ in the k domain can be written as follows:

$$T_f(k) = \int f(x) \cdot \Psi_k(x) \cdot dx \qquad [60]$$

where the function $\Psi_k(x)$ represents the transform function in the k domain. In a Fourier transform, this function is the complex exponential $e^{-j\omega t}$. Thus, the synthesis or inverse transform can be expressed as:

$$f(x) = \int T_f(k) \cdot \Psi_x(k) \cdot dk \qquad [61]$$

If x is the time variable and k the frequency variable, one can consider the function $\Psi_k(x)$ to be the impulse response of a filter. In this case, equation [60] is the convolution product between the signal $f(x)$ and the impulse response of a filter Thus, the result is the time response of this filter for the signal $f(x)$.

For any signal $x(t)$, one obtains its CWT from the following equation:

$$T_x(a,b) = \frac{1}{\sqrt{a}} \int_{-\infty}^{+\infty} x(t) \cdot \Psi\left(\frac{t-b}{a}\right) \cdot dt \qquad [62]$$

where b represents the time shift of the convolution, a is the scale of the wavelet (a function of the impulse response time of the considered filter) and Ψ is the wavelet function. One can perform as many convolutions as there are wavelet scales, where each wavelet scale, called a "daughter wavelet", is the impulse response of a specific filter for a particular frequency bandwidth defined by the scale. All the impulse responses, taken together, are called a "wavelet family". The typical impulse response that defines a specific filter is called the "mother wavelet".

If the Heisenberg–Gabor relation (i.e. the uncertainty principle) is taken into account, with the time responses (the results of convolutions) being functions of impulse response of the considered filter (i.e. of the *scale* of the wavelet), the time localization for low frequencies will be imprecise. In contrast, high frequencies (short impulse responses) will be localized with a good accuracy in the time domain.

The ratio between the central frequency and its bandwidth, which corresponds to the scale of the wavelet, is a constant.

After processing is complete, one has at one's disposal a whole series of filtered time responses. One can represent all the responses in a time–scale diagram, which by analogy corresponds to a time–frequency diagram. The scale of the frequencies corresponds to the central frequency of each filter corresponding to each daughter wavelet.

Among the most used wavelets for the analysis of non-stationary signals are Morlet's wavelet and the "Mexican hat", this latter being the second-order derivative of a Gaussian distribution. Morlet's wavelet is a complex wavelet that is the product of a complex exponential with a Gaussian distribution. This wavelet is shown in Figure 6.19(a), the real part as a continuous line, the imaginary part as a dashed line. Figure 6.19(b) shows the Fourier transform of this wavelet (i.e. the frequency bandwidth of the corresponding filter); the magnitude is normalized and the frequency scale is arbitrary.

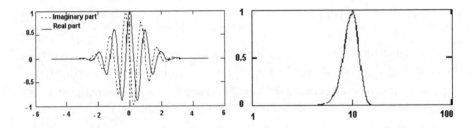

Figure 6.19. (*a*) *Normalized Morlet's wavelet;* (*b*) *modulus of the Fourier transform of the Morlet's wavelet*

A characteristic of this wavelet is that it is a complex wavelet. Thus, one can access the modulus and the phase of the transform. The central frequency, corresponding to the scale of the wavelet, is called the "scale frequency" and is given by the following relation:

$$f_{sc} = \frac{1}{\sigma}\frac{f_0}{a} \qquad [63]$$

where σ is the standard deviation of the Gaussian distribution and a is the scale of the wavelet, f_0 being the lowest frequency, which is defined as the inverse of the time duration of the signal $x\,(t)$.

6.4.3. The discrete wavelet transform

The CWT is very useful for the analysis of complex non-stationary signals. However, this process requires a great deal of computation time and generates many redundant terms. Various methods have been devised in order to reduce these disadvantages. Among these is the discrete wavelet transform (DWT).

Let us imagine a signal sampled with 2^N points. One can generate a family wavelet including n points, where n varies according to the following:

$$n_p = 2^{N-1}, 2^{N-2}, ..., 2^{N-p}$$
[64]

with p (the number of daughter wavelets generated), given by the relation $p = N - \log_2(M)$, where M is the next power of 2 above the number of points necessary to define the wavelet having the lowest scale. If one has a signal sampled with 1024 points, and if the smallest wavelet can be defined with 100 points (128 points will be considered), there will be a family of only three daughter wavelets. The process consists of projecting each of these daughter wavelets on the signal with a shift corresponding to the number of points of the considered daughter wavelet. One obtains a series of values for each wavelet scale; these are called "decomposition coefficients" and the number of coefficients obtained for a given scale is a function of the corresponding scale. Each of the series of coefficients gives the signal energy contained inside the frequency bandwidth corresponding to the scale of the wavelet. The DWT can be written as:

$$DWT_{m,k} = \int_0^\infty s(t) \gamma_{m,k}(t)\, dt$$
[65]

Each daughter wavelet $\gamma_{m,k}(t)$ is obtained by binary "expansion" or "contraction" m from the mother wavelet; each coefficient is obtained by projection of the considered daughter wavelet onto the signal, after dyadic translation k. The family of wavelets can be written as:

$$\gamma_{m,k}(t) = 2^{m/2} \gamma(2^m t - k)$$
[66]

In this case, the chosen wavelet is the impulse response of a low-pass filter having a finite impulse response. In practice, one uses two filters, the first of which is the low-pass filter and the second one in quadrature with the first (high-pass filter). Thus, one projects the signal onto an orthogonal basis of wavelets.

After processing (DWT), one can rebuild the signal at each level of decomposition m (i.e. each corresponding frequency bandwidth) with the following relation:

$$s_m(t) = \sum_k DWT_{m,k}\, \gamma_{m,k}(t) \tag{67}$$

the original signal $s(t)$ being the sum of all decomposition levels:

$$s(t) = \sum_m \sum_k DWT_{m,k}\, \gamma_{m,k}(t) \tag{68}$$

One of the most used wavelets is Daubechie's Wavelet, shown in Figure 6.20(a). This figure shows the "mother" wavelet (continuous line) and the "father" wavelet (dashed line), corresponding respectively to the impulse responses of the two filters; scales are arbitrary. Figure 6.20(b) shows the corresponding Fourier transforms of these wavelets; the scales are also arbitrary.

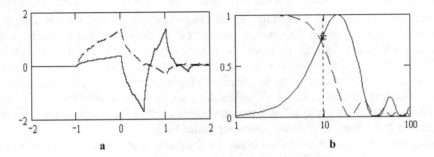

Figure 6.20. *(a) Daubechie's wavelet; (b) modulus of the corresponding Fourier transforms*

Daubechie's wavelet does not have an analytical expression in the time domain. However, one can uniquely express the corresponding filter in the frequency domain:

$$H(\omega) = \left(\frac{1+e^{j\omega}}{2}\right)^N P(\omega) \tag{69}$$

$P(\omega)$ being a trigonometric polynomial, 2π periodic. N represents the number of coefficients of the filter (of the wavelet). To respect the orthogonal conditions, this filter must have the following properties [TRU 98]:

$$|H(\omega)|^2 + |H(\omega+\pi)|^2 = 2$$

$$H(0) = \sqrt{2} \tag{70}$$

$$H(\pi) = 0$$

6.4.4. *Multiresolution*

The process consisting of decomposing a signal by DWT and rebuilding each of the decomposition levels is called "Multiresolution". Indeed, each rebuilding level corresponds to the time behavior of the signal in the considered frequency bandwidth. So, one obtains a signal "split" into successive half bandwidths. For a signal defined with a bandwidth B_p, decomposed into m levels, each rebuilt level L will include only the higher half bandwidth corresponding to the lower bandwidth of the preceding level. In this process, shown in Figure 6.21, the number m of the decomposition level is a function of the number of sampling points of the signal, which must be a power of 2:

$$m \le \frac{\log(N)}{\log(2)} \qquad [71]$$

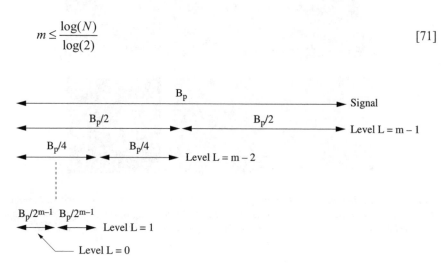

Figure 6.21. *Multiresolution process*

In Figure 6.21, one can see that the right-hand side of the diagram is empty. Indeed, this process does not allow the decomposition of the high-frequency parts obtained for each decomposition level. However, it is possible to perform a new decomposition for each high-frequency part and this process is called decomposition (or multiresolution) by wavelet packets. Nevertheless, this method requires as many wavelet families as decomposition levels.

Just like Fourier transforms and the DWT process, the Multiresolution process can be performed in two dimensions. Figures 6.22 show the 2-D Multiresolution process applied to an electric image obtained by the hybrid electromagnetic method (see section 6.3.3) on a carbon epoxy multilayer specimen including a hole. The multiresolution has been performed with eight decomposition levels using a Daubechie's wavelet having eight coefficients.

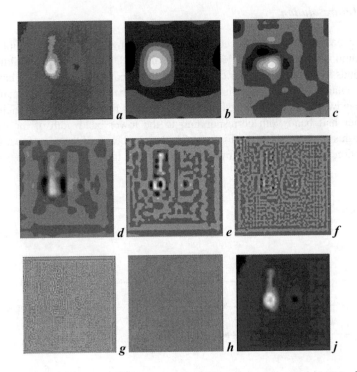

Figure 6.22. *2-D Multiresolution process applied to an electric image: (a) original image; (b) rebuilding at level 1; (c) rebuilding at level 2; (d) Rebuilding at level 3; (e) Rebuilding at level 4; (f) Rebuilding at level 5; (g) Rebuilding at level 6; (h) Rebuilding at level 7(j) Rebuilding at levels 2 + 3 + 4*

In these figures, the rebuilding at level 0 is not shown, as it only contains the very low frequencies of the image. Level 1 (Figure 6.22b) begins to show the searched-for characteristic (i.e. the hole) but with poor spatial definition due to the lowest frequency bandwidth. Levels 2, 3 and 4 clearly show the searched-for characteristic. Levels 5, 6 and 7 just contain noise. In fact, levels 2, 3 and 4 are sufficient to rebuild the image including relevant characteristics (see Figure 6.22j). If the original image is defined with a space frequency bandwidth Bp, the relevant characteristics are contained within $Bp/16 + Bp/32 + Bp/64$, about 1/8 of the full bandwidth of the original image. This method is therefore very useful for compressing any image or signal and for extracting the relevant characteristics, which can be invisible in the original data (e.g. in the case of a lower SNR).

6.4.5. *Denoising*

Various methods of denoising are applied to signals; and filtering or correlation can be used. These processes can give interesting results in many cases. However, the signal is often corrupted and some of the relevant information can be irreparably lost. David Donoho from the Statistics Department of the University of Stanford has developed a denoising algorithm based on DWT processing [DON 94]. This method allows the signal's corruption to be minimized; it is based on three hypotheses:

i) almost all of the energy contained in a signal can be represented by only some of the coefficients of the decomposition,

ii) the noise affects all coefficients of the decomposition,

iii) by decreasing the decomposition coefficients, one can rebuild the original signal without noise.

An example is given to illustrate these hypotheses.

Hypothesis 1. Figure 6.23(a) represents a time signal (the scales are arbitrary), Figure 6.23(b) represents the magnitude of the decomposition coefficients as a function of their position in the decomposition level. For a process having m levels of decomposition, the number of decomposition coefficients nb is given by:

decomposition at level $L = m$ -> $nb = 1$,

decomposition at level $L = m - 1$ -> $nb = 2$,

decomposition at level $L = m - 2$ -> $nb = 4$,

decomposition at level $L = m - 3$ -> $nb = 8$.

Generally:

$$nb_L = 2^L \qquad\qquad\qquad\qquad [72]$$

The total number of coefficients nt given by the decomposition is:

$$nt = \sum_0^L nb_L \qquad\qquad\qquad\qquad [73]$$

In the present case, $L = 9$ and $nt = 1023$.

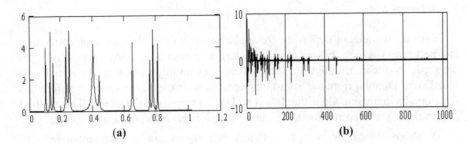

(a)   (b)

Figures 6.23. *(a) Time signal; (b) decomposition coefficients*

Since the amplitude of decomposition coefficients is representative of the energy contained in the signal, within the bandwidth corresponding to the decomposition level the whole coefficients represents the energy contained in the original signal. Figure 6.24 shows the sum of the decomposition coefficients (normalized in energy) as a function of their number. In this case, the limit is the first 100 coefficients.

A simple calculation shows that 90% of the energy of the signal is contained inside the first 45 coefficients and 99% of the energy is contained inside the first 100 coefficients. So, one can rebuild 99% of the original signal with only 100 coefficients (the total number being 1023).

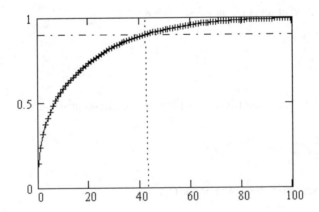

Figure 6.24. *Sum of the decomposition coefficients*

Hypothesis 2. Figure 6.25(a) represents the time signal from Figure 6.23(a) with some noise added, SNR = 4. Figure 6.25(b) shows the coefficients of the decomposition.

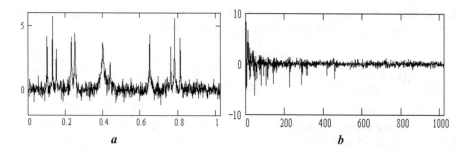

Figure 6.25. *(a) Time signal with noise added; (b) decomposition coefficients*

By comparing Figure 6.25(b) with Figure 6.23(b), one can see that the magnitude of all the coefficients is affected by noise.

Hypothesis 3. Figure 6.26(a) shows the time signal after the denoising process (i.e. the shrinking of the coefficients and rebuilding). This figure can be compared to Figure 6.23(a), reproduced in Figure 6.26(b) at the same scale. One can see that the noise has been completely removed, while, on the other hand, the original signal is practically unaffected by this processing.

The shrinking of coefficients can be carried out in two ways by use of two shrinking functions. The first one δ_h, called the "hard shrinking function", consists of setting to "zero" all coefficients having a value of k lower than a given threshold c and to rebuild the signal with the remaining coefficients.

$$\delta_h(k) = \begin{vmatrix} 0 & \text{if } |k| < c \\ k & \text{in other cases} \end{vmatrix} \qquad [74]$$

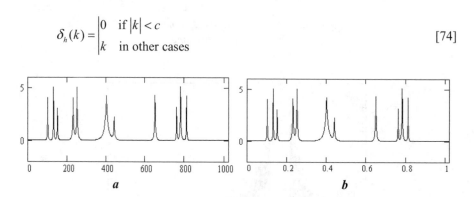

Figure 6.26. *(a) Denoised time signal; (b) Original time signal*

The second function δ_s called the "soft shrinking function" consists, like the first, of setting to zero all coefficients having a value k lower than c, but the value c

is also subtracted from each of the remaining coefficients. Thus, the signal will be rebuilt with the new values of the remaining coefficients:

$$\delta_s(k) = \begin{vmatrix} 0 & \text{if } |k| < c \\ \text{sign}(k) \cdot (|k| - c) & \text{in other cases} \end{vmatrix} \qquad [75]$$

The second method is preferable for statistical reasons. This is illustrated by Figures 6.27, 6.28 and 6.29, which represent respectively the histograms of the decomposition coefficients: before shrinking (Figure 6.27), after using the δ_h function (Figure 6.28) and after using the δ_s function (Figure 6.29). To make them easier to represent and understand, the histograms shown have been computed with a decomposition process performed with 15 levels. The total number of coefficients obtained is: $nt = 65535$.

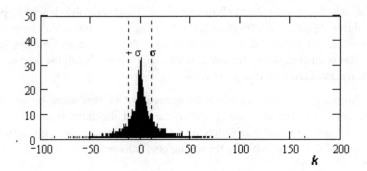

Figure 6.27. *Histogram of the decomposition coefficients before shrinking*

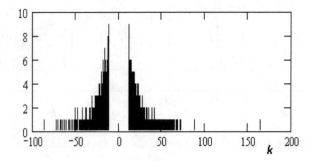

Figure 6.28. *Histogram of the decomposition coefficients after using the δ_h function*

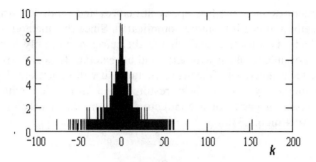

Figure 6.29. *Histogram of the decomposition coefficients*
after using the δ_s function

The "art" of denoising consists of choosing the right threshold value c in the shrinking function. D. Donoho [DON 94] recommends:

$$c = \sqrt{2 \cdot \log(nt)} \cdot \sigma \qquad\qquad [76]$$

σ being the standard deviation of the decomposition coefficients (see Figure 6.27).

6.5. Application to the SHM domain

6.5.1. *General principles*

In the domain of SHM, the goal is to determine at any given time, the state of health of the structure in use. So it is necessary to integrate multiple sensors inside the material during the manufacturing process in order to obtain what we might call a "Smart Structure". Generally, this process consists of integrating an active layer including various sensors (e.g. PZT sensors or optical fibers), between two layers of the composite structure.

In the case of electromagnetic methods, which are adaptations of methods used in the NDT/NDE field, the active layer is a flexible layer including wire networks, about 100–200 µm in thickness. This layer can be bonded onto the inner face of the structure or inserted between two layers of a multi-layer structure; see Figure 6.30.

As for NDT/NDE applications, there are three possible techniques: the magnetic technique, the electric technique and the hybrid technique. Which of these methods is most suitable depends on the type of material. The magnetic method is based on the variation of electrical conductivity so it will best be used with conductive materials and particularly carbon epoxy structures. The electric method, which is based on the variation of dielectric permittivity, will be best suited for use with dielectric structures such as glass epoxy structures and sandwich structures. For the hybrid method, it is a little more complicated. Since this method is based both on the conductivity variation and on the dielectric permittivity variation, it is theoretically possible to use it with any kind of material. However, as the electric method is better adapted for dielectric structures, it will be preferred in this case. The hybrid method gives the best results when it is used with composite structures consisting partly of a conductive medium and partly of a dielectric medium (see section 6.3.3).

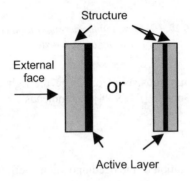

Figure 6.30. *Location of the active layer*

6.5.2. *Magnetic method*

The adaptation of the technique used in NDT/NDE domain involves the induction of eddy currents in the structure and measurement of the total resulting magnetic field (i.e. the incident field plus the reflected field; see section 6.3.2). Nevertheless, it is impossible to put the inductive loop and the measurement loop on the same face of the structure, because the distance r tends towards zero (see equation [56]) and the measured magnetic field becomes too weak. It is therefore necessary to put the inductive network on one face of the structure and the measurement network on the other face. This method cannot be used to build an SHM system, because it is impossible to put a sensor on the external face of a structure.

However, another solution is available. It consists of illuminating the structure with an antenna and measuring the transmitted magnetic field with a sensitive layer bonded on the inner face of the structure, as shown in Figure 6.31.

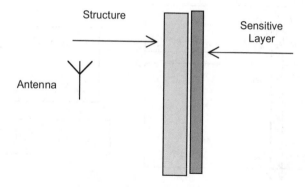

Figure 6.31. *Arrangement for an SHM system
using the magnetic method*

The sensitive layer is made of a double network of parallel wires short-circuited at one end, printed on a flexible substrate. The two networks are orthogonal, in order to form a matrix of squares. The geometry of this sensitive layer is shown in Figure 6.32. Each pair of wires can be considered as an elementary loop, which measures the transmitted magnetic flux Φm through the corresponding area. The sum of the two "loops" from each network gives the total magnetic flux ΦT through an elementary square. Thus, one can build a matrix of squares that represent an image of the structure including any damage that is present (see Figure 6.33). The distance between two wires in each network defines the spatial resolution of the system.

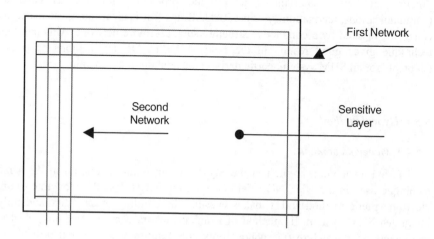

Figure 6.32. *Geometry of a sensitive layer*

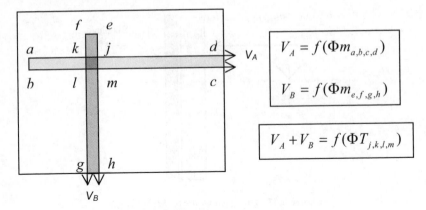

Figure 6.33. *Matrix building process*

This technique seems attractive, but it is not very interesting because the sensitive layer is a "passive" layer and the system needs an external arrangement (i.e. the antenna) to produce the induced electromagnetic field. In fact, this method cannot be considered as a fully integrated SHM system.

6.5.3. *Electric method*

This method can only be discussed theoretically, because it has not been developed. The reason for this lies in the domain of aeronautics, where purely dielectric materials are used precisely because they are dielectric and very often one function of these materials is to provide protection for various antennas (communications, navigation systems, RADAR, etc.). It is clear that these materials cannot be masked by a conductive structure such as a wire network. Although this technique gives good results in the field of NDT/NDE applications, it is not adaptable for an SHM system, particularly in the domain of aeronautics.

6.5.4. *Hybrid method*

6.5.4.1. *General comments*

The hybrid method applied in the SHM domain is an adaptation of the same technique used in the NDT/NDE field (see section 6.3.3). It is the most interesting electromagnetic method and it makes possible the building of an active and fully integrated SHM system. Its method of simultaneous measurement of both electric parameters of a structure (i.e. conductivity and permittivity) makes it possible to determine the origin of any damage. There are two versions of this technique. The first generation, which can be considered as an economical version, does not allow damage to be fully characterized. Nevertheless, it can be sufficient in many cases. The second generation, called the 2-D Hybrid method, allows full characterization of various types of damage and computation of the new values of local conductivity and local dielectric permittivity in the damaged zone.

6.5.4.2. *First generation*

The first generation of SHM system using the hybrid electromagnetic method is composed of two principal elements. The first element is an active layer called HELP Layer®-1 (Hybrid Electromagnetic Performing Layer) including a circuit inducing eddy currents inside the structure and a circuit for the measurement of the resulting electric field. The second element is an electronic system able to generate the excitation signal, to acquire the resulting signal, to perform the data reduction and to build the images of structures where the damage can be clearly seen [LEM 02].

As the sensitive layer of the magnetic method, the HELP Layer®-1 includes two orthogonal wire networks printed on a 100 μm thick flexible substrate. The main difference from the magnetic method is that one of these networks is not short-circuited at one end. The network that is short-circuited represents the induction loops, while the other network, which is open circuited, can be considered as a network of capacitances, which is sensitive to the electric field. The geometry of the HELP Layer®-1 is shown in Figure 6.34. Figure 6.35 represents an instrumented structure with the HELP Layer®-1.

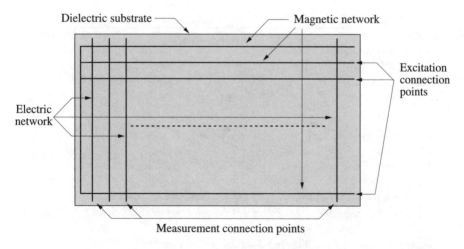

Figure 6.34. *Geometry of the HELP Layer®-1*

An example of an image obtained with this method is given in Figures 6.37 and 6.38. These images are obtained with a 16-ply orthotropic carbon epoxy plate $[0_2, 90_2]_{2s}$ of dimensions: $610 \times 305 \times 2$ mm. The plate has been damaged by six different defects: a 4 J impact (I2) inducing a severe delamination with fiber breakage, a 2 J impact (I1) inducing a light delamination, and four local burns produced by "high energy sparks" (30 V, 5 A) of various durations, with energies of 40 J (B2), 80 J (B3), 120 J (B1) and 400 J (B4) (see Figure 6.36). The magnetic field sensor network is excited by a frequency of 700 kHz, with a 100% amplitude modulation of 1 kHz.

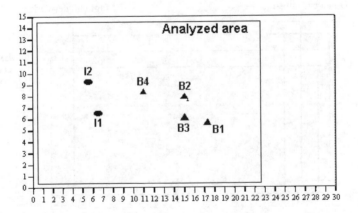

Figure 6.35. *An instrumented structure with the HELP Layer®-1*

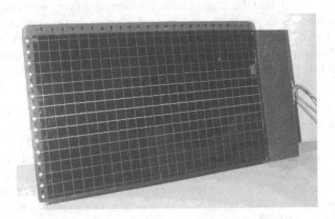

Figure 6.36. *Damaged structure*

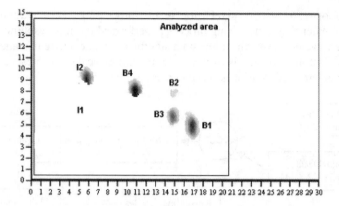

Figures 6.37. *Electromagnetic image of the damaged structure*

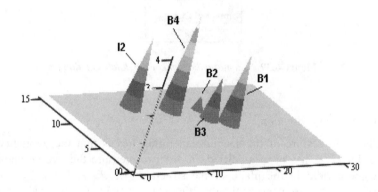

Figure 6.38. *3-D view of the electromagnetic image*

In Figures 6.37 and 6.38, one can see that all areas of damage are perfectly detected except the 2 J impact, probably because it involves no fiber breaking, so there is no variation of electric properties inside the structure under test. Figure 6.38 (the 3-D view of the electromagnetic image obtained) clearly shows the severity of the damage.

As explained in section 6.4 (signal processing) it is necessary to perform a process of data reduction to obtain understandable electromagnetic images. If the mesh of the networks is 20 mm, for an analyzed area of 440 × 280 mm, the experimental matrix has 308 points (22 × 14). The first step of the process consists of acquiring a matrix of the initial state of the structure under test (i.e. sound material). Each new matrix is compared with the initial matrix by subtraction. New data obtained after subtraction are re-sampled by linear interpolation, so as to build an image of 256 × 256 pixels for image processing considerations (matrix inversion). After that, a denoising process is performed by multiresolution (i.e. a 2-D discrete wavelet transform) to obtain a significant image where damage is clearly visible; at last this image is restored to its initial scale for presentation. The schematic diagram of the process is shown in Figure 6.39.

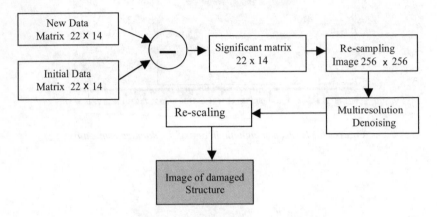

Figure 6.39. *Schematic diagram of the data reduction process*

6.5.4.3. *Second generation*

The main feature of the second generation lies is that the geometry of the HELP Layer®-1 has been modified, in order to measure the two components of the electric field in the plane of the structure (i.e. E_x and E_y), by using two crossed networks to measure the resulting electric field. This new HELP Layer®, called the 2-D HELP Layer®, includes three wire networks. The first one induces eddy currents in the structure, while the others are responsible for the measurement of the two components of the electric field. The new geometry of the 2-D HELP Layer® is shown in Figure 6.40. This feature makes possible the determination of the origin of the damage (see section 6.3.3) [LEM 04-2].

The first measuring network (the lower network in Figure 6.40) gives the term $E_{m1} = E_x - E_y$ (or $E_x + E_y$, depending on the direction of the wires), while the second measuring network (the upper network in Figure 6.40) gives the term $E_{m2} = E_x + E_y$ (or $E_x - E_y$), the difference $E_{m1} - E_{m2}$ being equal to $-2E_y$ (or $+2\ Ey$), while the sum $E_{m1} + E_{m2}$ is equal to $+2E_x$. This process, illustrated in Figure 6.41, allows each component of the electric field to be extracted.

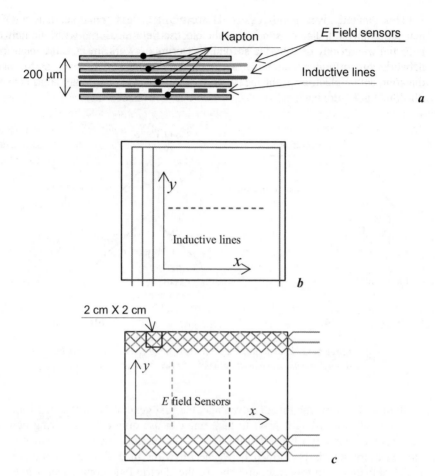

Figures 6.40. *The 2-D HELP Layer®: (a) Section view, (b) Inductive network, (c) Electric field measuring network*

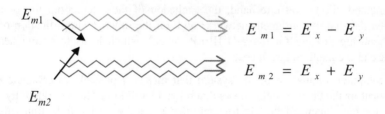

$$E_{m1} = E_x - E_y$$

$$E_{m2} = E_x + E_y$$

Figure 6.41. *The E-Field component extraction process*

This method gives good results. However, it is less sensitive than the first generation method. The reason is that the electric field measured with one network is partially short-circuited by the second measuring network. In fact, the measuring networks are orthogonal and the wires of the second network are in the same direction as the electric vector measured by the first network. This phenomenon is illustrated in Figure 6.42.

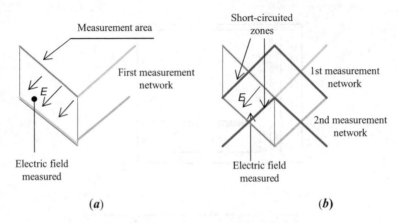

Figure 6.42. *Interaction between the two measurement networks: (a) only one measurement network, (b) two measurement networks*

In order to overcome this disadvantage, the first generation of the HELP Layer® can be used in a different way. In fact, one can use only one measuring network that has the geometry of the HELP Layer®-1 and perform two measurements. The first measurement is performed between two wires in differential mode (i.e. the classical manner); in this case, one obtains the electric field component *orthogonal* to the wires (a pair of wires being considered as an elementary capacitance). The second measurement will be performed by using only one wire in common mode; in this second case, one obtains the electric field component *parallel* to the wire, one wire being considered as an elementary capacitive antenna (see Figure 6.43). This method allows, on one hand, the exclusion of the interaction phenomenon between the two measurement networks and, on the other, a simplification of the manufacturing process for the 2-D HELP Layer®, which leads to a considerable decrease in the manufacturing cost.

A computer simulation of the system that makes it possible to calculate each component of the electric field shows that a local variation of the conductivity only affects the component of the electric field that is parallel to the induction wire. In contrast, a local variation of the dielectric permittivity affects both components. If one assumes that damage having a mechanical origin, such as an impact, induces a fiber break phenomenon that causes a variation of the conductivity (see section 6.3.3), then other kinds of damages such as burning or liquid ingress will cause a variation of the dielectric permittivity. Thus, it is possible to determine the origin of the damage by measurement of the two components of the resulting electric field. Figure 6.44 presents an algorithm enabling the solution of part of the inverse problem in order to determine the kind of damage and to compute the new values of conductivity and permittivity.

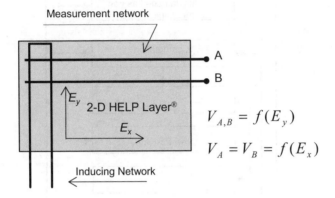

Figure 6.43. *The electric field component measurement process*

The process is as follows. First the measured E_x component is evaluated. If it is not significant (i.e. lower than 10^{-4}), the damage has a mechanical origin because there is only a variation in the conductivity. Then the new value of conductivity σ is computed by iteration up to the equality between E_y measured and E_y computed, in order to gain an idea of the severity of the damage. If the E_x component is significant (i.e. greater than 10^{-4}), the damage will have a different origin (e.g. a burn or liquid ingress), so the new values of ε and σ which are determining the damage severity can be computed by iteration.

A comparison between the results obtained and an ultrasonic C-Scan investigation makes it possible to determine the sensitivity of this method. Figures 6.45 and 6.46 show this comparison for "mechanical" damage and non-mechanical damage respectively. The first comparison (Figure 6.45) concerns a carbon epoxy plate $[45_2/0_2/-45_2/90_2]_s$ including various types of damage, produced by calibrated impacts of various energies. The energy of impact is plotted on the abscissa while the area of the damage forms the ordinate axis. The computed area of the electromagnetic images, which is considered as Gaussian, can be compared with the delamination area determined by an ultrasonic C-Scan, the limits being equal to 2σ. One can see that the sensitivity of the HELP Layer® is lower than that of the ultrasonic C-Scan.

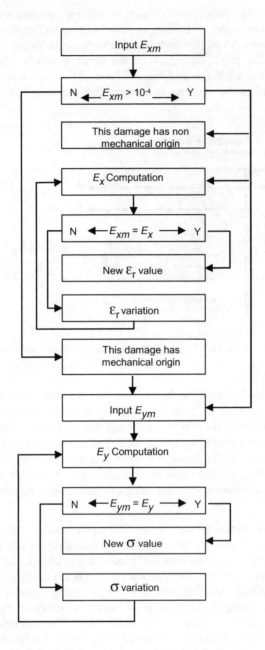

Figure 6.44. *Inverse problem resolution algorithm*

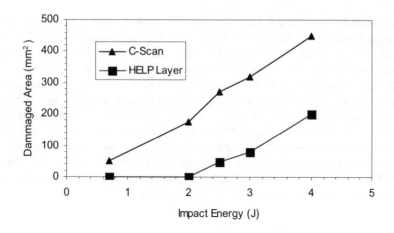

Figure 6.45. *Comparison between HELP Layer® and Ultrasonic C-Scan in the case of mechanical damage*

Figure 6.46 shows the same comparison in the case of thermal damage produced by an electric spark. The energy of the various damage events is calculated from the product $VI\Delta t$, V being the voltage, I the current and Δt the duration of the spark. In this figure one can see that the HELP Layer® is more sensitive than the ultrasonic method.

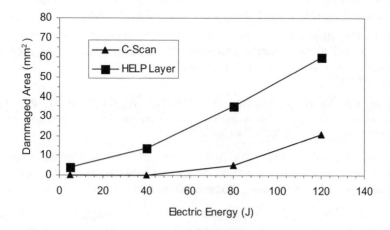

Figure 6.46. *Comparison between the HELP Layer® and Ultrasonic C-Scan in the case of non-mechanical damage*

To conclude about the HELP Layer®, and more generally about the hybrid electromagnetic method, one can say that this method detects practically all kinds of damage, and it is able to determine the kind of damage and its severity. Nevertheless, the sensitivity of this method is lower than that of the classical ultrasonic C-Scan method as far as damage having a mechanical origin, such as impacts, is concerned. Furthermore, the electromagnetic methods are unable to detect damage induced by a very low energy impact because it does not induce a fiber break, so that there is no variation in conductivity. Conversely, the hybrid electromagnetic method shows a greater sensitivity to other types of damage. In fact, this method is complementary to the ultrasonic methods such as the SMART Layer® from Stanford University. A tentative data fusion between the two has been performed and the results obtained are very interesting [LEM 03].

6.6. References

[BET 44] BETHE, H.A., "Theory of diffraction by small holes", *Physical Review*, vol. 7-8, 1944, pp. 367-377.

[CAS 81] CASEY K.F., "Low frequency electromagnetic penetration of loaded apertures", *IEEE Transaction on Electromagnetic Compatibility*, vol. EMC-23, no. 4, 1981, pp. 367-377.

[DON 94] DONOHO D., JOHNSTONE I., "Ideal denoising in an orthonormal basis chosen from a library of bases", *C.R. French Academy of Sciences*, Serie I, Paris 1994.

[FEY 86] FEYNMAN R.P. *Electromagnétisme*, vol. 1 and 2, InterEditions, Paris 1986.

[GOL 01] GOLDFINE N. *et al.*, "Surface mounted periodic field eddy currents sensors for Structural Health Monitoring", *Proceeding of SPIE*, vol. 4335, 2001, pp. 20-34.

[GRI 00] GRIMBERG R., SAVIN A., PREMEL D., MIHALACHE O., "Nondestructive evaluation of the severity of discontinuities in flat conductive materials using eddy currents transducer with orthogonal coils", *IEEE Transaction on Magnetics*, vol. 35, no. 1, 2000, pp. 299-331.

[GRI 01] GRIMBERG R., PREMEL D., LEMISTRE M., BALGEAS D., PLACKO D., "Compared NDE of damages in graphite epoxy composites by electromagnetics methods", *Proceeding of SPIE*, vol. 4336, 2001, pp. 65-72.

[LEM 01] LEMISTRE M.B., BALAGEAS D.L., "Electromagnetic Structural Health Monitoring for composite materials", *Structural Health Monitoring, The Demands and Challenges*, Ed. Fu-Kuo Chang, CRC Press, 2001, pp. 1281-1290.

[LEM 02] LEMISTRE M., BALAGEAS D., "A new concept for Structural Health Monitoring applied to composite materials", *Structural Health Monitoring*, DEStech publications, Ed. Daniel L. Balageas, 2002, pp. 493-507.

[LEM 03] LEMISTRE M., BALAGEAS D., "A hybrid electromagnetic acousto-ultrasonic method for SHM of carbon/epoxy structures", *SHM Journal*, vol. 2, no. 2, June 2003, pp. 153-160.

[LEM 04] LEMISTRE M., DEOM A., *Détection de brûlures dans les composites à base de carbone, Nouvelles méthodes d'instrumentation*, Ed. Hermes Lavoisier, vol. 2, 2004, pp. 305-312.

[LEM 04-2] LEMISTRE M., PLACKO D., "Evaluation of the performances of an electromagnetic SHM system for composite, comparison between numerical simulation", experimental data and ultrasonic investigation, *Proceedings of SPIE*, Ed. Tribikram Kundu, vol. 5394, pp. 148-156.

[LEM 98] LEMISTRE M., GOUYON R., BALAGEAS D., "Electromagnetic localization of defects in carbon epoxy materials", *Proceeding of SPIE*, Vol. 3399, 1998, pp. 89-96.

[MAD 99] MADAOUI N., SAVIN A., PREMEL D., VENARD O., GRIMBERG R., "An approach for Quantitative Nondestructive Evaluation of Discontinuities in Flat Conductive Materials Using Eddy Currents", 5th International Workshop on Electromagnetic Nondestructive Evaluation, Aug. 1999, Iowa, USA.

[TRU 98] TRUCHETET F. *Ondelettes pour le signal numérique*, Ed. Hermes, Paris 1998.

Chapter 7

Capacitive Methods for Structural Health Monitoring in Civil Engineering

7.1. Introduction

In most countries, the weight of new civil engineering constructions is constantly decreasing to the benefit of structure maintenance, basically for economic reasons. In the meantime, stakeholders are concerned about obtaining a precise knowledge of the health of the existing infrastructures, their preservation and the evolution of their safe-keeping.

Hence, monitoring and non-destructive testing (NDT) have become an important part of the operational systems which are being developed and are at the disposal of engineers for diagnosis [MAL 91, BUN 96]. The implemented techniques are related to the fields of Physics (seismics, gravimetry, electromagnetism, etc.) providing relevant information on the characteristics of the structures and their constituent materials.

In this context, electromagnetic techniques offer several areas of interest, among which are the possibility to propagate through voids (sometimes a contact between the sensor and the target is not necessary) and their sensitivity to water content, which is one of the major causes of the pathology of civil engineering structures.

A modern method has been studied for over 20 years in the French network of Public Works Laboratories: it is related to the assessment of the water content in civil engineering materials, such as concrete, masonry and earthwork materials used

Chapter written by Xavier DÉROBERT and Jean IAQUINTA.

under road structures. The so-called capacitive technique was used in a monitoring configuration using internal probes for a survey of historical buildings, as well as for non-destructive testing during periodic inspections. Because of the simplicity of the exploited principle, systems may be designed to fulfill particular requirements and/or specific geometries.

After a short description of the capacitive technique itself at the beginning of this chapter, we will present real implementations of such probes. A first (classical) application is related to the estimation of the water content of flat concrete structures, and a more recent one is for grouting control of external post-tensioned cables. The aim of the last part of the chapter is to illustrate the potentiality of the technique with an example of the monitoring of historic buildings.

7.2. The principle

There are applications of capacitive techniques in a large variety of disciplines. Such measurements are implemented on a regular basis to provide quantitative estimate of the water content of soils [FAR 02] or agricultural products [NEL 92], for example. The technique is also used to assess the porosity of volcanic rocks [RUS 99] and the void fraction in multi-phase flows [KEN 96]. Furthermore, capacitors are used for the qualitative differentiation of materials such as clear wood from knots and distorted grain [STE 96], to sound the snow packs in avalanche forecasts [LOU 98], characterize biological cells [ASA 02] and detect buried plastic landmines [MAM 99].

This technique has already been studied for several decades in the French network of Public Works Laboratories, initially for the measurement of the water content of soils [TRA 72, BAR 77, BLA 93], and afterwards on reinforced concrete before it is adapted to post-tensioned structures [IAQ 04].

The guiding principle consists of placing two (or more) electrodes on the outer surface of the samples and applying a voltage between them. This system forms a capacitor and changes in capacitance are indicative of the internal constituents (such as the nature of the materials or their moisture content).

If we were to apply the configuration of conventional parallel plate capacitors, it would require contact with the material under study from two opposite sides (Figure 7.1).

As a first approximation (the result is only valid when the spacing between the plates is much smaller than their dimensions), the value of the capacitance C is given by the following formula:

$$C = \varepsilon_0 \varepsilon_r \, S/e \qquad\qquad [1]$$

where S [m^2] is the surface area of the electrodes, e [m] the distance between them, ε [$\approx 8.854 \times 10^{-12}$ F m^{-1}] the absolute permittivity of vacuum, and ε_r the dielectric constant, which is an indicator of the medium's ability to store electric charge.

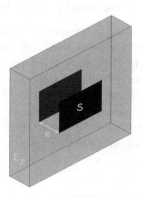

Figure 7.1. *The simplest type of capacitive probe is based on the so-called parallel plate capacitor. Two electrodes of area S, separated by the distance e, enclose a material of dielectric constant ε_r. A uniform electric field directed perpendicularly to the plates is created, and the capacitance of the system is proportional to the permittivity of the media*

For our applications related to measurement of the moisture content of cover concrete, the difference is that the electrodes of the sensor are placed next to one another (Figure 7.2), in order to provide a sufficient penetration depth of the electric field between sensing and driven devices on the same side of the material under study, and to allow measurements in situations where accessibility may be complicated [DIE 98].

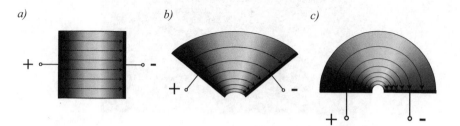

Figure 7.2. *The configuration of a capacitive sensor designed for the measurement of cover concrete moisture content can be seen as the result of gradually opening the angle between the two electrodes of a parallel plate capacitor from situation (a) to situation (c), so as to be able to apply it to plane objects*

However, for measurements carried out from the surface, neither the volume of investigation nor the penetration depths is precisely defined. In fact, the situation is complex, because the system appears as a heterogeneous mixing of dielectric materials and conductors. As a consequence, there is no analytical formulation (as there is for the parallel plate capacitor) giving the value of the capacitance as a function of the characteristics of the medium investigated.

The capacitance value is found by means of a (high-frequency) resonant circuit delivering an alternating voltage (note that a different kind of electronics, for instance a bridge, can be used). The resonant frequency shift is then obtained simply by using a frequency analyzer. In order to minimize the influences of temperature and ionic conduction, the oscillator is operated at 65 MHz. Practical reasons (related to the transmission of information over long distances) require the implementation of a frequency divider to modulate the signal towards lower frequencies (in the range of 5,000 – 40,000 Hz, according to the version).

7.3. Capacitance probe for cover concrete

7.3.1. *Layout*

The geometry of the sensor itself plays a very important role, and we have kept the possibility of employing different configurations of electrodes in order to reach various penetration depths, as shown in Figure 7.3.

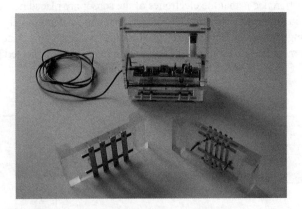

Figure 7.3. *Prototype of a capacitance probe for measuring the water content of flat concrete structures, with its connector and the resonant circuit (in the background). The shoe mounted at the bottom of the device employs a set of two large longitudinal electrodes. In the foreground, a different geometry using five small parallel electrodes (on the right), and a configuration based on four medium-sized electrodes (on the left)*

There are spaces on both sides of the electrodes, and the resulting layout does not constitute a single capacitor but a set of two capacitors in parallel. The first one (C_i) incorporates the object under examination and the second one (C_e) the surrounding environment (principally air and Plexiglas making up the probe), so that the total capacitance is the sum. The relationship between this quantity and the resonant frequency shift can be obtained, as the inductance of the circuit is known. C_e is a constant, provided that some precautions (shielding by the addition of a screen) are taken to prevent the "hands effect" and interference from external electrical field. Thus, since we are only interested in the changes in the capacitance, in the rest of this chapter we will only deal with C_i, the capacitance of the material inside.

7.3.2. Sensitivity

The principal reason for employing electrodes of varied dimensions and spacing is to reach different penetration depths in order to acquire information about the moisture gradient inside the cover concrete. Unfortunately, because of a strong dependency on the intrinsic material characteristics, it is rather difficult to delimit the investigation.

However, an indication of the properties of the inspected volume can be obtained by considering the three arrangements of electrodes placed in the center of the upper face of a (homogeneous) concrete slab, as shown in Figure 7.4. These numerical simulations were conducted with the commercial finite elements software package FEMLAB 2.3 [FEM 03].

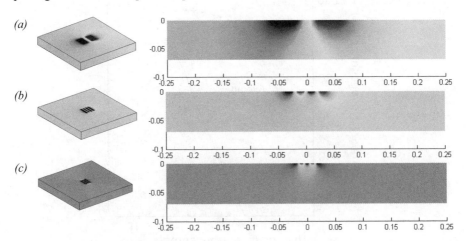

Figure 7.4. *Comparison of electric potential calculations for a 70 mm-thick slab of 500 × 500 mm. On the left-hand side of the figure are shown equipotential surfaces for: large plates (70 × 40 mm, 40 mm spacing); average size plates (70 × 10 mm, 10 mm spacing); small plates (70 × 5 mm, 5 mm spacing). On the right-hand side of the figure are shown the corresponding vertical cross-section plots in the center of the concrete slab*

Basically, with linear homogeneous isotropic materials, we use a three-dimensional electrostatic model of the system. Modeling the electric field is carried out using the electric potential V, calculated from the Laplace equation. Regarding the boundary conditions, the electrodes are equipotential surfaces, and the remainder is considered as electrically insulated.

From a qualitative point of view, the intensity of the electric field is extremely variable, depending on the configuration, in particular along the depth of the sample. Furthermore, the influence of the probe ranges from several centimeters (when a larger set of electrodes is used) to a few millimeters (when the device is mounted with smaller plates).

Additionally, the influence of the geometry and arrangement of the electrodes is confirmed quantitatively by capacitance calculations for several samples. The study of the sensitivity relates to the same concrete slab as before, still with the plates in the center of the upper face, but we are interested in thicknesses in the range of 5 to 150 mm (there are almost no changes beyond this depth). The values are normalized to unity when the sample is very thick. It turns out that approximately 95% of the maximal capacitance is obtained when the thickness τ_p is about 80 mm for the two larger plates (solid line), 20 mm in the case of the average sized plates (dashed line), and less than 9 mm for the smaller electrodes (see Figure 7.5).

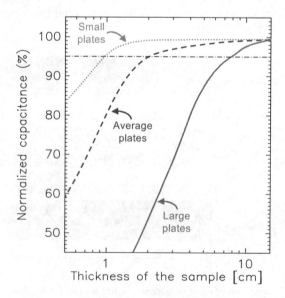

Figure 7.5. *Normalized capacitance values as a function of the thickness of the concrete sample for the different probe geometries (note that there is a log scale on the x axis). The investigated volume is much deeper when the large electrodes are used than with the average size or small plates*

This finding should be regarded only as an order of magnitude (it may depend on the moisture content), but it can be taken as a good indicator of the penetration depth of the sensors (the effective penetration depth being identified at a 5% reduction from the maximum capacitance). Indeed, it is practically impossible to find any modification of the capacitance when any material is added beneath a concrete slab of thickness τ_p. In addition, this means that operators must be careful when carrying out measurements with such a capacitance sensor on thin concrete slabs ($<<\tau_p$) because of the probable influence of materials located below.

7.3.3. *Example of measurements on the Empalot Bridge (Toulouse, France)*

In the context of a national project related to the evaluation of the cover concrete of reinforced structures by NDT, it was decided to perform both capacitive and GPR (ground penetrating radar) measurements in the same experiment. The site established at Empalot in Toulouse (France) was especially chosen because this bow-string bridge exhibits various typical pathologies. Most of these degradations are damage to the cover concrete (cracks, delaminations, etc.) that result from the corrosion of the reinforcement, structural faults or under-design.

A very useful quantity for the diagnosis of cover concrete, and in particular for the evaluation of corrosion, is the moisture content. Since both the radar and capacitive techniques are sensitive to this parameter, two areas of about 1.6×1.8 m were tested: a damaged one and a healthy one at the intrados of the deck (Figure 6a).

Radar investigations were performed through orthogonal profiles with a spacing of 20 cm, and capacitance measurements were made on a 20×20 cm grid.

One of the first steps during the radar investigation deals with the location of the reinforcement. By using specific processing of the parallel radar profiles, maps were constructed from chosen cross sections of the investigated medium.

The accuracy of the reinforcement map presented on the right-hand side of Figure 7.6b is noticeable. This result is due to a well-matched choice of the radar velocity during the computational process. As the map exhibits strong and thin reflected energy (in black) over the whole image, it is likely that the concrete is healthy. In contrast, in the other area the variations of radar velocity produce a worse reconstruction of the re-bars (as shown on the left-hand map of Figure 7.6b), the reason probably being a variation of the water content in the tested area.

Complementarily, our capacitive sensor was used with its three sets of electrodes in order to get an idea of moisture content of the cover concrete. The corresponding measurements are shown in Figure 7.6c (with points at intervals of 20 cm). Because it was not possible to carry out any calibration on that concrete, units are not presented and the interpretation will only be qualitative.

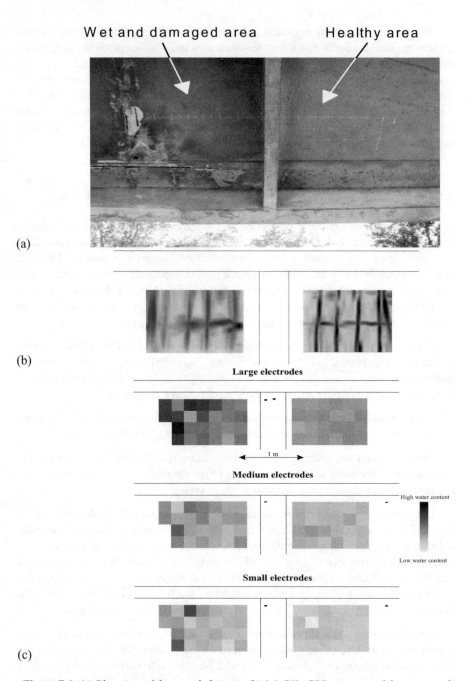

Figure 7.6. *(a) Plan view of the sounded areas; (b) 1.5 GHz GPR mapping of the areas and (c) capacitance maps obtained with the three sets of electrodes*

As already explained, each electrode set ensures a different penetration depth. For the large and average size sets, results have to be interpreted carefully because of the presence of the reinforcements. For the small electrodes, in spite of the large number of plates (so as to obtain a realistic averaged value), the estimation depends on the smoothness of the surface. In practice, because of the high roughness of the cover concrete, contacts were not perfect over the whole surface and measurements were problematic. Moreover, the interpretation must be carried out after interpretation of the two other capacitive maps (Figure 7.6c).

On the right-hand map, the moisture content is almost invariant: this had already been suspected from the radar result (the correlation between the two approaches can be explained by the fact that they are related to the same physical parameter, the dielectric constant of the concrete). This area will be considered as sound, with a low risk of corrosion.

In the left-hand maps of Figure 7.6(c), an area of significantly high moisture content appears on the left upper part. This is in good agreement with the visual inspection, which detected higher water content in this area, around a gargoyle.

The left-hand map corresponding to the large electrodes exhibits a gradient slope ($\geq 45°$) of water content decreasing from the top left corner. This feature suggests a similar gradient of corrosion risk for the reinforcement, dropping towards the right.

This example demonstrates the potential contribution of quasi-continuous NDT measurements for the assessment of water content over small areas. Knowing that large surfaces of concrete structures can be investigated by these techniques, such information is essential for the estimation of corrosion risk on a global scale.

7.4. Application for external post-tensioned cables

Many bridges and tunnels include external post-tensioned cables (they are called "external" because the sheaths are not hidden inside the concrete structure, but instead potentially accessible for measurement), either originally or after reinforcement of their structure (Figure 7.7).

These cables are positioned in High Density Poly-Ethylene (HDPE) ducts, where the residual space is filled under high pressure with a cement grout. The problem is that oxidation frequently spreads in zones not protected by this grout, in presence of "white paste" (that is to say, grout with a high water content that has not hardened) or air pockets (failure of injection).

A sketch of an external post-tensioned sheath containing a cable at the bottom is shown in Figure 7.8. Here, the volume of the HDPE envelope is imperfectly filled with cement grout, with an air pocket and a water-saturated layer of material.

Figure 7.7. *In the foreground, the long rectilinear black tubes are sheaths containing external post-tensioned cables. On the right-hand side of the figure, a capacitance sensor (first generation) is placed on one of the cables and connected to a portable computer (courtesy J.P. Sudret of the Public Works Regional Laboratory of Autun, France)*

The detection of such defects inside opaque ducts is cannot be carried out visually from the outside. Furthermore, the existing procedures often involve the damaging creation of apertures in the duct envelopes, so as to look inside or to introduce an endoscope. An alternative approach employing γ-rays is rather cumbersome and expensive, while acoustic methods did not prove to be precise enough. In this context auscultation with a capacitance probe was found to be promising in the laboratory [DUP 01].

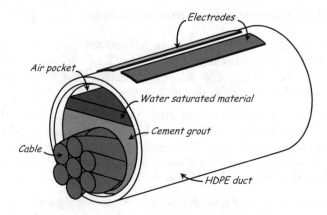

Figure 7.8. *Geometry of the system under test, with the sensing and driver electrodes outside the duct*

The experiments that were conducted used long transparent sample bodies, containing a steel cable and acting as part of the duct. A hole allowed the introduction of water or sand, and emptying as required. The method involved the acquisition of signals from electrodes placed on the periphery of the samples. In practice, the sensor was rotated around the sample, so as to provide information about the distribution of the material in the cross section under test, and then repeatedly moved along the axis.

The difference from the probe already employed for cover concrete was in the shape of the support of the electrodes (Figure 7.9).

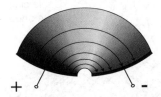

Figure 7.9. *After straightening the electrodes of the probe, as shown in Figure 7.2, the remaining operation just consists of slightly curving the plates in order to adapt their shape to the radius of curvature of the sheath*

As regards to the shape of the sensing devices, we employed, for the numerical computations, 150 mm long sections with an arc length of 20° and a 10° separation between them. In fact, using plates of smaller area results in a corresponding reduction in the capacitance makes it much more difficult to measure. Conversely, increasing the dimensions of the electrodes improves the signal-to-noise ratio, but it also increases the encumbrance and restricts the detectability of small features. This choice that was made was therefore a matter of compromise [XU 99].

7.4.1. *Influence of the location of the cable*

The example shown in Figure 7.10 relates to a configuration where the sheath (diameter 110 mm, thickness 5 mm) is entirely filled with cement grout (no air or "white paste"), and the steel cable is moved from the center towards the edge (at the bottom). The electrodes are slid around the axis of the duct, starting from the top (0°), passing the bottom (180°), and returning to the initial point, in order that the same cross-sectional area can be scrutinized.

If the cable is in the middle of the sheath, the capacitance obviously remains the same, whatever the position of the electrodes. A cable shift therefore produces an increase in the capacitance in the corresponding half of the duct; the surge affects the modeled values approximately in the zone between 90° and 270°. As the auscultation depth is about 20 mm in this construction, it turns out that phenomena occurring behind the cable may be invisible in certain configurations (when the cable is close to the edge) but the rest of the sheath should be properly sensed.

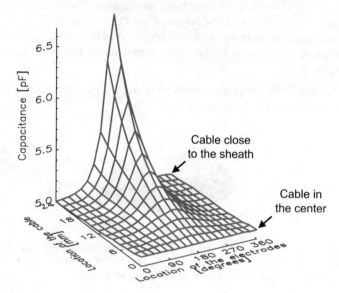

Figure 7.10. *Influence of the location of the cable on the value of the resulting capacitance of the system. The axes indicate the angular location of the sensor, and that of the cable in the sheath*

7.4.2. *Effect of air and water layers*

In this section we examine a situation with an HDPE sheath, a steel cable close to the bottom of the duct, with the cement grout filling all the gaps (Figure 7.11). On both sides of the plot (that is to say below 90° or above 270°), the capacitance is about 5 pF and, as long as it is in the sensing area of the electrodes, the influence of the cable dominates. The maximum value is obtained when the steel cable is near the outer surface of the duct (having its center located at 24 mm), with the electrodes very close to it (at an angle of 180°).

When a horizontal air layer is formed above the grout (the dashed curve on the plot), the capacitance drops off drastically: this change is a clear indication of the introduction of air within the detection volume [IAQ 04].

The third curve accounts for the presence of "white paste" with high moisture content, replacing the upper part of the grout, below the air region (maintained at the same thickness). Now, as the electrodes approach this layer, there is an increase in capacitance compared to the previous situations. This phenomenon especially affects the response of the device when it occurs exactly over the test zone, with little bumps on each side of the capacitance due to the cable.

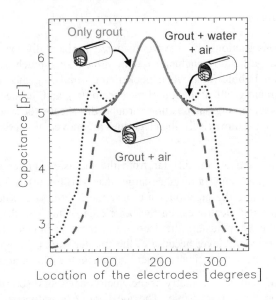

Figure 7.11. *Computed capacitance values for the following configurations:
cable + grout only (solid line); cable + air (dashed line);
cable + water + air (dotted line)*

The results can be summarized as follows: a capacitance value lower than that of the situation where there is only grout with a cable in the middle of the sheath is a distinguishing characteristic for air pockets inside. However, the signature of "white paste", producing a larger capacitance, is similar to that of the cable itself, but generally with a smaller extent. Accordingly, without any indication about the precise location of the cable, the interpretation of the measurements can be confusing.

Fortunately, there are indications that, between two spacers, the course of the cable within the sheath is rather regular, which is not true for randomly occurring injection defects of smaller size. Furthermore, because of the inclination of the ducts, development of air bubbles and "white paste" occurs in general towards the top [LER 00]. This *a priori* information will help discriminate between the cable and zones with a high moisture content, by considering capacitance changes as a function of the location along the sheath axis. Accordingly, smooth variation will be related to the cable, whereas more random changes are likely to correspond to defects having a much smaller extent.

7.4.3. *Small inclusions*

In the previous sections, we have considered horizontally extended layers. In this section, we will study inclusions with dimensions much smaller than the electrodes. An air bubble (or a water pocket, for simulating the "white paste") is created in the cement grout, at the top of the sheath (0°), and the steel cable is still in the center. For several angular positions around the sheath (between 0° and 180°, the other side being symmetrical), the electrodes are moved along the sheath (Figure 7.12).

As long as the sensor remains far from the inclusion, either angularly (from about 60°) or longitudinally (a distance larger than ≈ 180 mm), there is virtually no effect. However, when the electrodes are near the bubble (i.e. when the defect is located just below the electrodes, on the other side of the sheath) the response is noticeable. For the air inclusion, the decrease in the capacitance is smaller than in the case of the horizontal layer because of the smaller extent. Comparable behavior is observed for water, but this time with an increase in the capacitance.

It appears that, with the geometry of the electrodes used, discrimination between the two inclusion types is possible. The localization is rather fine, and a rough estimate of the dimension is also possible. A more detailed sensitivity investigation should bring additional information about the most suitable shape for the electrodes as a function of the minimum size of the defects to be investigated.

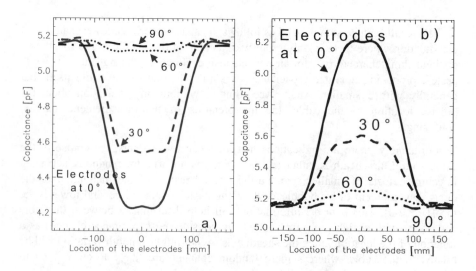

Figure 7.12. *Detection of inclusions for (a) air and (b) water*

7.4.4. *Example of an actual measurement*

A laboratory sample was specially prepared with an HDPE sheath, hardened grout at the bottom (no steel cable), an inclined layer of unhardened grout, and air at the top (see Figure 7.13).

In the actual probe, the electrodes were 165 mm long, with a 10° arc length and a 10° separation. We considered the same geometry for the numerical calculations.

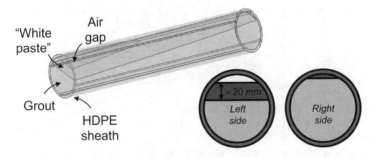

Figure 7.13. *Model used for the numerical study of the sample prepared for capacitance measurements, with the distribution of materials seen from the left- and right-hand sides. Grout is at the bottom of the sheath, with an inclined layer of "white paste" above, and an air gap at the top. The thickness of the "white paste" on the left-hand side is about 20 mm*

7.4.4.1. *Longitudinal inspection*

As a first stage, the probe was moved along the main axis, on the upper fiber (Figure 7.14). Because of the size of the electrodes, and the presence of stoppers plus seal plugs at both ends of the sample, it was not possible to position the probe everywhere on the sheath. Furthermore, as the probe has automatic compensation (which includes different calibrations and a readjustment), a point-to-point comparison is not possible. This method of processing of the data, through software and associated electronics, will be changed in the next release of the sensor (so as to get absolute values), but for the moment we are restricted to a qualitative comparison. In spite of this, modeled and experimental patterns exhibit a rather similar behavior.

As the maximum length of the air gap is close to 200 mm, when the electrodes are no longer above it (distance larger than about 285 mm) the variations are much less in terms of capacitance decrease or resonant frequency shift. After this, the "white paste" thickness becomes smaller (because of the inclination of the sample) and capacitance values start decreasing (this could not be measured experimentally, because of the settings constraints).

Note: for a similar analysis of defects located at the bottom of the sample, the sensitivity was really small (i.e. both the capacitance change and the corresponding measured frequency shift are almost negligible).

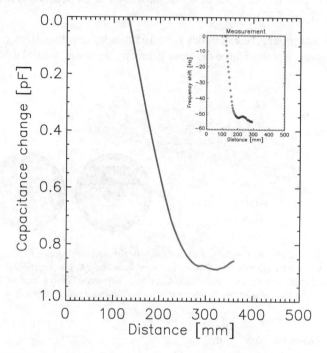

Figure 7.14. *Computed capacitance change along the upper fiber of the sample, and (in the insert on the right) the corresponding frequency shift measurements*

7.4.4.2. *Cross-sectional study*

For another series of tests the sensor was rotated around the sample so that different cross sections could be investigated. On the left-hand side the sensor was above the air gap and the "white paste" overlaying the grout (Figure 7.15a), while on the right-hand side of the sample (Figure 7.15b) it was only above "white paste" and the grout. Small icons on each graph show the related measurement of the resonant frequency shift for reference. The data are not very precise, but the behavior seems to be comparable for the model and the experiment.

It should be remembered that, for a sheath completely filled either with air, grout or "white paste", the capacitance measured when rotating the probe will remain the same (concentric circles on the graph). Moreover, as already shown, the presence in the sampled volume of a substance having a high or a low permittivity should respectively increase or decrease the corresponding capacitance values.

In the left-hand cross section (Figure 7.15a), the small bumps on both sides of the plot (at about ±45°) can be attributed to the "white paste" layer, and the depression at the top (0°) is due to the air gap (with a decrease in the capacitance value, compared to the adjacent points). On the right-hand cross section (Figure 7.15b), there is no longer any air and the presence of the material with a high moisture content is noticeable as the plot follows the "white paste" curve between approximately +20° and −20°. In the rest of the plot, the capacitance is close to the grout-only case.

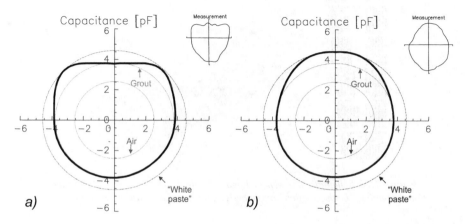

Figure 7.15. *Polar plots with the computed capacitance values for cross-sectional areas located on the left- (a) and right-hand (b) sides of the sample (thick line). The three circles correspond to situations where there would be only air (dashed line), grout (dotted line) or "white paste" (dotted–dashed) in the sheath. The small icons on each graph represent the corresponding measurements of the resonant frequency shift, for a qualitative comparison*

7.5. Future work

After conducting experiments with the capacitance sensors for applications involving the measurement of the moisture content of cover concrete and the auscultation of external post-tensioned cables, we are working on the design of new probes. The first point we have to address is the availability of various configurations of electrodes in terms of area and spacing, allowing different penetration depths for the electric field. For this purpose, employing a single device equipped with numerous plates and only activating some of them (through multiplexing) seems to be an answer. This solution is schematically shown in Figure 7.16 with an example of the sensing geometry.

In a sensor array, implementing two sets of seven contiguous plates is approximately equivalent to using two 74 × 74 mm single-sensor electrodes. The behavior is very similar although the two sets of seven contiguous plates are less efficient because of the difference in the effective surface of the electrodes Furthermore, this construction offers the possibility of adjusting the distance between the equivalent plates in the range of 3 to 80 mm (as shown on the left-hand side of the figure) without replacing the shoe of the probe.

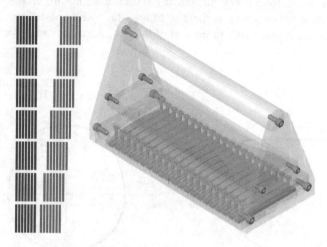

Figure 7.16. *Schematic of a sensor having 21 elements measuring 74 × 8 mm (3 mm spacing), each of which can be activated individually (as shown on the left)*

As well as making possible simultaneous measurements with variable volumes of investigation, this type of approach offers the possibility of making continuous NDT measurements. Such measurements, along with an onboard localization system, would enable the detection of significant variations in capacitance that are related to local defects. Accordingly, the typical signature of the figure (the shape of the curve) will provide qualitative information (detection of the defect itself) but also, when combined with modeling, more quantitative information (nature, dimension, etc.).

7.6. Monitoring historical buildings

In practice, monitoring techniques have not so far been widely applied in civil engineering, principally for two reasons. First, for most of the time, stakeholders' requirements are for global information regarding a structure, so that its sustainability can be assessed, whereas monitoring is more relevant for providing local knowledge. The second argument is related to the cost–effectiveness ratio of a monitoring approach for classical civil engineering structures, which can be seen as too high.

Historical buildings constructed of stone can certainly be considered to be within the frame of potential operations for monitoring systems. Indeed, because of the importance of the preservation of our heritage, and since common pathologies are due to moisture, a global indication of the health of the overall structure can be obtained from local measurements.

The propagation of moisture in masonry primarily proceeds from capillarity and presents a front that is more or less easily observable. Corresponding gradients may exhibit a horizontal profile, so that making water content measurements at various locations (for instance at increasing heights above the soil level) in representative areas can be convenient for monitoring.

During the last two decades, the French Regional Public Works Laboratory at Angers (Angers LRPC) has been working on capacitive monitoring techniques for the structure of historical buildings. The following section presents a summary of a part of their research and development, obtained from internal reports.

7.6.1. *Capacitance probe for moisture monitoring*

The physical principle implemented for the monitoring technique is very similar to the one previously introduced at the beginning of this chapter.

The basic idea is to have two metallic electrodes embedded in the material (Figure 7.17), and to apply an alternating electric current between them (with a nominal frequency of 35 MHz). This system forms a capacitor, and changes in capacitance are indicative of the internal constituents (including the nature of the materials or their moisture content). The oscillator is protected by a watertight plastic box before being connected to the electrodes.

For a given material, a linear variation of frequency is the consequence of a linear change of the water content. A careful calibration enables the operators to find the relationship between these two quantities. The sensitivity of measurements depends on the geometry of the electrodes (see Figure 7.17). As an example, for the monitoring of limestone (stone from Lavoux, France) with 6 mm diameter cylindrical electrodes (80 mm length and 20 mm spacing), a 1% variation in the water content is responsible for a 50 Hz variation in the oscillator frequency.

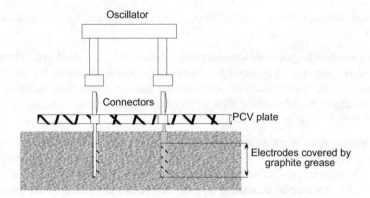

Figure 7.17. *Schematic of the monitoring device,*
including the positioning of the electrodes

A laboratory calibration of the capacitive sensor, carried out on the Lavoux Stones, led to the following linear relationship:

$$F = -50W + 5000 \tag{2}$$

where F is the measured frequency in Hz and W the weight percentage of the water content.

7.6.2. *Environmental conditions*

Preliminary experimentation on masonry stones showed a significant influence of temperature and humidity on the measurements. Accordingly, after numerous laboratory tests with dry stones and types of concrete, the following compensation terms were introduced for each sensor:

$$C_T = 1.02 - 10^{-3} * T \tag{3}$$

$$C_H = 1.004 - 1.2 * 10^{-4} * H \tag{4}$$

where C_T and C_H are corrective factors for temperature and humidity respectively, T is the temperature (in °C) and H the relative humidity (in %).

These adjustments are necessary if comparisons of the measurements are to be made over long periods of time, as well as over daily or seasonal cycles.

7.6.3. *Study on a stone wall test site*

In response to a request from the LRMH (the acronym for the French Research Laboratory of Historical Buildings), the LRPC at Angers built in its premises a stone wall intended to optimize a methodology suitable for monitoring the storage and evaporation of water in certain types of stone (Figure 7.18).

Two different types of stone were used: freestone and Lavoux stone, which is a fine limestone from the western part of France, both commonly used in historic buildings. These materials have contrasting properties regarding permeability to water (liquid and vapor phases), capillarity and porosity. Thermal and watertight insulations were added to avoid a lateral gradient of water content, keeping in mind that the objective of the study was an assessment of the diffusion of water along the vertical and horizontal directions.

Meteorological parameters were monitored over a period of nine months during 1998 at the meteorological station at Angers-Avrillé, 10 km away from the test site. Dominant winds are west/southwest.

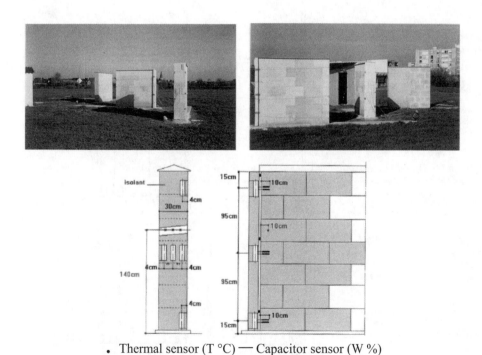

• Thermal sensor (T °C) — Capacitor sensor (W %)

Figure 7.18. *Wall stone test sites, with a schematic of the implantation of the capacitive and thermal sensors (courtesy LRPC at Angers)*

Some results of water content measurement are shown in Figure 7.19. Periodic rainfall had been observed since September produced an accumulation of water, which diffused into the wall. It was observed that the progression of the moisture front inside the Lavoux stone wall is more significant than in the freestone, although the capillarity and permeability coefficients of the latter are higher than those of the former.

It seems that Lavoux stone is not able to keep out water which has accumulated on the exposed side in successive rainfall.

Nevertheless, this study on controlled materials, conducted in actual conditions, made it possible to define an effective procedure for monitoring masonry structures, yielding information about the range of water content variations inside the stones.

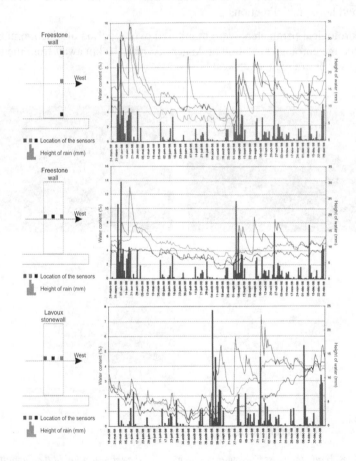

Figure 7.19. *Capacitive monitoring over a period of nine months on freestone and Lavoux stone. Influence of rain on the water content diffusion (courtesy LRPC at Angers)*

7.6.4. *Water content monitoring of part of the masonry of Notre-Dame La Grande church (Poitiers, France)*

Notre-Dame La Grande church, built during the 11th and 12th centuries, is located in Poitiers (France). This Roman art monument is renowned for its completely carved front (Figure 7.20).

Since the 19th century, the frontage has suffered from changes due to fishmongers and salt-makers working against the walls of the church. As a result of the saturation of the soil, salt has migrated into the limestone walls by capillarity.

Figure 7.20. *Frontage of Notre-Dame La Grande church (Poitiers, France)*

Recent restoration was carried out to repair the stones and prevent new water infiltration into the walls, principally by adding a watertight seal and a drain in front of the western side.

LRMH then commissioned Angers LRPC to install on the western front of the church capacitive sensors in order to monitor the progress of water in the inside facing walls, and to detect any vertical gradient of the water content and capillary phenomena over a period of several years.

The sensors chosen were similar to those already tested on the laboratory stone wall (Figure 7.17), and they were implanted in a vertical line at heights of 0.5, 1, 2, 3, 4 and 5 m, in the center of the wall. The internal facing wall is made of Lavoux stones sealed with lime mortar.

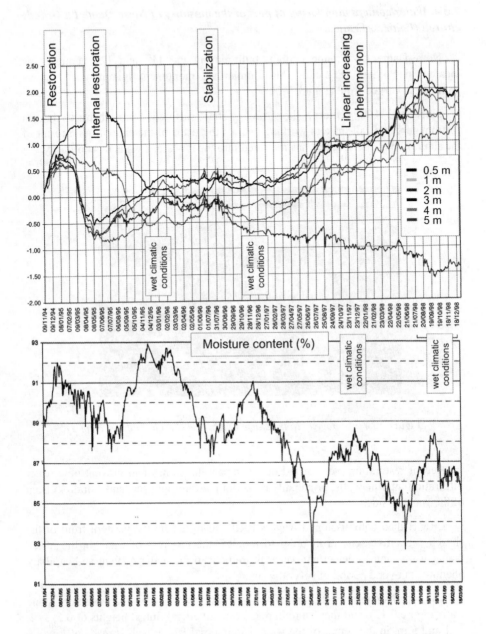

Figure 7.21. *Monitoring of the water content and moisture of the west wall of the church during the period from November 1994 to December 1998 (courtesy LRPC at Angers)*

In addition, thermal and vapor–moisture sensors were placed at about 4 m from the soil level, and general meteorological information (temperature, relative humidity, pluviometry, velocity and direction of the wind) came from a station sited in Poitiers.

After a calibration phase (using the sensors at 4 m), measurements were taken every 90 minutes and converted into water content. Then, a daily average was obtained and compared with that of the first day taken as a reference (November 1994).

The results from the water content and the vapor–moisture sensors in the wall are given in Figure 7.21. The lower curve shows the environmental conditions prevailing at the local meteorological station, and mainly relates to the pluviometry and outdoor moisture. As a matter of interest, the seasonal variation of the climate (rainy winters and sunny summers) can be clearly seen in the graphs.

The beginning of the monitoring corresponded to the end of the restoration work, in autumn 1994, allowing a survey of the water content in the wall at that time. The initial increase due to the fresh mortar was followed by a decrease of water content during the 1995 winter season, corresponding to the mortar setting. Then, an internal coat, fitted on the wall (which completed the restoration in March 1995), induced an increase in water content as measured by all the sensors.

During 1996, the migration of water seemed to stabilize, and no further capillarity effects were observed. It is likely that the weather conditions have a very low impact on the water content in the heart of the stone wall.

During 1997, although that year was not especially rainy, the water content increased noticeably at all heights from 0.5 to 4 m. This phenomenon does not appear to be the consequence of capillarity effects. Similar results were obtained in 1998, when there was an increase in the water content.

The sensor located at 5 m from the soil level exhibited a quite different profile with a continuous decrease of the water content. The reason for this is probably that, at this altitude, the wall is not affected by either capillarity effects or the rain.

In contrast, the other capacitive sensors proved to be very sensitive to rainy conditions (with a few months' delay). The interpretation given by the architects is this was caused by water infiltration coming from a part of roof at a height between 4 and 5 m.

Eventually, even if though water infiltration biased some of the measurements, the differences in the values between the lowest sensors (0.5, 1 and 2 m heights) have shown a very small effect of capillarity on the stones, confirming, if necessary, the efficiency of the building restoration.

7.7. Conclusion

Evaluating the civil engineering heritage is a constant priority of infrastructure administrators, because of the high cost of construction compared to that involved in maintenance. And monitoring and non-destructive techniques take an increasingly important role in providing a reliable diagnosis. The emerging approaches (and this is not an exhaustive list) have the following features: they are contact-less; they implement mobile devices (energetically autonomous, lightweight, wireless instruments); and they enable high-efficiency dynamic measurements in real time.

Monitoring devices (like the capacitive probes) and NDT (i.e. radar) appear to be complementary tools, as shown in section 7.3.3. The former are helpful for the calibration of NDT sensors, whereas the latter add value on core information, when producing evidences for how representative homogeneous areas are. Moreover, such techniques may lead to optimized positioning of destructive soundings when necessary.

A key feature of the capacitive sensing system (which has been the topic of this chapter) lies in that it is indicative of the presence of water, known to be at the origin of most common pathologies of concrete and masonry structures. Seen as atypical, this approach remained marginal for years in the field of civil engineering because of the difficulty in interpreting the measurements. This lack of knowledge is now being filled with the help of a modeling approach.

Technical developments in the near future will focus on the implementation of array sensors (combinations of sensing elements of the same type in a single device), for instance multi-plate capacitors, and on data processing suited to each application. Advances in this post-processing will necessarily benefit from the resolution of inverse problems and from a better knowledge of the materials. As civil engineering structures constitute a rather heterogeneous mix, they have to be EM characterized over a wide frequency range. This study is currently being undertaken in several research laboratories.

7.8. Acknowledgements

The authors would like to express their gratitude to the network of Public Works Regional Laboratories (Autun and Angers) and in particular to: J.P. Sudret and A. Dupas for their extensive involvement during the capacitive experiments on the external post-tensioned ducts; J. Godin, A. Briffault and M. Pithon, from LRPC at Angers for the monitoring studies; and the people involved in the National Project RGC&U on the evaluation of concrete pathologies by NDT, for the measurements performed on Empalot Bridge. They would also like to thank L.M. Cottineau from the French Public Works Research Laboratory (Nantes) for his invaluable help regarding the technical aspects of probe development.

7.9. References

[ASA 02] ASAMI K., "Characterization of heterogeneous system by dielectric spectroscopy", *Progress in Polymer Science*, vol. 27, 2002, pp. 1617-1659.

[BAR 77] BARON J.P., TRAN N.L., "Méthodes de mesure et de contrôle des teneurs en eau de matériaux dans les LPC", *Bull. des Ponts et Chaussées*, no. 87, 1977, pp. 85-96.

[BLA 93] BLASZCZYK F., BLASZCZYK R., TROCHET B., BIGORRE M., DUPAS A., "Mesure de la teneur en eau en continu d'un matériau granulaire: TRITON II. Mesure à la jetée d'un transporteur", *Bull. des Ponts et Chaussées*, no. 186, 1993, pp. 85-87.

[BUN 96] BUNGEY J.H., MILLARD S.G., *Testing of Concrete in Structures*, Blackie Acad. & Prof., 3rd edition, 1996.

[DIE 98] DIEFENDERFER B.K., Al-QADI I.L., YOHO J.J., RIAD S.M., LOULIZI A., "Development of a capacitor probe to detect subsurface deterioration in concrete", *Materials Research Society Symposium Proceedings*, vol. 503, 1998, pp. 231-236.

[DUP 01] DUPAS A., SUDRET J.P., CHABERT A., "Méthode de diagnostic de câbles de précontrainte externe contenus dans des gaines", French Patent no. 0107719, National institute for industrial property, Paris, 2001.

[FAR 02] FARES A., ALVA A.K., "Soil water components based on capacitance probes in a sandy soil", *Soil Science Society of America Journal*, vol. 64, 2002, pp. 311-318.

[FEM 03] FEMLAB, "Reference manual – Version 2.3", by COMSOL, http://www.comsol.com, 2003.

[IAQ 04] IAQUINTA J., "Contribution of capacitance probes for the inspection of external prestressing ducts", Proceedings of the 16th World Conference on Nondestructive Testing, Montréal (Canada), 2004.

[KEN 96] KENDOUSH A.A., SARKIS Z.A., "A non-intrusive auto-transformer technique for the measurement of void fraction", *Experimental and Thermal Fluid Science*, vol. 13, 1996, pp. 92-97.

[LER 00] LE ROY R., WILLAERT M., MIRMAND H., ROUANET D., "Identifying the parameters that encourage the formation of air pockets and water pockets in cement grouts for pre-stressing ducts", *Bulletin des Laboratoires des Ponts et Chaussées*, vol. 229, 2000, pp. 53-70.

[LOU 98] LOUGE M.Y., FORSTER R.L., JENSEN N., PATTERSON R., "A portable capacitance snow sounding instrument", *Cold Regions Science and Technology*, vol. 28, 1998, pp. 73-81.

[MAL 91] MALHOTRA V.M., CARINO N.J., *Handbook on Nondestructive Testing of Concrete*, CRC Press, 1991, 341 pp.

[MAM 99] MAMAISHEV A.V., "Interdigital dielectrometry sensor design and parameter estimation algorithms for non-destructive material evaluation", PhD dissertation, Massachusetts Institute of Technology, U.S.A, 1999.

[NEL 92] NELSON S.O., "Measurement and applications of dielectric properties of agricultural products", *IEEE Transactions on Instrumentation and Measurement*, vol. 41, 1992, pp. 116-122.

[RUS 99] RUST A.C., RUSSEL J.K., KNIGHT R.J., "Dielectric constant as a predictor of porosity in dry volcanic rocks", *Journal of Volcanology and Geothermal Research*, vol. 91, 1999, pp. 79-96.

[SAA 00] SAARENKETO T., SCULLION T., "Road evaluation with ground penetrating radar", *J. Appl. Geophys.*, vol. 43, 2000, pp. 119-138.

[STE 96] STEELE P.H., KUMAR L., "Detector for heterogeneous material", US Patent no. 5,585,732, US Patent Office, Washington, DC, 1996.

[TRA 72] TRAN N.L., AMBROSINO R., "Mesure de la teneur en eau des sols et des matériaux par une méthode capacitive: 3- Applications de la méthode capacitive de mesure de la teneur en eau", *Bull. des Ponts et Chaussées*, no. 60, 1972.

[XU 99] XU H., YAN H., WANG S., "Optimum design of capacitance sensor array for imaging the solid distribution of fluidized bed", Proceedings of the 15th International Conference on Fluidized Bed Combustion, Savannah (Georgia), 1999.

Short Biographies of the Contributors

Jean-Christophe Abry received a PhD degree from the University of Provence and Ecole Centrale de Lyon (ECL) in 1998, with a thesis on the "*In situ* health monitoring of CFRP using electrical properties measurement". At present, he is working in the field of Tribology as a research engineer at the Laboratory of Tribology and Dynamics of Systems at ECL.

Daniel Balageas, Deputy Head of the Structures and Damage Mechanics Department (DMSE) of ONERA and part-time Associate Professor at ENS-Cachan, has 40 years of experience in aerospace research. He is author of more than 80 and 30 publications/communications in the NDE and SHM fields respectively, and in 2002 established the European Workshop on SHM.

Xavier Dérobert, senior scientist at LCPC, has been a specialist in Ground Penetrating Radar (GPR) for over 10 years in civil engineering. He is author of more than 30 publications/communications in NDT and has organized two national GPR conferences in 2001 and 2003.

Claus-Peter Fritzen, Professor of Mechanics at the Institute of Mechanics and Control-Mechatronics of the University of Siegen (Germany), is a researcher and teacher in the fields of vibrations, structural identification, smart structures and SHM. He has published more than 100 journal and conference papers, 50 of them related to damage identification and SHM.

Alfredo Güemes, full Professor of Material Science and Director of the Aerospace Materials & Processes Department at the Polytechnic University of Madrid, and Co-Chairman of the Stanford International Workshops on SHM since 1997, has led three research projects on Fiber Optic Sensors, sponsored by the European Commission and ESA. He has authored 14 papers on these topics and given more than 50 communications to international conferences.

Philippe Guy is Associate Professor at the National Institute of Applied Sciences (INSA-Lyon), Head of Eurinsa (the European First Cycle Cursus of INSA), and researcher in the Physical Metallurgy and Material Physics Research Group (GEMPPM) at INSA. He was Chairman of the 4th Japan–France Seminar on Intelligent Materials and Structures held in Lyon (July 2002).

Jean Iaquinta has been working as a researcher at the French Public Works Research Laboratory of Paris (France) in the Division of Metrology and Instrumentation since 2001. He is author of more than 40 peer-reviewed papers and communications to international conferences in the fields of remote sensing and non-destructive evaluation of civil engineering infrastructures.

Michel Lemistre is senior scientist at the Structures and Damage Mechanics Department of ONERA and associate researcher at the SATIE Laboratory of ENS-Cachan. He gives courses at the University of Evry (France). He is author of more than 50 publications and communications in the field of signal processing and electromagnetic techniques for SHM and NDE.

Jose Manuel Menendez works as a specialist for Smart Composites Development and Applications at Airbus-Spain. He obtained a PhD degree at the University of Madrid in 1999 with a thesis on FBG characterization. He has written several papers and two patents related to fiber-optic sensors.

Thomas Monnier holds a PhD in material engineering from INSA Lyon (France) and is currently a research fellow in the INSA Group on Metallurgy and Materials Science. His research interests include ultrasonic material characterization, ultrasound propagation and detection using embedded transducers, SHM and NDE of composites.

Michelle Salvia, Associate Professor at the Ecole Centrale of Lyon (France) and researcher in the field of materials science (polymers, composites and smart materials) at the LTDS Laboratory (CNRS), has authored several papers on durability and on electrical methods for SHM of composite materials. She was editor of the special issue (December 2004) of *Annales de Chimie, Sciences des Matériaux* on "Recent advances in intelligent materials".

Index

W, X, Y, Z